Probability Theory and Mathematical Statistics for Engineers

This book is essential reading for practising engineers who need a sound background knowledge of probabilistic and statistical concepts and methods of analysis for their everyday work. It is also a useful guide for graduate engineering students and researchers.

The theoretical aspects of modern probability theory is the subject of the first part of the book. In the second part, it is shown how these concepts relate to the more practical aspects of statistical analyses. Although separated, the two parts of the book are presented in a unified style and form a well-structured unity. Moreover, besides discussing a number of fundamental ideas in detail, the author's approach to the subject matter is particularly useful for the interested reader who wishes to pursue the study of more advanced topics.

This book has an unusual combination of topics based on the author's education as a Nuclear Physicist and his many years of professional activity as a consultant in different fields of Engineering.

Paolo L. Gatti formerly Head of the Vibration Testing and Data Acquisition Division of Tecniter s.r.l., Cassina de' Pecchi, Milan, Italy, works now as an independent consultant in the fields of Engineering Vibrations, Statistical Data Analysis and Data Acquisition Systems.

Spon's Structural Engineering: Mechanics and Design series

Innovative structural engineering enhances the functionality, serviceability and life-cycle performance of our civil infrastructure systems. As a result, it contributes significantly to the improvement of efficiency, productivity and quality of life.

Whilst many books on structural engineering exist, they are widely variable in approach, quality and availability. With the *Structural Engineering: Mechanics and Design* series, Spon Press is building up an authoritative and comprehensive library of reference books presenting the state-of-the-art in structural engineering, for industry and academia.

Topics under consideration for the series include:

- Structural Mechanics
- Structural Dynamics
- Structural Stability
- Structural Reliability
- Structural Durability
- Structural Assessment
- Structural Renewal
- Numerical Methods
- Model-based Structural Simulation
- Composite Structures
- Intelligent Structures and Materials

Books published so far in the series:

1. Approximate Solution Methods in Engineering Mechanics
   A.P. Boresi and K.P. Chong 1–85166–572–2
2. Composite Structures for Civil and Architectural Engineering
   D.-H. Kim 0–419–19170–4
3. Earthquake Engineering
   Y.-X. Hu, S.-C. Liu and W. Dong 0–419–20590–X
4. Structural Stability in Engineering Practice
   Edited by L. Kollar 0–419–23790–9
5. Innovative Shear Design
   H. Stamenkovic 0–415–25836–7
6. Modeling and Simulation-based Life-cycle Engineering
   Edited by K. Chong, S. Saigal, S. Thynell and H. Morgan 0–415–26644–0

Potential authors are invited to contact the Series Editors, directly:

Series Editors Professor Ken P. Chong National Science Foundation, 4201 Wilson Blvd, Room 545, Arlington, VA 22230, USA e-mail: kchong@nsf.gov

Professor John E. Harding Department of Civil Engineering, University of Surrey, Guildford, GU2 5X11, UK e-mail: j.harding@surrey.ac.uk

Probability Theory and Mathematical Statistics for Engineers

Paolo L. Gatti

CRC Press
Taylor & Francis Group
Boca Raton London New York

CRC Press is an imprint of the
Taylor & Francis Group, an **informa** business
A SPON PRESS BOOK

CRC Press
Taylor & Francis Group
6000 Broken Sound Parkway NW, Suite 300
Boca Raton, FL 33487-2742

First issued in paperback 2019

ISBN-13: 978-0-415-25172-3 (hbk)
ISBN-13: 978-0-367-44652-9 (pbk)

British Library Cataloguing in Publication Data
A catalogue record for this book is available
from the British Library

Library of Congress Cataloging in Publication Data
Gatti, Paolo L., 1959–
Probability theory and mathematical statistics for engineers / Paolo L. Gatti. – 1st ed.
p. cm. – (Structural engineering: mechanics and design)
Includes bibliographical references and index.
1. Engineering – Statistical methods. 2. Probabilities.
I. Title. II. Series.

TA240.G38 2004
620′.001′5192–dc22 2004006776

Visit the Taylor & Francis Web site at
http://www.taylorandfrancis.com

and the CRC Press Web site at
http://www.crcpress.com

To the dearest friends of many years,
no matter how far they may be.
To my parents Paola e Remo and in
loving memory of my
grandmother Maria Margherita.

Paolo L. Gatti

Contents

Preface

Tanta animorum imbecillitas est, ubi ratio decessit

L.A. Seneca (De Constantia Sapientis, XVII)

A common attitude among engineers and physicists is, loosely speaking, to consider statistics as a tool in their toolbox. They know it is 'there', they are well aware of the fact that it can be of great help in a number of cases and, having a general idea of how it works, they 'dust it off' and use it whenever the problem under study requires it.

A minor disadvantage of this pragmatic, and in many ways justifiable (after all, statistics is not their main field of study) point of view is that one often fails to fully appreciate its potential, its richness and the complexity of some of its developments. A more serious disadvantage is the risk of improper use, although it is fair to say that this is rarely the case in science when compared with other fields of activity such as, for instance, politics, advertising and journalism (even assuming the good faith of the individuals involved).

These general considerations aside, in the author's mind the typical reader of this book (whatever the term 'typical reader' may mean) is an engineer or physicist who has a particular curiosity – personal and/or professional – for probability and statistics. Although this typical reader is surely interested in statistical techniques and methods of practical use, his/her focus is on 'understanding' rather than 'information' on this or that specific method, and in this light he/she is willing to tackle some mathematical difficulties in order to, hopefully, achieve this understanding. Since I found myself in this same situation a few years ago and it is now my opinion that the reward is well-worth the effort (this, however, in no way implies that I have reached a full understanding of the subject-matter – unfortunately, I feel that it is not so – it simply means that after a few years of study the general picture is much clearer and many details are much sharper now than then), I decided to write a book which would have fulfilled my needs at that time. It goes without saying that there are many good books on the subject – a good number of them are on my shelf and I have often referred to them either for work or in writing this book – and this is why I included a rather detailed list of references at the end of each chapter.

The book is divided into two main parts: Part 1 (Chapters 1–4) on probability theory and Part 2 (Chapters 5–7) on mathematical statistics. In addition, three appendices (A, B and C) complement the book with some extra material relevant to the ideas and concepts presented in the main text.

With regard to Part 1 on probability, some mathematical difficulties arise from the circumstance that the reader may not be familiar with measure theory and Lebesgue integration, but I believe that it would have been unfair to the 'typical reader' to pretend to ignore that modern probability theory relies heavily on this branch of mathematics. In Part 2, on the other hand, I tried as much as possible to show the way in which, in essence, this part is a logical – even if more application-oriented – continuation of the first; a fact that, although obvious in general, is sometimes not clear in its details.

In all, my main goal has been to give a unified treatment in the hope of providing the reader with a clear wide-angle picture where, in addition, some important details are in good focus. On this basis, in fact, he/she will be able to pursue the study of more advanced topics and understand the main ideas behind the specific statistical techniques and methods – some of which are rather sophisticated indeed – that he/she will encounter in this and/or other texts.

Also, it is evident that in writing a book like this some compromise must be made on the level of mathematical exposition and selection of topics. In regard to the former I have striven for clarity rather than mathematical rigor; in fact, there exist many excellent books written by mathematicians (some of them are included in the references) where rigor is paramount and all the necessary proofs are given in detail. For the latter, it is only fair to say that many important topics, including probably at least one of everyone's favourites, have been omitted. Out of necessity, in fact, some choices had to be made (a few of them have been made painfully along the way, leaving some doubts that still surface now and then) and I tried to do so with the intention of writing a not-too-long book without sacrificing the spirit of the original idea that had me started in the first place. Only the readers will be able to tell if I have been successful and faithful to this idea.

Finally, it is possible that, despite the attention paid to reviewing all the material, this book will contain errors, omissions, oversights or misprints. I will be grateful to the readers who spot any of the above or who have any comments for improving the book. Any suggestion will be received and considered.

Paolo L. Gatti
Milano
September 2004

Part I

Probability theory

1 The concept of probability

1.1 Different approaches to the idea of probability

Probabilistic concepts, directly or indirectly, pervade many aspects of human activities, from everyday situations to more advanced and specific applications in natural sciences, engineering, economy and politics. It is the scope of this introductory chapter to discuss the fundamental idea of probability which, as we will see, is not so obvious and straightforward as it may seem. In fact – in order to deal with practical problems in a first stage and to arrive at a sound mathematical theory later – this concept has evolved through the centuries, changing the theory of probability from an almost esoteric discipline to a well-established branch of mathematics.

From a strict historical point of view, despite the fact that some general notions have been common knowledge long before the seventeenth century (e.g. Cardano's treatise 'Libel de Ludo Aleæ' (Book of Dice Games) was published in 1663 but written more than a century earlier), the official birth of the theory dates back to the middle of the seventeenth century and its early developments owe much to great scientists such as Pascal (1623–1662), Fermat (1601–1665), Huygens (1629–1695), J. Bernoulli (1654–1705), de Moivre (1667–1754), Laplace (1749–1827) and Gauss (1777–1855).

Broadly speaking, probability is a loosely defined term employed in everyday conversation to indicate the measure of one's belief in the occurrence of a future event when this event may or may not occur. Moreover, we use this word by indirectly making some common assumptions: probabilities near 1 (100%) indicate that the event is extremely likely to occur, probabilities near zero indicate that the event is almost not likely to occur and probabilities near 0.5 (50%) indicate a 'fair chance', that is, that the event is just as likely to occur as not.

If we try to be more specific, we can consider the way in which we assign probabilities to events and note that three main approaches have developed through the centuries. Following the common terminology, we call them

(1) the classical approach,
(2) the relative frequency approach,
(3) the subjective approach.

This order agrees with the historical sequence of facts. In fact, the classical definition of probability was the first to be given, followed by the relative frequency definition and – not long before Kolmogorov's axiomatic approach was introduced in 1931 – by the subjective definition. Let us examine them more closely.

1.2 The classical definition

The first two viewpoints mentioned in Section 1.1, namely the classical and the relative frequency approaches, date back to a few centuries ago and originate from practical problems such as games of chance and life insurance policies, respectively. Let us consider the classical approach first.

In a typical gambling scheme, the game is set up so that there exists a number of possible outcomes which are mutually exclusive and equally likely and the gambler bets against the House on the realization of one of these outcomes. The tossing of a balanced coin is the simplest example: there are two equally likely possible outcomes, head or tail, which are mutually exclusive (that is both faces cannot turn up simultaneously) and the bet is, say, the appearance of a head.

More specifically, the *classical* (or the gambler's) definition of probability can be used whenever it can be reasonably assumed that the possible outcomes of the 'experiment' are mutually exclusive and equally likely so that one calculates the probability of a particular outcome A as

$$P(A) = \frac{n(A)}{n(S)} \tag{1.1}$$

where $n(A)$ is the number of ways in which outcome A can occur and $n(S)$ is the total number of ways in which the experiment can proceed. Note that with this definition we do not need to actually perform the experiment because eq. (1.1) defines an '*a priori*' probability. In tossing a fair coin, for instance, this means that without even trying we can say that $n(S) = 2$ (head or tail) and the probability of a head is $P(A) \equiv P(\text{head}) = 1/2$. Also, in rolling a fair die – where six outcomes are possible, that is, $n(S) = 6$ – the appearance of any one particular number can be calculated by means of eq. (1.1) and gives 1/6 while, on the other hand, the appearance of, say, an even number is 1/2.

1.2.1 Properties of probability on the basis of the classical definition

In the light of the simple examples given in Section 1.2, the classical definition can be taken as a starting point to give some initial definitions and determine a number of properties which we expect a 'probability function' to have. This will be of great help in organizing some intuitive notions in a more systematic manner – although, for the moment, in a rather informal way and on the basis of heuristic considerations (the term 'probability function' itself is here used informally just to point out that a probability is something that assigns a real number to each possible outcome of an experiment). In order to do so we must turn to the mathematical theory of sets (the reader may refer to Appendix A for some basic aspects of this theory). First of all, we give some definitions:

(a) we call *event* a possible outcome of a given experiment;
(b) among events, we distinguish between *simple events*, which can happen only in one way, are mutually exclusive and equally likely;
(c) *compound events*, which can happen in more than one way.

Then,

(d) we call *sample space* (or event space) the set of all possible simple events.

Note that this definition justifies the fact that simple events are also often called *sample points*. In the die-rolling experiment, for example, the sample space is the set $\{1, 2, 3, 4, 5, 6\}$, a simple event is the observation of a six and a compound event is the observation of an even number (2, 4, or 6).

Adopting the notations of set theory, we can view the sample space as a set W whose elements E_j are the sample points. Then, any compound event A is a subset of W and can be viewed as a collection of two or more sample points, that is, as the union of two or more simple events. In the die-rolling experiment above, for example, we can write

$$A = E_2 \cup E_4 \cup E_6 \tag{1.2}$$

where we called A the event 'observation of an even number', E_2 the sample point 'observation of a 2' and so on. In this case, it is evident that $P(E_2) = P(E_4) = P(E_6) = 1/6$ and, since E_2, E_4 and E_6 are mutually exclusive we expect an 'additivity property' of the form

$$P(A) = P(E_2 \cup E_4 \cup E_6) = P(E_2) + P(E_4) + P(E_6) = 1/2 \tag{1.3a}$$

An immediate consequence of eq. (1.3a) is that

$$P(W) = P\left(\bigcup_{j=1}^{6} E_j\right) = \sum_{j=1}^{6} P(E_j) = 1 \tag{1.3b}$$

because it is clear that one of the six faces must necessarily show up.

Moreover, if we denote by A^C the complement of set A (clearly $W = A \cup A^C$: for example, in the die experiment if A is the appearance of an even number then the event A^C represents the non-occurrence of A, that is, the appearance of an odd number; therefore $A^C = E_1 \cup E_3 \cup E_5$), we have

$$P(A^C) = 1 - P(A) \tag{1.4}$$

If, on the other hand, we consider two events, say B and C, which are not mutually exclusive, a little thought leads to

$$P(B \cup C) = P(B) + P(C) - P(B \cap C) \tag{1.5a}$$

where $P(B \cap C)$ is called the *compound probability* of events B and C, that is, the probability that B and C occur simultaneously.

An example will help clarify this idea: returning to our die-rolling experiment, let, for example, $B = E_2 \cup E_3$ and $C = E_1 \cup E_3 \cup E_6$, then $B \cap C = E_3$ and, as expected, $P(B \cup C) = (2/6) + (3/6) - (1/6) = (4/6)$.

For three non-mutually exclusive events, say B, C and D, eq. (1.5a) becomes

$$\begin{aligned} P(B \cup C \cup D) = P(B) + P(C) + P(D) \; - \; P(B \cap C) \\ - P(B \cap D) - P(C \cap D) + P(B \cap C \cap D) \end{aligned} \tag{1.5b}$$

as the reader is invited to verify. In general, the extension of eq. (1.5a) to n events $A_1, A_2, \ldots, A_n$ leads to the rather cumbersome expression

$$\begin{aligned} P\left(\bigcup_{k=1}^{n} A_k\right) = \sum_{k=1}^{n} P(A_k) - \sum_{k_1<k_2} P(A_{k_1} \cap A_{k_2}) + \cdots + (-1)^{m+1} \\ \times \sum_{k_1<k_2<\cdots<k_m} P(A_{k_1} \cap A_{k_2} \cap \cdots \cap A_{k_m}) \\ + \cdots + (-1)^{n+1} P\left(\bigcap_{k=1}^{n} A_k\right) \end{aligned} \tag{1.5c}$$

of which eq. (1.5b) is just the special case $n = 3$. Also note that a similar relation applies for the intersection of n events, that is,

$$\begin{aligned} P\left(\bigcap_{k=1}^{n} A_k\right) = \sum_{k=1}^{n} P(A_k) - \sum_{k_1<k_2} P(A_{k_1} \cup A_{k_2}) + \cdots + (-1)^{m+1} \\ \times \sum_{k_1<k_2<\cdots<k_m} P(A_{k_1} \cup A_{k_2} \cup \cdots \cup A_{k_m}) \\ + \cdots + (-1)^{n+1} P\left(\bigcup_{k=1}^{n} A_k\right) \end{aligned} \tag{1.5d}$$

which, for instance, in the case $n = 3$ (we call them again events B, C, D) becomes

$$P(B \cap C \cap D) = P(B) + P(C) + P(D) - P(B \cup C) \\ - P(B \cup D) - P(C \cup D) + P(B \cup C \cup D) \tag{1.5e}$$

Now, the careful reader has probably noticed that so far we have not yet expressed the notion of mutually exclusive events in set language. This is done by writing that two events B and C are mutually exclusive if $B \cap C = \emptyset$, where $\emptyset$ is the empty set. In order to deal with this minor complication and make complete sense of the equations above, we need to include the empty set in the sample space and require

$$P(\emptyset) = 0 \tag{1.6}$$

In probability terminology, $\emptyset$ is called the *impossible event.*

Proceeding in our discussion, let us now introduce two other definitions of practical importance – namely conditional probability and independent events – together with the properties that follow from these definitions.

Intuitively, we can argue that the probability of an event can vary depending upon the occurrence or non-occurrence of one or more related events: in fact, for instance, in the die-rolling experiment it is different to ask 'what is the probability of a 6?' or 'what is the probability of a 6 given that an even number has fallen?'. The answer to the first question is 1/6 while the answer to the second question is 1/3. This is the concept of *conditional probability*, that is, the probability of an event A given that an event B has already occurred. The symbol for conditional probability is $P(A|B)$ and its definition is

$$P(A|B) = \frac{P(A \cap B)}{P(B)} \tag{1.7}$$

provided that $P(B) \neq 0$. (As a side note on $P(A \cap B)$, the reader is invited to verify that $P(A \cap B) \geq P(A) + P(B) - 1$, which is known as Bonferroni's inequality.)

Equation (1.7) yields immediately the multiplication rule for probabilities, that is,

$$P(A \cap B) = P(B)P(A|B) = P(A)P(B|A) \tag{1.8a}$$

which can be generalized to a number of events $A_1, A_2, \ldots, A_n$ as follows

$$P\left(\bigcap_{j=1}^{n} A_j\right) = P(A_1)P(A_2|A_1)P(A_3|A_1 \cap A_2)\cdots P\left(A_n|\bigcap_{j=1}^{n-1} A_j\right) \tag{1.8b}$$

If the occurrence of event B has no effect on the probability of event A, then A and B are said to be *independent* and we can express this fact in terms of conditional probability as

$$P(A|B) = P(A) \tag{1.9a}$$

or, equivalently, since we expect symmetry (if A is independent of B then B is independent of A)

$$P(B|A) = P(B) \tag{1.9b}$$

Clearly, two mutually exclusive events are not independent because, from eqs (1.7) and (1.6), we have $P(A|B) = 0$ when $A \cap B = \emptyset$. Also, if A and B are two independent events, we get from eqs (1.7) and (1.9)

$$P(A \cap B) = P(A)P(B) \tag{1.10a}$$

which is referred to as the *multiplication rule* for independent events and can be assumed as the definition of independent events. For n mutually (or collectively) independent events $A_1, A_2, \ldots, A_n$ one would expect the relation

$$P\left(\bigcap_{j=1}^{n} A_j\right) = P(A_1)P(A_2)\cdots P(A_n) = \prod_{j=1}^{n} P(A_j) \tag{1.10b}$$

but it will be shown (Section 2.2.2) that eq. (1.10b) is not enough to define collective independence. For the moment, we simply point out that three (or more) random events can be independent in pairs without being mutually independent. This is illustrated by the following example.

Example 1.1 Consider a lottery with 8 numbers (from 1 to 8) and let $E_1, E_2, \ldots, E_8$, respectively, be the simple events of extraction of 1, extraction of 2, etc. Let

$$A_1 = E_1 \cup E_2 \cup E_3 \cup E_4$$
$$A_2 = E_3 \cup E_4 \cup E_5 \cup E_8$$
$$A_3 = E_1 \cup E_2 \cup E_3 \cup E_5 \cup E_6 \cup E_8$$

Now, $P(A_1) = P(A_2) = 1/2$ and $P(A_3) = 3/4$. It is then easy to verify that $P(A_1 \cap A_2) = 1/4 = P(A_1)P(A_2), P(A_2 \cap A_3) = 3/8 = P(A_2)P(A_3)$ and $P(A_3 \cap A_1) = 3/8 = P(A_3)P(A_1)$, which means that the events are pairwise independent. However, $P(A_1 \cap A_2 \cap A_3) = 1/8 \neq P(A_1)P(A_2)P(A_3) = 3/16$ meaning that the three events are not mutually, or collectively, independent.

Another important result is known as the total probability formula. Let $A_1, A_2, \ldots, A_n$ be n mutually exclusive events such that $\cup_{j=1}^{n} A_j = W$, where W is the sample space (in set language this concept is expressed by saying that the sets A_j form a finite partition of W; see Appendix A, Section A.1). This means that exactly one of the A_j will occur. Then, a generic event $B \subset W$ can be expressed as

$$B = \bigcup_{j=1}^{n} (B \cap A_j) \tag{1.11}$$

where the n events $(B \cap A_j)$ are mutually exclusive because so are the A_j. Owing to eq. (1.5c) we get

$$P(B) = P\left(\bigcup_{j=1}^{n} (B \cap A_j)\right) = \sum_{j=1}^{n} P(B \cap A_j)$$

so that using eq. (1.8a), we obtain the *total probability formula*

$$P(B) = \sum_{j=1}^{n} P(A_j) P(B|A_j) \tag{1.12}$$

which can be interpreted by saying that $P(B)$ is a weighted average of the conditional probabilities $P(B|A_j)$, each term being weighted by the probability of the event on which it is conditioned. Also, due to its importance and in view of future developments, we anticipate here that eq. (1.12) remains true for $n \to \infty$, that is, when the sets $A_1, A_2, \ldots$ form a countable partition of W.

With the same assumptions as above on the events $A_j (j = 1, 2, \ldots, n)$, let us now consider a particular event A_k; the definition of conditional probability yields

$$P(A_k|B) = \frac{P(A_k \cap B)}{P(B)} = \frac{P(A_k \cap B)}{\sum_{j=1}^{n} P(A_j) P(B|A_j)} \tag{1.13}$$

where eq. (1.12) has been taken into account. By virtue of eq. (1.8a) we can write $P(A_k \cap B) = P(A_k) P(B|A_k)$ so that substitution in eq. (1.13) yields

$$P(A_k|B) = \frac{P(A_k) P(B|A_k)}{\sum_{j=1}^{n} P(A_j) P(B|A_j)} \tag{1.14}$$

which is known as *Bayes' formula* and deserves some comments.

First, as for the total probability formula, eq. (1.14) is true if $n \to \infty$. Second, eq. (1.14) is particularly useful for experiments consisting of stages. Typically, the A_js are events defined in terms of a first stage (or, otherwise, the $P(A_j)$ are known for some reason) while B is an event defined in terms of the whole experiment including a second stage; asking for $P(A_k|B)$ is then, in a sense, 'backward', we ask for the probability of an event defined at the first stage conditioned on what happens in a later stage. In Bayes' formula this probability is given in terms of the 'natural' conditioning, that is, conditioning on what happens at the first stage of the experiment. This is why the $P(A_j)$ are called the '*a priori*' (or prior) probabilities whereas $P(A_k|B)$ is called '*a posteriori*' (posterior or inverse) probability. The advantage of this approach is to be able to modify the original predictions by incorporating new data. Obviously, the initial hypotheses play an important role in this case, if the initial assumptions are based on insufficient knowledge of the process, the prior probabilities are no better than reasonable guesses.
Two examples will help clarify the use of Bayes' formula.

Example 1.2 Among voters in a certain area, 40% support party 1 and 60% support party 2. Additional research indicates that a certain election issue is favoured by 30% of supporters of party 1 and by 70% of supporters of party 2. One person at random from that area – when asked – says that he favours the issue in question. What is the probability that he/she is a supporter of party 2? Now, let

- A_1 be the event that a person supports party 1, so that $P(A_1) = 0.4$;
- A_2 be the event that a person supports party 2, so that $P(A_2) = 0.6$;
- B be the event that a person at random in the area favours the issue in question.

Prior knowledge (the results of the research) indicate that $P(B|A_1) = 0.3$ and $P(B|A_2) = 0.7$. The problem asks for the '*a posteriori*' probability $P(A_2|B)$, that is, the probability that the person who was asked supports party 2 given the fact that he/she favours that specific election issue. From Bayes' formula we get

$$P(A_2|B) = \frac{P(A_2)P(B|A_2)}{P(A_1)P(B|A_1) + P(A_2)P(B|A_2)} = 0.778$$

Then, obviously, we can also infer that $P(A_1|B) = 1 - P(A_2|B) = 0.222$.

Example 1.3 In order to detect a particular disease, assume that there is a medical test which is 98% accurate, that is, if someone has the disease the test will be positive 98% of the time and, conversely, if someone has not the disease the test will be negative 98% of the time. Also, assume that medical research has shown that – at any given time in a certain area – 0.5% of the population has the disease. Now, you live in that area, imagine that you take the test and the test comes out positive. The question is: what is the probability that you have the disease given the fact that you tested positive? or, in less mathematical terms: how worried should you be? Surprisingly, the answer is that you should be cautiously optimistic.

In fact, let A_1 be the event that you have the disease and A_2 be the event that you do not have the disease. This implies that the prior probabilities are $P(A_1) = 0.005$ and $P(A_2) = 0.995$. In addition to this, we must consider the conditional probabilities on the accuracy of the test, namely

(a) $P(p|A_1) = 0.98$, the probability of testing positive given that you actually have the disease (the lowercase letter p is for positive test), and
(b) $P(p|A_2) = 0.02$, the probability of testing positive given that you do not have the disease.

Then, according to Bayes' formula (1.14), the probability of having the disease given a positive result of the test is

$$
\begin{aligned}
P(A_1|p) &= \frac{P(A_1)P(p|A_1)}{P(A_1)P(p|A_1) + P(A_2)P(p|A_2)} \\
&= \frac{(0.005)(0.98)}{(0.005)(0.98) + (0.995)(0.02)} = 0.198 \cong 20\%
\end{aligned}
$$

The result is less surprising if we think that out of n administered tests we will obtain $0.0248n$ positive results (the denominator of Bayes' formula) which, for the most part ($0.0199n$) are false positives. Then, since only $0.0049n$ are real positive tests (the numerator of Bayes' formula), the probability we are looking for is precisely $0.0049/0.0248 = 0.198$.

1.2.2 *More on the classical definition: a short digression on combinatorials*

With the classical definition of probability in mind, one observation of practical nature is in order. Definition (1.1) requires counting, that is, in order to determine the probability of an event A we need to enumerate the favorable outcomes which contribute to A and the possible outcomes of the experiment. Sometimes the counting may be easy, but for large sample spaces it may not be an easy task. In fact, suppose we are given the following problem: if we choose three cards at random from a deck of 52 cards, what is the probability of extracting at least one ace? Problems such as this one can be answered perfectly well with definition (1.1), but it is evident that the counting becomes soon impracticable. Before turning to this problem we need some definitions.

From combinatorial analysis we know that the number of *permutations* (i.e. ordered arrangements) of n distinct objects taken r at a time ($r \le n$) is given by

$$P_{n,r} = n(n-1)\cdots(n-r+1) = \frac{n!}{(n-r)!} \tag{1.15}$$

(We recall that $n! \equiv n(n-1)(n-2)\cdots 1$ so that, for example, $3! = 6, 4! = 24$, $5! = 120$ etc. Also, by definition, $0! = 1$). Clearly, the number of permutations of n distinct objects taken n at a time is simply the number of ordered arrangements of the n objects and is $P_{n,n} = n!$. If, however, the n objects are not all distinct but there are, say, k_1 objects of a type, k_2 objects of a second type different from the first, ... and k_m of the mth type different from the first, second,..., $(m-1)$th type, we surely do not expect to have $n!$ permutations because some of these $n!$ possibilities will not be distinguishable. It is not difficult to determine that in this case we have

$$P_n(k_1, k_2, \ldots, k_m) = \frac{n!}{k_1!k_2!\cdots k_m!} \tag{1.16}$$

permutations (clearly $k_1 + k_2 + \cdots + k_m = n$). A simple example will clarify this assertion. Suppose we have the three letters a, b, b; this means that out of these letters $k_1 = 1$ are of one type (the letter a) and $k_2 = 2$ are of a second type (the two b's, which are indistinguishable). Then we only have three permutations, that is, the arrangements $\{a, b, b\}$, $\{b, b, a\}$ and $\{b, a, b\}$, in agreement with eq. (1.16) which yields $3!/(2!1!) = 3$.

When the order of the objects is not important we speak of combinations. Specifically, the *combinations* of n distinct objects taken r at a time ($r \le n$) is

$$C_{n,r} = \binom{n}{r} \equiv \frac{n!}{r!(n-r)!} = \frac{P_{n,r}}{r!} \tag{1.17}$$

where we have $C_{n,r} \le P_{n,r}$ because order is now irrelevant. For example, if $n = 3$ (objects a, b and c) and $r = 2$, the fact that the number of combination is less than the number of permutations is evident if one thinks that in a permutation the arrangement $\{a, b\}$ is considered different from the arrangement $\{b, a\}$, whereas in a combination they count as one single arrangement.

Incidentally, it should be noted that the calculations of factorials can be often made easier by using Stirling's formula, that is, $n! \cong n^n e^{-n}\sqrt{2\pi n}$, which results in relative errors smaller that 1% for $n \ge 10$ (note, however, that absolute errors increase as n increases).

Now, returning to the card problem, we can consider the event of extracting at least one ace – event A – as the sum of three mutually exclusive events:

- event A_1 = extraction of one ace;
- event A_2 = extraction of two aces;
- event A_3 = extraction of three aces.

and we have, according to property (1.3a)

$$P(A) = P(A_1) + P(A_2) + P(A_3)$$

The probability $P(A_1)$ can be calculated as follows: the possible combinations of three cards out of a 52 card deck is $C_{52,3}$. One ace can be extracted in $C_{4,1}$ different ways and other two cards (which are not aces) can be chosen in $C_{48,2}$ different ways. In other words, we have $(C_{4,1})(C_{48,2})$ favourable cases out of a total of $C_{52,3}$ cases. Similarly, in obtaining $P(A_2)$ we have $(C_{4,2})(C_{48,1})$ favourable cases out of $C_{52,3}$ cases and in calculating $P(A_3)$ we have $(C_{4,3})(C_{48,0})$ favourable cases out of $C_{52,3}$ cases. Therefore

$$\begin{aligned} P(A) &= \frac{(C_{4,1})(C_{48,2})}{C_{52,3}} + \frac{(C_{4,2})(C_{48,1})}{C_{52,3}} + \frac{(C_{4,3})(C_{48,0})}{C_{52,3}} \\ &= 0.20416 + 0.01303 + 0.00018 \cong 0.217 \end{aligned}$$

Finally, let us see what happens when repetitions are allowed. If we denote by $\hat{P}_{n,r}$ the permutations with repetitions of n objects taken r at a time (note that now it can be $r > n$) it is easy to determine that

$$\hat{P}_{n,r} = n^r \tag{1.18}$$

because all of the r objects to be taken can be chosen in n ways. As an example we can determine how many two digits numbers we can form with the three numbers 1, 2 and 3. Equation (1.18) yields $\hat{P}_{3,2} = 3^2 = 9$; in fact, in addition to the six arrangements of $P_{3,2}$ we have the three arrangements $\{1,1\}$, $\{2,2\}$ and $\{3,3\}$.

If now we denote by $\hat{C}_{n,r}$ the combinations with repetitions of n objects taken r at a time (and here also it can be $r > n$) we have

$$\hat{C}_{n,r} = \binom{n+r-1}{r} = \frac{n(n+1)\cdots(n+r-1)}{r!} = C_{n+r-1,r} \tag{1.19}$$

because two arrangements are now considered distinct if (i) they differ by at least one object or (ii) they differ by the number of times that a given object appears in the arrangement. So, for example, with the three numbers

1, 2 and 3 we have, according to eq. (1.19), $\hat{C}_{3,2} = 12/2! = 6$. These are precisely the arrangements $\{1,1\}$, $\{1,2\}$, $\{1,3\}$, $\{2,2\}$, $\{2,3\}$, $\{3,3\}$ (and the arrangements $\{2,1\}$, $\{3,1\}$ and $\{3,2\}$ do not appear because – being order irrelevant in a combination – they are the same as $\{1,2\}$, $\{1,3\}$ and $\{2,3\}$).

1.3 The relative frequency approach to probability

The classical definition (1.1) is frequently used and works well in many circumstances, but has its limitations as well. First of all, even when the basic assumptions of mutually exclusive and equally likely sample points is valid, what do we do if the number of possible outcomes is infinite? For example, we may ask: what is the probability that an integer drawn at random from the set of all positive integers is even? The answer that comes natural is 1/2 because, intuitively, in the set of all positive integers there are just as many even numbers as odd numbers and it could also be argued that any sufficiently large set of the first N positive integers contains $N/2$ even numbers so that the ratio (1.1) 'tends' to $1/2$ as $N \to \infty$. This sounds reasonable, but it seems to depend on the natural ordering of integers, for example, if we order the integers as 1, 3, 5, 2, 7, 9, 11, 4, ... (i.e. three odd numbers followed by an even number) the 'limiting argument' would lead us to believe that the probability of drawing an even number is 1/4. Also, it is possible to order the positive integers in such a way that the ratio (1.1) keeps oscillating and never approaches a definite number as N increases indefinitely. Indeed, some researchers in the past had tried to extend the classical definition of probability to deal with an infinite number of events. Their approach led to the concept of geometrical probability in which, typically, the probability is calculated as the ratio of the measures (lengths, areas, etc.) of two regions of space. This definition, however – although useful in many cases – ran into some paradoxical results and was criticized to the point that some authors were convinced that it was not possible to determine objectively – that is, in a way independent on the method used to calculate it – a value of probability in the case of an infinite number of outcomes (in the light of Kolmogorov's approach we will see that, in essence, the problem lies in the fact that not all 'regions of space' – intended as subsets of the real line $\mathbb{R}$, the plane $\mathbb{R}^2$, etc. – are 'measurable'). The following 'meeting problem' is a typical (and not paradoxical) application of geometric probability. Two persons A and B decide to meet at a given place between 2 and 3 p.m. The first to arrive must wait 20 min and then leave. If the two arrival times t_A and t_B are independent and random (between 2 p.m. and 3 p.m.) what is the probability that they actually meet? We briefly sketch the solution inviting the reader to work out the details. The meeting takes place if $|t_A - t_B| \le 20$; if we consider t_A and t_B as the x–y coordinates in a plane, this condition delimitates an area S_1 within a square of total area $S_2 = 60^2$ (we assume the minute as our basic unit). Then, the meeting probability is determined by the ratio S_1/S_2 and yields $S_1/S_2 = 5/9$.

Another serious drawback of the classical definition arises when its basic assumptions are no longer valid; for example, we cannot say anything about the probability of a head in the case of an unbalanced coin or the probability of a six with a biased die; the mutually exclusive events are still there but they are definitely not equally likely. In addition to this, we have no answer to common questions such as: what is the probability of more than five defective pieces out of a lot of 1000 pieces manufactured by such and such company? or, what is the probability that an Italian male will live longer than the age of 80? Intuition suggests that there must be an answer to questions like these but the point is here that in many interesting cases it may not even be possible to determine the set of mutually exclusive and equally likely outcomes.

The consequence of these observations is that we must somehow extend – or change altogether – our definition of probability if we want a number of interesting problems to fit into the theory. In this regard, in fact, it is known that as early as the end of the seventeenth century insurance companies were faced with the problem of determining the probability of death of their clients according to age, gender, health condition, etc. Their solution was to start keeping records, year after year, of the number of people in different age groups and on the mortality rate in each group. Then, dividing this latter quantity by the former, they were able to obtain a reliable estimate of the probability they were looking for.

This procedure is nothing but an early example of the *relative frequency approach* to probability, in which the probability of interest is obtained as a ratio between two numbers: n_A – the number of times an event A occurs – and n, that is the total number of cases at our disposal. In formula we write

$$\widetilde{P}(A) = \frac{n_A}{n} \tag{1.20}$$

and we call $\widetilde{P}(A)$ the *statistical probability* of occurrence of the event A. All readers are probably well aware of the fact that this is a common way to calculate probabilities in natural sciences and technological problems.

In this form, the relative frequency approach applies to cases in which we have a 'population' or a 'sample' (these terms will become clearer in future chapters) of n elements and we observe the occurrence of the event A in n_A elements out of the n at our disposal. Equivalently, we can say that eq. (1.20) pertains to cases in which an 'experiment' can be repeated many times under a given set of conditions and we observe the result.

At this point, a few remarks are in order. First of all, it is evident that eq. (1.20) defines an *a posteriori* probability because we must actually perform the experiment in order to calculate the value of $\widetilde{P}(A)$. Second, this calculated quantity $\widetilde{P}(A)$, in turn, is generally considered a reliable approximation of the 'whole picture' – that is, the experiment of interest – on the basis of the fact that past and present experience show that – for a given experiment – $\widetilde{P}(A)$ is almost constant for sufficiently large values of n. In fact, for example, if we have the patience to roll a balanced die over and over again (1000, 2000, 5000 times or more) we will obtain a value of, say, $\widetilde{P}(six)$ close to 1/6, with smaller and smaller fluctuations as n increases.

Now, the facts that definition (1.20) can be calculated also in a number of cases in which eq. (1.1) applies and that experimental evidence shows that $\widetilde{P}(A) \cong P(A)$ for sufficiently large 'samples' has led researchers to postulate that – even in cases in which (1.1) does not apply – there exists an entity $P(A)$ called the 'probability of event A' with the following properties

(i) it does not depend on the observer,
(ii) it can be approximated with a higher and higher degree accuracy by eq. (1.20) as we increase the number n at the denominator.

This last property deserves some comments. However, before doing this, let us return for a moment to definition (1.20) itself. It is clear that this definition overcomes some of the limitations of the classical definition because, for instance, it is now no longer necessary for the possible outcomes of the experiment to be equally likely. So, for example, we can apply definition (1.20) to determine the probability of a head in the case of an unbalanced coin; after a sufficiently large number of trials we would find that the ratio on the r.h.s. of eq. (1.20) is significantly different from 1/2. Conversely, a value of the ratio on r.h.s. significantly different from 1/2 after a large number of trials would lead us to suspect that the coin is unbalanced. In addition, with a sufficiently large sample at our disposal, we can now answer the question on the probability of an Italian male to live longer than eighty or the probability of a defective item out of a production lot.

Let us now return to property (ii) above and note that the steps which have brought us there are, in essence:

(a) the observation of long-term regularities (which have been known for centuries) on the value of the ratio (1.20) when n becomes large (in other words, this ratio 'tends' to become constant as n increases);
(b) the fact that, in cases when both definitions (1.1) and (1.20) apply, the statistical probability gets closer and closer to the classical probability as n increases.

These two facts support our intuitive beliefs that; first, there exists a definite value of probability even for events for which the classical definition does not apply and, second, that this (unknown) value can be approximated reasonably well by the ratio (1.20) after a large number of trials.

At this point, anyone familiar with elementary calculus would be tempted to define the probability of an event A – once postulated that this quantity exists – as

$$P(A) = \lim_{n\to\infty} \widetilde{P}(A) = \lim_{n\to\infty} \frac{n_A}{n} \tag{1.21}$$

which is called the *Von Mises' definition* of probability (his point of view was that any '*a priori*' definition was meaningless and that only an empirical definition could be useful in the field of natural sciences). However, if we recall the notion of convergence which, in strict mathematical terms, reads: the sequence of real numbers $\{x_n\}_{n=1}^{\infty}$ is said to converge to the real number x if $\forall \varepsilon > 0 \; \exists \; N \in \mathbb{N} : |x_n - x| < \varepsilon \; \forall n \geq N$ (or, in words: a sequence of real numbers x_n converges to a limit x whenever, for sufficiently large values of n, the quantity $|x_n - x|$ becomes smaller than an arbitrarily chosen positive number ε), we note that the limiting statement of eq. (1.21) is quite strong.

All we can say for the moment is that we probably need a different notion of convergence which, in the light of the above considerations, will be weaker than the usual definition of basic analysis.

We delay these mathematically issues to later chapters and close this section with two final remarks. First, in Section 1.3.1 we have obtained a number of properties satisfied by a 'probability function' (eq. (1.3a) onwards) in the light of the classical definition; however, these same properties can be obtained by starting from the relative frequency definition. In fact, after noting that the concept of mutually exclusive outcomes makes perfect sense even in cases when the classical definition does not apply, let, for example, A and B be two mutually exclusive outcomes of a given experiment which is repeated n times. Then – out of the n repetitions – if event A has occurred n_A times and event B has occurred n_B times, the event $A \cup B$ has occurred $n_A + n_B$ times. Therefore, from eq. (1.20) we get

$$\widetilde{P}(A \cup B) = \frac{n_A + n_B}{n} = \widetilde{P}(A) + \widetilde{P}(B)$$

which is the additivity property for exclusive events. As another example, consider n repetitions of an experiment where, given an event B with non-zero probability, another event A will occur only if $A \cap B$ occurs. Then, the relative frequency of occurrence of event A among those times in which B has occurred is, by definition, the conditional probability $\widetilde{P}(A|B)$. This can be obtained from (1.20) as

$$P(A|B) = \frac{n_{A \cap B}}{n_B} = \frac{n_{A \cap B}/n}{n_B/n} = \frac{\widetilde{P}(A \cap B)}{\widetilde{P}(B)}$$

which is just the relative frequency counterpart of eq. (1.7). Clearly, the same applies to all other properties given in Section 1.2.1.

The second remark considers the main limitation of this approach to probability, namely the requirement of a sufficiently large sample. In many real world situations – for a number of reasons varying from the nature of the problem under study to cost of the experiment, etc. – it is simply not possible to fulfill this request. In addition, one is often faced with cases in which it is not even feasible to repeat the experiment a second time. Think, for example, to the publisher of this book who may ask the perfectly reasonable question: what is the probability of selling more than, say, 500 copies within the first year? He/she is not at all interested in repeating the experiment many times – which would be impossible anyway – but only in the result of the first, and unique, trial.

It is in response to questions like these that the subjective approach was introduced.

1.4 The subjective viewpoint

If one takes its name too literally, the *subjective* (or personal) *approach* could be quickly dismissed by saying that, since it reflects a personal opinion, it is always applicable because anyone can have a personal opinion about anything. As it often happens, however, things are not so clear-cut. This approach arose in the light of the practical difficulties – and the inapplicability altogether in many cases of interest – encountered by the classical and relative frequency approaches. An example has been given above, but it is not difficult to find many others. For instance, after an oil spill an expert could be asked about the probability to contain the spill before it causes widespread damage. Clearly, the answer can only be an informed personal opinion because the factors into play – amount of oil spilled, sea and wind conditions during clean-up, etc. – make this spill unique. So, according to the subjective point of view, the probability of an event may vary from person to person depending not only on the available information but also on the importance given to this information. As a consequence, even two people with the same amount of information may assign different probabilities to a given event.

On a more practical side, those who favour this point of view also add that in order to determine a value of probability we must somehow force the interested person to take action – for instance, by betting a sum of money – thus inferring his/her degree of belief by his/her actions. Without getting into much detail, we may note that, in all, there exist a number of reasonable logical justifications at the basis of this approach; however, there are also two main serious drawbacks. First, in order to arrive at meaningful results, one of the basic assumptions of the scientific community is that the probability of an event is an 'existing entity' which must not depend on the person who determines it. Second, a number of studies have shown that personal intuition and judgment on specific probabilistic problems – compared to exact statistical calculations when these are possible – often lead to wrong answers, either because in complex situations reasonableness is sometimes misleading or because of some form of bias which, consciously or unconsciously, may have a significant influence on our judgment. Despite these disadvantages, however, one point should be made: even when we adopt a subjective approach, we necessarily imply, implicitly or explicitly, some basic rules that our probability must obey. A little thought on the problem shows that these rules, once organized and cast in mathematical form – are the same rules given in Section 1.2.1. Moreover, since we drew a similar conclusion in Section 1.3, it seems that the rules and properties of a 'probability function' are more important than the way in which we assign probabilities to events. This final remark prepares the way to Kolmogorov's axiomatic approach which is at the basis of the modern mathematical theory of probability.

1.5 Summary

Although the term 'probability' is frequently used by each one of us in everyday conversations, the idea of probability is not so obvious as it may seem. In fact, the basic concept has evolved through the centuries – starting officially from the seventeenth century, but, unofficially, much longer before this – leading to three main viewpoints. Following the common terminology, we called these, the classical approach, the relative frequency approach and the subjective approach to probability. Historically, the *classical definition* (Section 1.2) was the first to be given, originating mainly from the typical scheme of games of chance (the tossing of a coin, the rolling of one or more dice, the roulette, etc.) in which the gambler bets on the occurrence of a particular event chosen among a finite number of possible outcomes. The basic assumption is that it must be possible to identify a finite set of 'simple events' which are mutually exclusive and equally likely. Only in this way, by means of eq. (1.1), one can obtain an '*a priori*' value for the probability of the event of interest - where the term '*a priori*' refers to the fact that this probability, if the game is fair, can be known beforehand without even carrying out the experiment (i.e. tossing a coin, rolling a die, etc.). On the basis of this definition – perhaps the most intuitive for all of us – and with the help of basic mathematical notions borrowed from set theory, Section 1.2.1 is devoted to an informal discussion of some fundamental properties which we expect from a 'probability function', also introducing some important definitions such as conditional probability and independent events. In addition, a short digression on combinatorial analysis (Section 1.2.2) is given as an aid in the calculation of classical probabilities when the sample space is large, that is, when counting both the number of favorable cases and the total number of possibilities is not an easy task.

The classical approach to probability applies to a variety of cases but leaves many interesting questions unanswered. For example, what do we do if the number of possible outcomes is not finite? or, what is the probability of a head with an unbalanced coin? or else, what is the probability that my uncle will live longer than 80 given the facts that he is a male, lives in a certain state, is 68 and in good health conditions? This and similar problems, to which we all believe there is an answer, do not fit in the classical scheme because its basic assumption is no longer valid. We must then introduce (Section 1.3) the concept of an '*a posteriori*' probability, that is, a probability calculated on the basis of the results of an experiment. This leads to the *relative frequency definition* of probability in which an experiment is repeated many times – or, stated differently, we collect a large sample – and we record the results counting the number of times that a given outcome has actually occurred. The rationale behind this approach, once postulated that a definite value probability exists for these kinds of problems, lies in the long-term regularities of the relative frequency for a large number of repetitions. These regularities have been known for centuries and lead to the conclusion that, broadly speaking, there is some sort of 'tendency' of the calculated relative frequency to the actual probability of the event of interest. The term 'tendency', in turn, implies a limit – von Mises' hypothesis - but not necessarily the usual limit of sequences of elementary calculus that von Mises had considered. An important point, however, is that the expected properties of probability given in Section 1.2.1 can be re-obtained on the basis of this strictly experimentally-based definition of probability.

One main problem with the relative frequency approach is the requirement of a large sample because in real world problems, more often than not, we will not be able to collect a large sample. Moreover, in some cases we will not even be able to repeat the experiment a second time.

A tentative answer to these problems is given by the concept of *subjective probability*, a value assigned to an event by individuals interested in this event on the basis of the risk they are willing to accept by taking action in favor of its occurrence. Clearly, this value reflects the belief of the individual, his/her judgement being based, in general, on a number of factors: past experience in similar situations, knowledge of the problem and, last but not least, personal 'gut-feelings'. This approach to probability has the advantage of being practically applicable to any problem but its main drawback is the lack of objectivity. Therefore, since we generally assume that the probability of an event should not depend on the person who determines it, this is not a widely accepted definition among the scientific community. Nonetheless, even probabilities assigned on a subjective basis must have some properties, and a little thought on the problem shows that, basically, they should be the same as the ones given in Section 1.2.1.

On the basis of these considerations the conclusion of this chapter is that, from a mathematical viewpoint, the properties of a 'probability function' seem to be more important than the way in which we assign probabilities to this or that event. It is precisely this line of thinking that leads to the formal axiomatic approach (due to the Russian mathematician Kolmogorov) considered in the following chapters.

References and further reading

[1] Costantini, D., '*La Probabilità e le Sue Regole*', Proceeding from the Seminar 'Metodologie Statistiche per il Trattamento delle Misure', CISM, Udine (Italy) 9–11 Oct. (1991).

[2] Gnedenko, B.V., '*Teoria della Probabilità*', Editori Riuniti, Roma (1987) (also available in English, '*The Theory of Probability*', 4th edn., Chelsea, New York, 1968).
[3] Kolmogorov, A.N., '*Foundations of the Theory of Probability*', reprinted by the American Mathematical Society, Providence, Rhode Island (2000).
[4] Milton, J.S., Arnold, J.C., '*Introduction to Probability and Statistics, Principles and Applications for Engineering and the Computing Sciences*', 2nd edn., McGraw-Hill, New York (1990).
[5] Ross, S.M., '*Introduction to Probability Models*', 7th edn., Harcourt Academic Press, San Diego (2000).
[6] Solnes, J., '*Stochastic Processes and Random Vibrations, Theory and Practice*', John Wiley & Sons, Chichester (1997).
[7] Ventsel, E.S., '*Teoria delle Probabilità*', Edizioni Mir, Mosca (Russia) (1983).

Lighter reading

[8] Paulos, J.A., '*Innumeracy, Mathematical Illiteracy and its Consequences*', Vintage Books, New York (1988).
[9] Weaver, W., '*Lady Luck, the Theory of Probability*', Anchor Books, Doubleday & Company, Inc., New York (1963).

2 Probability: the axiomatic approach

2.1 Introduction

The key point of the previous chapter is that the properties of a 'probability function' seem to play a more important role than the definition of probability itself. More specifically, from a mathematical point of view we are led to the idea that the way in which we assign probabilities to events is almost secondary with respect to the fact that these probabilities must satisfy a number of well-defined properties. Therefore, by temporarily ignoring the way in which we assign probabilities to events – but, at the same time, by defining what exactly is meant by 'event' – we can simply define a 'probability function' as something that satisfies a given set of rules. By so doing, each one of the definitions given in Chapter 1, the classical definition, the relative frequency definition, etc., turns out to be just a special case of 'probability function' which works perfectly well in the appropriate context. So, in fair games of chance (dice, roulette, lotteries, etc.) we adopt the classical probability, in repeated experiments where the classical definition cannot be used (mortality rates, measurement of a physical quantity, etc.), we turn to the relative frequency definition of probability, etc.

This, in essence, is the idea at the basis of the axiomatic approach to probability due to Kolmogorov: events are special subsets of a 'universal' set and a probability is a non-negative, real-valued set function (see Appendix A) which 'measures' events. We intentionally use the word 'measure' here because the developments of this axiomatic approach parallel closely many aspects of the branch of Mathematics known as 'theory of measure and integration' in which an important role is played by the so-called Lebesgue measure and Lebesgue integral. For a detailed discussion of this subject the reader is referred, for example, to [4, 6–8, 10] or, for a more probabilistic oriented treatment to [1, 2, 9, 12].

2.2 Probability spaces

In order to make more mathematically precise the concepts of Chapter 1 and extend them to a broader class of problems, our first step is to introduce the definition of elementary probability space.

Definition 2.1 We call *elementary probability space* a triplet (W, R, P) where W is a set, R is an algebra of subsets of W and P is a set function defined on R with values in the real interval $[0, 1]$ – that is, $P : R \to [0, 1]$ – which satisfies the following relations (called 'elementary probability axioms'):

(EP1) $P(W) = 1$

(EP2) If $A_1, A_2, \ldots, A_n \in R$ and $A_i \cap A_j = \emptyset$ for $i \neq j$ then

$$P\left(\bigcup_{k=1}^{n} A_k\right) = \sum_{k=1}^{n} P(A_k) \tag{2.1}$$

The members of the algebra R (the definition of algebra of sets is given in Appendix A) are called *events* and axiom (EP2), in words, is phrased by saying that the probability function P is *finitely additive.* For this reason elementary probability spaces are sometimes called finitely-addititive probability spaces.

With Definition 2.1 we are able to deal with all the situations which involve the probabilities of only a finite number of events and it is important to note that all the properties given in Chapter 1 (Section 1.2.1) descend from (EP1) and (EP2). So, for instance, it is easy to show that

(a) $P(\emptyset) = 0$;
(b) $P(A^C) = 1 - P(A)$ for every $A \in R$.

Less immediate are the properties

(c) $P(A \cup B) = P(A) + P(B) - P(A \cap B)$ for all $A, B \in R$;
(d) *monotonicity*: $A, B \in R$ and $B \subset A$ implies $P(B) \leq P(A)$;
(e) *finite subadditivity*: if $A_1, A_2, \ldots, A_n \in R$, then

$$P\left(\bigcup_{k=1}^{n} A_k\right) \leq \sum_{k=1}^{n} P(A_k) \tag{2.2}$$

The proofs of (c) and (d) are left to the reader (hint for (c): $A = (A \cap B) \cup (A - B)$; hint for (d): $A = B \cup (A - B)$) while (e) can be proven by writing

$$\bigcup_{k=1}^{n} A_k = A_1 \cup \left(A_2 \cap A_1^C\right) \cup \left(A_3 \cap A_2^C \cap A_1^C\right) \\ \times \cup \cdots \cup \left(A_n \cap A_{n-1}^C \cap \cdots \cap A_1^C\right) \tag{2.3}$$

and then noting that the sets on the r.h.s. of eq. (2.3) are disjoint (i.e. their probabilities add according to eq. (2.1)). Property (e) then follows from the inequalities $P(A_2 \cap A_1^C) \leq P(A_2), P(A_3 \cap A_2^C \cap A_1^C) \leq P(A_3)$, etc.

Also, if the events $A, B \in R$ and $P(B) > 0$ the conditional probability $P(A|B)$ of event A given B is defined by eq. (1.7), with the consequence that all the considerations of Section 1.2.1 (specifically, the total probability formula and Bayes' rule) still apply. However, since the introduction of conditional probability (with respect to an event B) implies the definition of the set function $P_B(A) = P(A|B)$ for every $A \in R$, the question arises if, mathematically speaking, the triplet (W, R, P_B) is itself a probability space. The answer is yes because properties (EP1) and (EP2) are satisfied. In fact. $P_B(W) = P(W \cap B)/P(B) = P(B)/P(B) = 1$ and if $A_1, A_2, \dots, A_n$ are mutually disjoint sets of R then

$$P_B\left(\bigcup_k A_k\right) = \frac{1}{P(B)} P\left[\left(\bigcup_k A_k\right) \cap B\right] = \frac{1}{P(B)} P\left[\bigcup_k (A_k \cap B)\right]$$

$$= \sum_k \frac{P(A_k \cap B)}{P(B)} = \sum_k P_B(A_k)$$

Elementary probability spaces are the mathematical setting in which experiments with a finite number of outcomes are formulated. In general, W is taken as the set of all possible outcomes of the experiment and the algebra R is the power set $\mathbb{P}(W)$, that is, the collection of all subsets of W which is, indeed, an algebra of sets. Within this context, if – by any appropriate means – we assign definite values of probability to the elements of W (which form a finite partition of W and are often called the simple events) it is possible to determine the probability of any event, that is of any member of $\mathbb{P}(W)$. This possibility of 'extending' the probability from a smaller set to a larger set – that is from W to $\mathbb{P}(W)$ in this case – is worthy of notice and we will have more to say about it later (see Section 2.1.1). For the moment an example will help clarify these ideas.

Example 2.1 In the light of Definition 2.1, let us reconsider the experiment of rolling a fair die. Here W is the set $\{\{1\}, \{2\}, \dots, \{6\}\}$ (although it may be redundant in this case, for reasons that will become clearer as we progress it is desirable to distinguish between the events $\{1\}, \dots, \{6\}$ and the real numbers $1, \dots, 6$) and probabilities are assigned according to the classical definition on the basis of symmetry considerations, that is,

$$P(\{1\}) = P(\{2\}) = \cdots = P(\{6\}) = 1/6 \tag{2.4}$$

However, the algebra $R = \mathbb{P}(W)$ is a collection of $2^6 = 64$ sets and we can determine the probabilities of all these events simply on the basis of eq. (2.4). So, for example, one may ask for the probability of obtaining an odd number, which we call, say, event A. Since $A = \{1\} \cup \{3\} \cup \{5\}$

and these three events belong to R it follows that $A \in R$; then, noting that the three events $\{1\}, \{3\}, \{5\}$ are mutually disjoint, we use the finite additivity (EP2) to get $P(A) = 1/6 + 1/6 + 1/6 = 1/2$. By the same token it is immediate to determine, for instance, the probability of a number less than five, or the probability that a 6 or a 1 show up, etc., because all these events belong to the algebra R. Also, note that the same line of reasoning applies even if the die is not fair; the elementary probability space is now $(W, \mathbb{P}(W), P_{\mathrm{U}})$ where the only difference is the function P_{U} (the subscript U is for 'unfair'). Clearly, $P_{\mathrm{U}} \neq P$ and we would probably have to adopt a relative frequency approach in order to determine the basic probabilities $P_{\mathrm{U}}(1), P_{\mathrm{U}}(2), \ldots, P_{\mathrm{U}}(6)$ which, nonetheless, must satisfy the condition (EP1), that is, $P_{\mathrm{U}}(1) + \cdots + P_{\mathrm{U}}(6) = 1$.

The example above is intentionally simple because it shows clearly the main idea behind the mathematical symbolism. Along the same line of reasoning, it will not be difficult for the reader to find many other examples. In order to progress further, however, the notion of elementary probability space may turn out to be inadequate. In fact, some applications require a mathematical setting where it should be possible – at least in principle – to perform an infinite number of operations on the elements of W which, in turn, may not be a finite set itself. These considerations lead to another definition:

Definition 2.2 We call *probability space* a triplet (W, S, P) where W is a set, S is a σ-algebra of subsets of W (the definition of σ-algebra is given in Appendix A) and P is a set function defined on S with values in the real interval $[0, 1]$ which satisfies the probability axioms

(P1) $P(W) = 1$

(P2) If $A_1, A_2, \ldots, \in S$ and $A_i \cap A_j = \emptyset$ for $i \neq j$ then

$$P\left(\bigcup_{n=1}^{\infty} A_n\right) = \sum_{n=1}^{\infty} P(A_n) \tag{2.5}$$

Turning to terminology, property (P2) – which is clearly more general than (EP2) – is called σ-additivity (or, often, countable additivity; note that the fact that S is a σ-algebra implies that the countable union on the l.h.s. of eq. (2.5) belongs to S) and now we call events all the members of S.

Starting from the probability axioms (P1) and (P2) it can be shown that properties (a), (b) and (c) still hold while (d) and (e) can be written for a countable collection of sets. The form of this generalization is straightforward for (e) but less obvious for (d) which now reads: (d′) if $A_1, A_2, \ldots, \in S$ and $A \subset \cup_n A_n$ then $P(A) \leq \sum_n P(A_n)$. This property is called *countable monotonicity*.

In addition, we also have an important *continuity* property of the σ-additive function P:

Proposition 2.1 *(a) If the sets $A_1, A_2, \ldots, \in S$, $A_1 \supset A_2 \supset \cdots$ and $A_n \downarrow A$ (i.e. $\{A_n\}$ is a decreasing sequence and $A = \cap_n A_n$) then*

$$P(A) = \lim_{n\to\infty} P(A_n) \tag{2.6a}$$

(b) If $A_1, A_2, \ldots, \in S$, $A_1 \subset A_2 \subset \cdots$ and $A_n \uparrow A$ (i.e. $\{A_n\}$ is an increasing sequence and $A = \cup_n A_n$) then

$$P(A) = \lim_{n\to\infty} P(A_n) \tag{2.6b}$$

As an incidental remark note the following: since the hypotheses of Proposition 2.1 (a) and (b) can be written in set notation as $A = \lim_{n\to\infty} A_n$ (see Appendix A), eqs (2.6a) and (2.6b) amount to the equality

$$P(\lim_{n\to\infty} A_n) = \lim_{n\to\infty} P(A_n) \tag{2.6c}$$

thus meaning that the limit operation can be moved inside and outside the probability sign (with the obvious understanding that we have a limit of sets on the l.h.s. of (2.6c) and a limit of real numbers on the r.h.s.).

Another important point to be made is that σ-additivity implies finite additivity but the reverse, in general, is not true. In this regard, however, it can be shown that finite additivity plus continuity imply σ-additivity and, in fact, some authors (see, for instance, [5]) replace axiom (P2) by (EP2) plus continuity which, in turn, is sometimes stated as: (c) if $\{A_n\} \in S$ is a decreasing sequence such that $A_n \downarrow \emptyset$, then $P(\lim_{n\to\infty} A_n) = 0$.

2.2.1 *A digression on measure theory: the Lebesgue measure and Caratheodory extension theorem*

At this point the reader familiar with measure theory has already noted the similarities between probability spaces and measure spaces, the former being a special case of the latter because, in mathematical terms, the probability function P is just a finite (meaning that $P(W)$ is finite, that is, $P(W) < \infty$) measure defined on a σ-algebra of sets (see Appendix B). In this light, an important results in measure theory has to do with the possibility of extending a σ-additive measure from a limited collection of sets to a σ-algebra of sets.

Without getting into strict mathematical details (which can be found in the references), the general procedure can be outlined as follows. One starts with a real, non-negative set function m defined on a semialgebra G of subsets of a set W (i.e. $m: G \to \mathbb{R}_+ \cup \{0\}$). Then, a first theorem states that there is a unique extension of m to the algebra $R(G)$, that is, the algebra generated by G. Let us denote this extension by the symbol m'. Clearly, for every set $A \in G$ we have $m'(A) = m(A)$ but m' is considered different from m because $R(G) \supset G$ and therefore its domain is different. Now, if the original measure m is σ-additive it turns out that it can be extended to a σ-additive measure $\overline{\mu}$ whose domain is a collection of sets M which, in turn

(a) is a σ-algebra,
(b) is much larger than $R(G)$, that is, $M \supset R(G)$.

The construction of this 'larger' extension is accomplished by two intermediate stages in which one first introduces the so-called 'outer measure' μ^* defined on the power set $\mathbb{P}(W)$ and then restricts this domain to all sets A satisfying the following property: given any $\varepsilon > 0$ there is a set $B \in R(G)$ such that

$$\mu^*(A \Delta B) < \varepsilon \tag{2.7}$$

The final result is that condition (2.7) defines a collection of sets $M \subset \mathbb{P}(W)$ with the properties (a) and (b) above. The members of M are called *measurable* sets (clearly, all sets of G and of $R(G)$ turn out to be measurable) and the set function μ^* restricted to the domain M – and denoted by the symbol $\overline{\mu}$ – is called the *Lebesgue* extension of m. Then, the last step is the proof that $\overline{\mu}$ is, indeed, σ-additive on its domain M, that is, it is a measure.

In regard to this last statement it should be noted that the process of extension cannot be terminated at the stage of the outer measure because it is shown that μ^* is not σ-additive on $\mathbb{P}(W)$. Therefore, since σ-additivity is the key property we want to maintain in the extension, the outer measure does not fit our needs until we take a further step and restrict its domain to the collection M of measurable sets.

At this point the question could be asked if M coincides with $S(G)$, where $S(G)$ is the σ-algebra generated by the original collection of sets G.

The answer is negative and it turns out that $M \supset S(G)$. However, the following result holds: if $A \in M$ then A can be expressed as $B \cup N$ where $B \in S(G)$ and N is a subset of a set $N' \in S(G)$ with $\mu^*(N') = 0$. Mathematically speaking – see Appendix B (proposition B.1) and the references for further details – this means that the Lebesgue measure is the 'completion' of the outer measure restricted to $S(G)$, but for our purposes it simply means two things

(a) M and $S(G)$ are only slightly different,
(b) there is little loss of generality in saying that the original measure m can be extended to the σ-algebra $S(G)$.

Now, since a probability is a special case of measure, the preceding discussion on extension of measures has an important counterpart in probability theory. This is the so-called *Caratheodory extension theorem* which can be stated as follows:

Let G be an algebra (or, more generally, a semialgebra) of subsets of a set W and let $\overline{P}$ be a set function $\overline{P} : G \to [0, 1]$ satisfying (P1) and (P2) (clearly, when (P2) makes sense, that is, if the union on the l.h.s. belongs to G). Then there exists a unique σ-additive extension P of $\overline{P}$ to the σ-algebra $S(G)$ such that $P(A) = \overline{P}(A)$ for every set $A \in G$.

So, two points can be made at the end of this mathematical digression:

(1) the Caratheodory extension theorem motivates the fact that in Definition 2.2 the domain of the probability function P is, from the start, chosen to be a σ-algebra of sets;
(2) we have the possibility of calculating the probability of any event (i.e. any set of the σ-algebra) once the values of probability have been assigned to a limited number of events.

Once again, however, we note that the theory does not say how to determine the 'basic' probabilities – except for the special cases $P(W) = 1$ and $P(\emptyset) = 0$. This is in no way a limitation of the theory, but, on the contrary, is a circumstance which allows a high degree of flexibility and gives us the possibility to deal with a large number of real-world situations.

Finally, the discussion of this section justifies the fact that from now on we will often speak of 'probability measures', where this term refers to σ-additive, finite measures with the property $P(W) = 1$.

2.2.2 *Stochastic independence*

In Chapter 1, we briefly introduced the notion of independent events. Since this concept plays a central role in probability and is frequently used in applications, it is useful to discuss it in more detail now that we have a precise notion of event at our disposal.

As a preliminary remark, it is the author's opinion that it is desirable to distinguish between 'physical independence' and 'stochastic independence', where the first term refers to the real-world situation and the second to its mathematical counterpart in probability theory. As a matter of fact, we generally formulate the concept of independence on the basis of intuition: events are considered independent when they seem to have no causal relation. This idea, however, relies ultimately on our experience of the real world and represents a 'physical (or logical) independence' which – before being incorporated in the theory – needs to be translated in probabilistic language. In order to do so, we consider two events A and B and argue that we can call them mutually independent if the occurrence of one does not 'condition' the probability of occurrence of the other. With this in mind we write $P(A|B) = P(A)$ and

$P(B|A) = P(B)$ and using the definition of conditional probability (eq. (1.7)) we get the multiplication rule

$$P(A \cap B) = P(A)P(B) \tag{2.8}$$

with the assumption of symmetry, that is, if A is independent of B, then B is independent of A.

Equation (2.8) expresses a so-called condition of 'stochastic independence', where the term 'stochastic' emphasizes the mathematical concept rather than a real-world situation of no causal relation between the two events. Now, since, *a priori*, there is no necessary strict logical connection between the mathematical model and its real-world counterpart, we should check eq. (2.8) against practical examples in which the classical and/or relative frequency definition of probability applies; if the agreement is satisfactory we can assume eq. (2.8) as a valid statement of independence. We will not do it here and leave it to the reader, but a little thought shows that this is, indeed, the case. In this light, we therefore accept the fact that in order to establish the independence of two events in the probabilistic or stochastic sense we must show that eq. (2.8) – or some mathematically equivalent rule – holds. This is why it is worth investigating some consequences and extensions of eq. (2.8).

Proposition 2.2 *If any one of the pairs* $\{A, B\}$, $\{A, B^C\}$, $\{A^C, B\}$, $\{A^C, B^C\}$ *is an independent pair, then all the pairs are independent pairs.*

We only prove one of the statements and leave the rest to the reader. Noting, for instance, that A *can be written as the union of two disjoint sets as* $A = (A \cap B) \cup (A \cap B^C)$, *from the independence of events* A *and* B *it follows*

$$\begin{aligned} P(A \cap B^C) &= P(A) - P(A \cap B) = P(A) - P(A)P(B) \\ &= P(A)[1 - P(B)] = P(A)P(B^C) \end{aligned}$$

meaning that events A, B^C *form an independent pair.*

Proposition 2.3 *Any event* A *is independent of the impossible event* $\emptyset$ *and the sure event* W *(the proof is immediate).*

The following proposition is given because the notions of independent events and mutually exclusive (i.e. disjoint) events can sometimes be confused.

Proposition 2.4 *If two events* A *and* B *have positive probabilities and are mutually exclusive, they cannot be independent. Conversely, if* A *and* B *have positive probabilities and are independent, they cannot be mutually exclusive.*

The mathematical proof is immediate but the following remark may be more appropriate: if two events A and B are mutually exclusive the occurrence of A implies the occurrence of B^C and the occurrence of B implies the occurrence of A^C. Therefore, since one event 'conditions' the other, they cannot be independent. On the other hand, if A and B are a stochastically independent pair then (Proposition 2.2) A, B^C and A^C, B are independent pairs and the occurrence of one event must have no effect on the occurrence of the other. Since this is not true for mutually exclusive events, it follows that A and B cannot be mutually exclusive.

We are now in a position to extend the definition of stochastic independence to more than two events. A collection of events $\mathbb{A} = \{A_n\}$, finite or not, is called a collectively (or mutually) independent class if the product rule holds for all finite subcollections of $\mathbb{A}$, that is, if

$$P(A_{k_1} \cap A_{k_2} \cap \cdots \cap A_{k_m}) = P(A_{k_1})P(A_{k_2}) \cdots P(A_{k_m}) \tag{2.9}$$

for all collections of indices $\{k_1, k_2, \ldots, k_m\}$, with m finite (and clearly $m \geq 2$).

A few remarks on this definition are in order. First, we generalize Proposition 2.2 and then we consider the relation – or lack thereof – between independence of an entire class and independence of its subclasses. In the spirit of the book not all proofs will be given but the interested reader can find them in more mathematically oriented texts (for instance, see the references at the end of this chapter).

Proposition 2.5 *If the events A_n are collectively independent and one or more events are replaced by their complement (or by $\emptyset$ or W), independence is maintained.*

Proposition 2.6 *If $A_1, A_2, \ldots, A_n$ are such that $A_{k_1}, A_{k_2}, \ldots, A_{k_m}$ are independent for all distinct indices $\{k_1, k_2, \ldots, k_m\}$ with $m = 2, 3, \ldots, n-1$ it does not follow that $A_1, A_2, \ldots, A_n$ are collectively independent. Stated differently, Proposition 2.6 says that $(n-1)$-wise collective independence does not imply n-wise independence. Note that this is a generalization of what has been said in Chapter 1 (Example 1.1), namely that for $n > 2$ pairwise independence does not imply collective independence.*

Proposition 2.7 *If $P(A_1 \cap A_2 \cap \cdots \cap A_n) = P(A_1)P(A_2) \cdots P(A_n)$ it does not follow that $P(A_{k_1} \cap A_{k_2} \cap \cdots \cap A_{k_m}) = P(A_{k_1})P(A_{k_2}) \cdots P(A_{k_m})$ for $m < n$;*

As an example illustrating Proposition 2.7, consider tossing two dice simultaneously and let:

- A be the event that the second die shows 1, 2 or 3;
- B be the event that the second die shows 3, 4 or 5;
- C be the event that the sum of the faces is 9.

Now, $P(A) = P(B) = 1/2$ and $P(C) = 4/36 = 1/9$ and also – noting that the event $A \cap B \cap C = \{6, 3\}$ – we have

$$P(A \cap B \cap C) = 1/36 = P(A)P(B)P(C)$$

but the events are not pairwise independent; in fact

$$P(A \cap B) = 1/6 \neq P(A)P(B) = 1/4$$
$$P(B \cap C) = 1/12 \neq P(B)P(C) = 1/18$$
$$P(A \cap C) = 1/36 \neq P(A)P(C) = 1/18$$

showing that the validity of the product rule for an entire collection of events does not imply that it holds for its subcollections.

Proposition 2.8 *Let the events $A_1, A_2, \ldots, A_n$ be collectively independent. If we divide them in groups and for each group we form a new event by means of unions, intersections and complementation, then these newly formed events are mutually independent.*

As an example of Proposition 2.8 consider three collectively independent events A, B, C and consider, for instance, the two groups $\{A, B\}$ and $\{C\}$. Then we can prove that the two events $A \cup B$ and C are independent. In fact

$$\begin{aligned} P[(A \cup B) \cap C] &= P[(A \cap C) \cup (B \cap C)] \\ &= P(A \cap C) + P(B \cap C) - P(A \cap B \cap C) \\ &= P(A)P(C) + P(B)P(C) - P(A)P(B)P(C) \\ &= P(C)[P(A) + P(B) - P(A)P(B)] = P(C)P(A \cup B) \end{aligned}$$

Similarly, it can be shown that, say, $A \cup B$ and C^C are independent, that $A \cup B^C$ and C are independent, etc.

An important point to be made is that independence of events depends on the probability function and is not an intrinsic characteristic of events themselves. Therefore, a given collection of events $\{A_n\}$ can be independent with respect to a probability function P without being independent with respect to $P' \neq P$. For example, having shown that the conditional probability (given an event B with $P(B) > 0$) P_B is a probability function in its own right, it is evident that we can define two events A_1, A_2 to be conditionally independent if

$$P(A_1 \cap A_2 | B) = P(A_1 | B)P(A_2 | B) \tag{2.10}$$

or, equivalently, $P_B(A_1 \cap A_2) = P_B(A_1)P_B(A_2)$ where this notation emphasizes that independence is meant with respect to the probability P_B. However, in general, conditional independence does not imply ordinary independence (i.e. with respect to the original function P), nor does ordinary independence imply conditional independence.

In closing this section we make two final observations on independence, one of practical nature and the other more theoretical.

The first is that one application (among many others) of independence comes from reliability theory when one considers the probability of failure of an engineering system. The usual procedure is to (ideally) break up the system under study in a number of, say, n subsystems with known reliabilities $r_1, r_2, \ldots, r_n$ ($r_k = 1 - p_k$ where p_k is the probability of failure of the kth subsystem). Then, if we call A the event that the system succeeds, A_k the event that the kth subsystem succeeds and make the basic assumption that $A_1, A_2, \ldots, A_n$ is a collectively independent class, the two basic cases are as follows:

(a) The subsystems are connected in series (i.e. the failure of any one subsystem causes the whole system to fail). In this case $A = \cap_{k=1}^{n} A_k$ and the reliability $P(A)$ of the entire system is given by

$$P(A) = \prod_{k=1}^{n} P(A_k) = r_1 r_2 \cdots r_n \tag{2.11}$$

so that, for example, if we have three subsystems with reliability 0.85 each, the reliability of the whole system is $P(A) = (0.85)^3 \cong 0.614$.

(b) The subsystems are connected in parallel (i.e. the system works as long as at least one of its subsystems works). Then $A = \cup_{k=1}^{n} A_k$ and since by de Morgan's law

$$A = \bigcup_{k=1}^{n} A_k = \left(\bigcap_{k=1}^{n} A_k^{\mathrm{C}} \right)^{\mathrm{C}}$$

we can use the independence assumption to get

$$\begin{aligned} P(A) &= P\left(\bigcap_{k=1}^{n} A_k^{\mathrm{C}} \right)^{\mathrm{C}} = 1 - P\left(\bigcap_{k=1}^{n} A_k^{\mathrm{C}} \right) = 1 - \prod_{k=1}^{n} P\left(A_k^{\mathrm{C}}\right) \\ &= 1 - \prod_{k=1}^{n} [1 - P(A_k)] = 1 - (1 - r_1)(1 - r_2)\ldots(1 - r_n) \end{aligned} \tag{2.12}$$

so that, for example, a parallel system of three components with $r_k = 0.85$ each has a total reliability $P(A) = 0.997$.

The second and final observation goes back to the beginning of this section, and precisely to the reason why we said that the distinction between physical and stochastic independence is desirable. In the light of the product rule we note that two or more events – simply because of the numerical values of their probabilities – may turn out to be stochastically independent even when physical independence may not seem fully justified. In general, this occurrence has no consequences in applications but is worthy of mention because it represents a debated point in the philosophy of all scientific disciplines where probability plays a part.

2.3 Random variables and distribution functions

In many experiments involving elements of randomness, we are often interested in some numerical quantity associated with the possible outcomes rather than in the outcomes themselves. So, to each element in the sample space W we assign – in some convenient way – a real number with the intention of making probability statements on this or that numerical value or set of values. Mathematically, this process corresponds to defining a real-valued function $X : W \to \mathbb{R}$ which, when certain properties are satisfied, is called a random variable. Before turning to these properties it is useful to give some examples.

(i) In tossing two dice, for instance, a gambler may not be interested in the individual outcomes $\{i, j\}$, $i, j = 1, 2, \ldots, 6$, but in the sum $i + j$. So, to each element in the sample space – that is, the 36 ordered pairs $\{1, 1\}, \{1, 2\}, \ldots, \{6, 6\}$ – he/she assigns a number between 2 and 12 thus defining the function $X(\{i, j\}) = i + j$.

(ii) In a sequence of N shots to a target of diameter D the ballistic department of the Army may be interested in the distance d between the impact point and the bull's eye. The function in this experiment is defined by $X(k\text{th shot}) = d_k$ where $1 \le k \le N$ and $0 \le d_k \le D$ (assuming that each shot hits the target).

(iii) In the daily production of 500 lots of 100 pieces each, a company is interested in the number of defective pieces in each lot, so that $X(k\text{th lot}) = n_k$ where $1 \le k \le 500$ and $0 \le n_k \le 100$.

Many other examples can be made, but the point is that in any case of interest an appropriate real-valued function X is defined and we are faced with the problem of extending the mathematical model to include numerical-valued phenomena subjected to chance. In the light of the definitions of Section 2.2, this means that two questions need to be answered: (a) what kind of functions, mathematically speaking, qualify as random variables? and (b) since the probability P is defined for events (i.e. subsets of W), how do we make probability statements on the values belonging to the range $\text{Rg}(X)$ – which is a subset of the real numbers $\mathbb{R}$ or $\mathbb{R}$ itself – of the function X?

In regard to the second question we first make a preliminary remark. If we denote by w a point element of the sample space W, then $X(w) \in \text{Rg}(X) \subset \mathbb{R}$

and, typically, we will be interested in the probability that:

(1) $X(w)$ has a particular value a ($a \in \mathbb{R}$), or
(2) $X(w)$ lies in the range of values $a < X(w) \leq b$ ($a, b \in \mathbb{R}; a < b$), or
(3) $X(w) \leq a$.

Noting that the definition of X implies an 'inverse' relation between Rg(X) and the original space W, conditions (1)–(3) are equivalent to saying that we are interested in assigning probabilities to the subsets of W

(1′) $X^{-1}(\{a\}) = \{w \in W : X(w) = a\}$
(2′) $X^{-1}(a, b] = \{w \in W : X(w) \in (a, b]\}$
(3′) $X^{-1}(-\infty, a] = \{w \in W : X(w) \in (-\infty, a]\}$

respectively. Clearly, this can be done if and only if these sets are events, that is, they are members of the σ-algebra S (or the algebra R if we are dealing with an elementary probability space). So, in practice, the procedure works if $X^{-1}\{a\}, X^{-1}(a, b], X^{-1}(-\infty, a] \in S$ and only in this case it makes sense to speak of the probabilities $P(X^{-1}\{a\})$, $P(X^{-1}(a, b])$ and $P(X^{-1}(-\infty, a])$.

If, in addition, we consider that $\mathbb{R}$ is equipped with the 'natural' σ-algebra **B** of Borel sets (see Appendix A, Section A3) and that the subsets $\{a\}$, $(a, b]$ and $(-\infty, a]$ are just special cases of Borel sets, we arrive at the formal definition of random variable:

Definition 2.3 Given the probability space (W, S, P) we call *random variable* (r.v. for short) a real-valued function $X : W \to \mathbb{R}$ such that $X^{-1}(B) \in S$ for every Borel set $B \in \mathbb{B}$.

Definition 2.3 is the answer to the two questions (a) and (b) above; it says which kind of functions are random variables and, at the same time, automatically guarantees the possibility of making probability statements by using subsets of Rg(X). (Remark: at this point one could ask the following question: when $W \subseteq \mathbb{R}$ why not use the direct image of the function instead of its inverse image in order to define a random variable? The reason lies in the fact that the inverse image preserves set operations – see Appendix A, eq. (A.14) and Proposition A.8 – while the direct image, in general, does not. So, for example, given a function f, it is true that $f^{-1}(A^C) = [f^{-1}(A)]^C$ while, in general, $f(A^C) \neq [f(A)]^C$.)

Now, besides the unfortunate terminology of calling 'variable' a function (a usage, however, so widespread in literature that one has no choice but to adhere to it), we may once again turn to the theory of measure and integration and note that in mathematical terms a r.v. is a so-called P-measurable function. In fact (see, for example, [1] or [7]) the general definition is as follows: if W is set with a σ-additive measure ν defined on a σ-algebra S of subsets of W, a real-valued function $f : W \to \mathbb{R}$ is called ν-measurable (or, for some authors, 'S-measurable' or simply 'measurable') if $f^{-1}(B) \in S$ for every Borel set $B \subset \mathbb{R}$. This fact is worthy of mention because it means that all theorems and results on measurable functions can immediately be taken over to probability theory. We will state and use these theorems if and whenever needed in the course of the discussion; for the moment, however, only one proposition will suffice.

Proposition 2.9 *Let (W, S, P) be a probability space and let X be a real valued function on W, that is, $X : W \to \mathbb{R}$. Then the following statements are equivalent:*

(a) *X is a random variable*
(b) $X^{-1}(-\infty, a] \in S$ *for every* $a \in \mathbb{R}$
(c) $X^{-1}(-\infty, a) \in S$ *for every* $a \in \mathbb{R}$
(d) $X^{-1}[a, +\infty) \in S$ *for every* $a \in \mathbb{R}$
(e) $X^{-1}(a, +\infty) \in S$ *for every* $a \in \mathbb{R}$

In particular, Proposition 2.9 justifies the fact that one often finds the following statement: a function $X : W \to \mathbb{R}$ is a r.v. if and only if the set $X^{-1}(-\infty, a] = X^{-1}\{w \in W : X(w) \leq a\}$ belongs to the σ-algebra S (i.e. is an event). Also, it should be added that one rarely needs to worry about questions of measurability in applications because the definition of random variable – although fundamental in the construction of a coherent picture – is sufficiently general to cover most cases of practical interest.

With the notions given above we can now give the definition of probability distribution function (of a r.v.). Before doing so, however, we make some preliminary considerations.

Given a random variable X on a probability space (W, S, P) the preceding discussion shows that we can make probability statements by considering the inverse images (through X) of Borel sets because the quantity $P(X^{-1}(B))$ is well-defined for every $B \in \mathbb{B}$. Then, if we define the set function $P_X : \mathbb{B} \to [0, 1]$ by means of the relation

$$P_X(B) \equiv P(X^{-1}(B)) = P\{w \in W : X(w) \in B\} \tag{2.13}$$

it can be shown that P_X is a σ-additive probability measure, that is, it satisfies properties (P1) and (P2) given in Section 2.2 – as the reader is invited to verify.

This fact implies that the original probability space (W, S, P) 'induces' – through X – a real probability space $(\mathbb{R}, \mathbb{B}, P_X)$ which, in turn, is completely determined by (W, S, P). The reverse statement is not, in general, true, meaning that the space $(\mathbb{R}, \mathbb{B}, P_X)$ does not determine uniquely the space (W, S, P). However, in applications this is not a problem because the original space is generally a tacitly implied notion of theoretical interest and $(\mathbb{R}, \mathbb{B}, P_X)$ is all we need for most practical situations. In this light, therefore, it is common usage to write $P(X \in B)$ in place of $P_X(B)$ with the implicit understanding that the meaning of this notation – which refers to the 'induced' space – is expressed by eq. (2.13).

Following this line of reasoning we note, in particular, that the quantity $P(X^{-1}(-\infty, a])$ is well-defined for every real number a. Then, since $P(X^{-1}(-\infty, a]) = P_X(-\infty, a] = P_X(\{w \in W : X \leq a\})$ is written as $P(X \leq a)$ in the notation introduced above, we can define the real function $F_X : \mathbb{R} \to [0, 1]$

$$F_X(a) \equiv P(X \leq a) \tag{2.14}$$

which is called the *probability distribution function* (PDF for short) of the random variable X. Also, since a is any real number we can replace it by x and write $F_X(x)$ as it is customary in ordinary calculus.

The function $F_X(x)$ satisfies the following properties:

(D1) F_X is non-decreasing, that is, $a \leq b$ implies $F_X(a) \leq F_X(b)$
(D2) $\lim_{x \to -\infty} F_X(x) = 0$ and $\lim_{x \to +\infty} F_X(x) = 1$
(D3) F_X is right-continuous at every point, that is,

$$F_X(x+) \equiv \lim_{h \to 0+} F_X(x + h) = F_X(x)$$

We only prove (D1) here but it is worth noting that the properties above hold because P_X is a σ-additive probability measure in its own right.

In fact, for example, the proof of (D1) is as follows: if $a \leq b$ then $(-\infty, a] \subset (-\infty, b]$, therefore from the monotonicity property of probability measures it follows $P_X(-\infty, a] \leq P_X(-\infty, b]$ and hence $F_X(a) \leq F_X(b)$.

A direct consequence of the definition of PDF is that

$$P(a, b] = F_X(b) - F_X(a) \tag{2.15}$$

where once again we point out that by $P(a, b]$ we mean the probability $P_X(a, b] = P_X(a < X \leq b)$. So, since the interval $(-\infty, b]$ is the union of the two disjoint sets $(-\infty, a]$ and $(a, b]$, from the additivity of P_X we get $P_X(-\infty, b] = P_X(-\infty, a] + P_X(a, b]$. Rearranging terms, eq. (2.15) follows.

In regard to property (D3) we may ask about the left limit. This limit exists because F_X is a monotone function (property (D1)) and therefore it can only have discontinuities of the first kind (i.e. finite jumps, see Ref. [10] for details). Left-continuity, however, is not in general guaranteed. In fact, consider a point $x = x_0$ on the real line; the set $\{w \in W : X = x_0\}$ or, for short, $\{X = x_0\}$ can be expressed as

$$\{X = x_0\} = \bigcap_{n=1}^{\infty} \{x_0 - 1/n < X \leq x_0\}$$

where the sets $\{x_0 - 1/n < X \le x_0\} = (x_0 - 1/n, x_0]$ form a decreasing sequence whose limit is $\{X = x_0\}$. Then, from the continuity property (a) of Proposition 2.1 we get

$$\begin{aligned} P_X(X = x_0) &= \lim_{n\to\infty} P_X(x_0 - 1/n, x_0] = \lim_{n\to\infty} (F_X(x_0) - F_X(x_0 - 1/n)) \\ &= F_X(x_0) - \lim_{n\to\infty} F_X(x_0 - 1/n) = F_X(x_0+) - F_X(x_0-) \end{aligned} \tag{2.16}$$

where in the last expression $F_X(x_0+) = F_X(x_0)$ is the right limit and $F_X(x_0-)$ is the left limit. Equation (2.16), in essence, means that if $P(X = x_0) \neq 0$ then F_X has a jump at $x = x_0$, the magnitude of the jump being precisely $P(X = x_0)$; if, on the other hand $P(X = x_0) = 0$ then F_X is continuous at $x = x_0$ because $F_X(x_0+) = F_X(x_0-)$.

One word of caution is in order at this point: some authors define the distribution function as $F_X(a) \equiv P(X < a)$ instead of (2.14). As a consequence, the roles of right- and left-continuity are interchanged, that is, property (D3) is replaced by left-continuity and F_X may not be right-continuous at some points. Owing to Proposition 2.9, this is not a problem; nonetheless, some attention should be paid because – once a definition is given – consistency must be maintained throughout.

Now, in regard to random variables and their PDFs, the preceding discussion can be summarized as follows: starting from a probability space (W, S, P) and given a r.v. X we can consider the 'induced' probability space $(\mathbb{R}, \mathbb{B}, P_X)$ and determine the PDF of X by means of eq. (2.14). The distribution function F_X, in turn, satisfies properties (D1)–(D3). However, since in many practical cases one specifies a r.v. X by simply giving its distribution function F_X, the question can be asked if this procedure is justified. We anticipate here that the answer is yes, but before tackling this problem we give a preliminary definition:

Definition 2.4 We call *probability distribution function* (PDF) any function $F : \mathbb{R} \to [0, 1]$ satisfying properties (D1)–(D3).

With this definition the question above can be reformulated in more general terms: given a PDF F, is there a probability measure $\hat{P}$ defined on the σ-algebra of Borel sets such that $\hat{P}(a, b] = F(b) - F(a)$?

The answer is yes because for every $a, b \in \mathbb{R}$, we can define – on the set of all right-semiclosed intervals of $\mathbb{R}$ – the function $\overline{P}(a, b] \equiv F(b) - F(a)$. Then, since $\overline{P}$ can be shown to be σ-additive on its domain, we are under the hypothesis of Caratheodory theorem (Section 2.2.1). Therefore $\overline{P}$ can be extended to a probability measure $\hat{P}$ whose domain is the σ-algebra of Borel sets $\mathbb{B}$ and, by construction, satisfies the requirement $\hat{P}(a, b] = F(b) - F(a)$. Also, it is not difficult to show that $\hat{P}(-\infty, a] = F(a)$ for every $a \in \mathbb{R}$.

This function $\hat{P}$, in turn, is the probability measure of the 'induced' space (i.e. the measurable space $(\mathbb{R}, \mathbb{B})$) and one may ask if there is at least a probability space (W, S, P) on which a r.v. X with distribution function F can be defined so that $\hat{P} = P_X$ for some r.v. X. The answer is again affirmative because we can always supply the probability space in a canonical way by setting $W = \mathbb{R}$, $S = \mathbb{B}$ and using the identity map as our random variable, that is $X(w) = w$ $(w \in W)$ (as a matter of fact, this canonical way of constructing the 'original' space (W, S, P) is the implicit assumption made in almost all practical cases). Then, since $P_X(B) \equiv P\{w : X(w) \in B\} = \hat{P}(B)$, X has induced (working, so to speak, backwards) the probability measure P and therefore the PDF F. The conclusion is that, summarizing, if F is a PDF (in the sense of Definition 2.4) then it is the distribution function of some random variable X. It should be noted, however, that while it is true that a random variable defines uniquely its distribution function, the reverse statement, in general, does not hold and a given PDF can correspond to many different random variables. Nonetheless, as far as probability statements are concerned, the PDF F_X provides a complete description of the r.v. X. In other words, given F_X, we can calculate the probability that X takes on values in B where B is any Borel set of the real line.

As a final remark to this section we can once again turn to the terminology of measure theory and observe that, in mathematical terminology, P_X is a so-called Lebesgue–Stieltjes measure on the real line (see Appendix B). Probably, this remark does not say much to the reader who is not familiar with measure theory but it may be helpful if one refers to more mathematically oriented literature. As a matter of fact, any non-negative, σ-additive and finite Lebesgue–Stieltjes measure μ_F on $\mathbb{R}$ can be defined by means of an appropriate non-decreasing, right-continuous function F which is bounded below and above by the quantities $F(-\infty) \equiv \lim_{x\to-\infty} F(x)$ and $F(\infty) \equiv \lim_{x\to\infty} F(x)$, respectively, where the two limits are assumed to be finite. Then μ_F is called the Lebesgue–Stieltjes measure corresponding to F and F is said to be the generating function of μ_F (incidentally, it is worth pointing out that if we relax the assumption of finite limits and choose $F(x) = x$ then $\mu_F = \mu_x$ is the Lebesgue measure on the real line).

Returning to probability theory, for any PDF F_X we have, clearly, $F_X(-\infty) = 0$ and $F_X(\infty) = 1$ and P_X is the Lebesgue–Stieltjes measure corresponding to F_X. Conversely, F_X is the generating function of P_X. This fact sets up a one-to-one correspondence between PDFs and the Lebesgue–Stieltjes measures on $\mathbb{R}$ satisfying $\mu_F(\mathbb{R}) = 1$.

2.3.1 *Types of random variables and their distribution functions*

Random variables can be classified as discrete or continuous depending on the mathematical structure of their range (as a subset of $\mathbb{R}$) or, alternatively, on the continuity properties of their distribution functions. This second classification is generally preferable for two reason: first, from a probabilistic (and statistical) point of view, the actual random variable is often less important than its PDF and, second, the PDF reflects the properties of the probability measure P_X which, in turn, plays an important role in the calculation of the (Lebesgue–Stieltjes) integrals used to obtain parameters such as the expected value, the variance and, in general, the so-called 'moments' of a random variable.

From the preceding section we know that a PDF is a real-valued function defined on the real line and satisfying the properties (D1)–(D3). In this regard we may recall two theorems from analysis stating that:

(i) any monotone function on the real line has at most a countable number of discontinuities (and these are only discontinuities of the first kind);
(ii) any monotone, right-continuous function can be written as the sum of a continuous monotone function and a right-continuous jump function. Moreover, this decomposition is unique.

So, by virtue of theorem (ii) we have a first classification of random variables in discrete, continuous and mixed, the mixed case being when both the continuous and the jump function of the decomposition are different from zero. If the continuous function of the decomposition is identically zero we have the discrete case, that is, a PDF which increases only by finite 'steps' (the discontinuities of the first kind) at the points $x_1, x_2, \ldots, \in \mathbb{R}$ which, clearly, form at most a countable set of points (theorem (i)) and are the values taken on by the random variable. Therefore, we can call *discrete* all random variables which have an at most countably infinite set of possible values. Among these, one sometimes distinguishes the subclass of *simple* random variables, which can take on only finitely many possible values.

Example 2.2 Consider the experiment of rolling a fair die. The elementary probability space (W, R, P) is $W = \{\{1\}, \{2\}, \ldots, \{6\}\}$, $R = \mathbb{P}(W)$ and $P(\{j\}) = 1/6$ for every $\{j\} = \{1\}, \{2\}, \ldots, \{6\}$. A 'natural' random variable on this space is defined by $X(\{j\}) = j$ (note that we distinguish between the event $\{j\}$ – that is, the outcome of the experiment – and the real number j) and it is clearly a simple random variable because its range is the set of real numbers $\mathrm{Rg}(X) = \{1, 2, \ldots, 6\}$. Also, from the definition above we have $X^{-1}(j) = \{j\}$ and $P_X(j) = P\{X^{-1}(j)\} = 1/6$ so that the PDF F_X is the jump function $F_X(j) = j/6$ with discontinuities of magnitude 1/6 at the points $x_1 = 1, x_2 = 2, \ldots, x_6 = 6$.

(Note that we have tacitly implied the passage to the 'induced' probability space whose elements are: (a) the subset of $\mathbb{R}$ $\overline{W} = \{1, 2, \ldots, 6\}$, (b) the algebra $\overline{R} = \mathbb{P}(\overline{W})$ and (c) the probability measure $P_X(j) = P\{X^{-1}(j)\} = 1/6$ for $j = 1, 2, \ldots, 6$.)

This example illustrates what has been said above, if $A_X = \{x_1, x_2, \ldots\} \subset \mathbb{R}$ is the set of possible values of the discrete random variable X, then its PDF F_X is a jump function with a discontinuity at each x_n of magnitude

$$p_n \equiv P\{X^{-1}(x_n)\} = F_X(x_n) - F_X(x_n-) \tag{2.17}$$

(see eq. (2.16)). Moreover, F_X is constant between any two neighbouring points x_n and x_{n+1} (the reader is invited to prove it) and takes the upper value at each discontinuity. From their definition, the quantities p_n are clearly non-negative and must satisfy the so-called 'normalization condition'

$$\sum_n p_n = 1 \tag{2.18}$$

Thus, in terms of probabilities, a discrete random variable can be completely specified by the (finite or countable) set of real numbers A_X and the probabilities p_n. Moreover, if we introduce the Heaviside (or step) function

$$H(x) \equiv \begin{cases} 0, & x < 0 \\ 1, & x \geq 0 \end{cases} \tag{2.19}$$

we note that F_X can be written as

$$F_X(x) = \sum_n p_n H(x - x_n) \tag{2.20}$$

so that, in the case of Example 2.2, eq. (2.20) is $F_X(x) = \sum_{j=1}^{6}(1/6)H(x-j)$ which, as expected, gives the value $F_X(j) = j/6$ for each $j = 1, 2, \ldots, 6$, is zero for $x < 1$ and unity for $x \geq 6$. If, for notational convenience, we want to emphasize the fact that the quantities p_n are relative to the random variable X, we can formally introduce the function $p_X : A_X \to [0, 1]$ such that $p_X(x_n) \equiv p_n$ for $n = 1, 2, \ldots$. In this symbolism, the function p_X is often called the *probability mass function* (pmf) of the random variable X.

Example 2.3 Given a probability space (W, S, P) and a subset $A \subset \mathbb{P}(W)$ the *indicator* function of A, denoted by I_A (or by χ_A) is defined as

$$I_A(w) \equiv \begin{cases} 0, & w \notin A \\ 1, & w \in A \end{cases} \tag{2.21}$$

It is worth noting in passing that some authors call I_A the characteristic function of A. We will not follow this terminology because in probability and statistics the term 'characteristic function' is widely used to denote a different concept (see Section 2.4).

The reader is invited to prove the following statements:

(a) I_A is a (discrete) random variable if and only if $A \in S$, that is, if A is an event (and therefore the quantities $P(A)$ and $P(A^C) = 1 - P(A)$ are well-defined);
(b) when A is an event, the distribution function F_I of the random variable I_A can be written as

$$F_I(x) = [1 - P(A)]H(x) + P(A)H(x-1) \tag{2.22}$$

where H is the Heaviside function of eq. (2.19). Other examples of discrete random variables will be given later.

Returning to theorem (ii) at the beginning of this section we note that a PDF F_X is continuous on the real line if the jump function of the decomposition is identically zero. This means that there are no points $x_n \in \mathbb{R}$, $n = 1, 2, \ldots$, such that $P(X = x_n) > 0$ and we have $P(X = x) = 0$ for all $x \in \mathbb{R}$ (which, clearly, implies $F_X(x) = F_X(x-)$ for all $x \in \mathbb{R}$). In this case the random variable X is also called *continuous*. At this point, an interesting question could be asked: since F_X is continuous and on the real line we have the notion of Lebesgue integral at our disposal (which generalizes and extends the notion of Riemann integral known from elementary analysis, see Appendix B), is it possible to represent F_X as a Lebesgue integral of an appropriate Lebesgue-integrable function f_X? In other words, is it possible to write

$$F_X(x) = \int_{-\infty}^{x} f_X(t)\,\mathrm{d}t \tag{2.23}$$

(where the integral in intended in the Lebesgue sense) for some f_X? Moreover, is it true that the derivative of F_X equals f_X? The answer, in general, is no but it is yes if F_X is *absolutely continuous* (the definition can be found in Appendix B). In this regard two comments are worthy of mention:

(1) the class of absolutely continuous functions is a proper subclass of continuous functions and
(2) they are precisely the class of functions for which the second fundamental theorem of calculus (eq. (2.24), also known as Newton-Leibnitz formula) applies: that is,

Proposition 2.10 *If F is absolutely continuous on the interval $[a, b]$ then $F' = \mathrm{d}F/\mathrm{d}x$ is Lebesgue-integrable on $[a, b]$ and*

$$\int_{a}^{x} F'(t)\,\mathrm{d}t = F(x) - F(a) \tag{2.24}$$

for all $x \in [a,b]$. In other words, only within the class of absolutely continuous functions we can restore a function by integrating its derivative.

Therefore, when a PDF F_X is absolutely continuous eq. (2.23) holds and

$$\frac{dF_X}{dx} = f_X \tag{2.25a}$$

In this case the random variable associated with F_X is also called absolutely continuous and the derivative of F_X – that is, the function f_X – is called the *probability density function* (pdf) of the random variable X. Also, from the general properties of F_X it is evident that the pdf f_X must satisfy the 'normalization' condition

$$\int_{-\infty}^{+\infty} f_X(t)\, dt = 1 \tag{2.25b}$$

which is the absolutely continuous counterpart of eq. (2.18).

Now, although continuous PDFs which are not absolutely continuous are very seldom encountered in applications, it is however worthwhile spending a word on what happens in the general case. If a PDF F is continuous but not absolutely continuous then mathematical analysis shows that F can be represented as a sum $F(x) = g(x) + s(x)$ where g is an absolutely continuous function and s is a 'singular' function, where by 'singular' we mean a continuous function whose derivative is zero a.e. (almost everywhere in the sense of the Lebesgue measure on the real line, see Appendix B). Then, $F'(x) = g'(x)$ and integration of F' does not restore F, but only its absolutely continuous component.

At this point we can return to theorem (ii) at the beginning of this section and see that the most general kind of PDF can be represented as the sum of three components: a jump function, an absolutely continuous function and a singular function (these two latter functions, when considered together, form the continuous part of the decomposition of theorem (ii)). Integrating the derivative of the PDF leaves only the absolutely continuous component, while the other two functions 'disappear without a trace' (note that the derivative of a jump function is zero a.e. where, again, a.e. is intended in the sense of the Lebesgue measure on $\mathbb{R}$). Now, as far as PDFs and their classification are concerned, these comments are sufficient for our purposes. However, the argument can be taken further when we note from the preceding discussion that any PDF is defined by means of a probability measure which, in turn, is a finite measure defined on all Borel sets of the real line. In this light, the above decomposition of a general PDF turns out to be a particular case of a result of analysis on the decomposition of measures (see Appendix B). This result can be adapted to our present purposes and stated as in the following Proposition 2.11 by first giving the preliminary definitions:

(a) a measure m on $\mathbb{R}$ is continuous if $m(\{x\}) = 0$ for all $x \in \mathbb{R}$ and absolutely continuous (with respect to the Lebesgue measure μ on $\mathbb{R}$) if $\mu(A) = 0$ implies $m(A) = 0$;
(b) a measure m on $\mathbb{R}$ is singular (with respect to the Lebesgue measure μ on $\mathbb{R}$) if there is a Borel set B with $\mu(B) = 0$ and $m(B^C) = 0$;
(c) a measure m on $\mathbb{R}$ is discrete if there is a countable (Borel) set D with $m(D^C) = 0$.

Proposition 2.11 *Given a probability measure P on $\mathbb{R}$ there are three unique probabilities P_{ac}, P_{sc}, P_d and three positive numbers a, b, c with $a+b+c = 1$ such that*

$$P = aP_{ac} + bP_{sc} + cP_d \tag{2.26}$$

where P_{ac} is absolutely continuous (with respect to the Lebesgue measure μ on $\mathbb{R}$), P_{sc} is singular (with respect to the Lebesgue measure on $\mathbb{R}$) and continuous and P_d is discrete. It is then a consequence of the Radon–Nikodym theorem (see Appendix B) that P_{ac} can be expressed as the Lebesgue integral of a non-negative integrable function $f : \mathbb{R} \to \mathbb{R}$, that is,

$$P_{ac}(B) = \int_B f \, d\mu \tag{2.27}$$

for all Borel sets $B \subset \mathbb{R}$.

The connection between a probability measure P and the corresponding PDF F is such that if we can decompose P as in eq. (2.26) then

$$F = aF_{ac} + bF_{sc} + cF_d \tag{2.28}$$

where $F_{ac}(x)$ is an absolutely continuous function corresponding to P_{ac}, $F_{sc}(x)$ is a singular continuous function corresponding to P_{sc} and $F_d(x)$ is a jump (discrete) function corresponding to P_d. Clearly, the definition of absolute continuity for measures and for functions are two distinct concepts, but it can be shown that if P_{ac} is an absolutely continuous probability measure on $\mathbb{R}$ then $F_{ac}(x) = P_{ac}(-\infty, x]$ is an absolutely continuous function and conversely. Also, with the appropriate definitions in mind, the same relation exists between P_{sc} and $F_{sc}(x)$ and between P_d and $F_d(x)$.

Example 2.4 Perhaps the most famous type of absolutely continuous random variable is a random variable whose pdf is the so-called *normal* (or

Gaussian) probability law: this is

$$f_X(x) = \frac{1}{\sigma\sqrt{2\pi}} e^{-(x-\bar{x})^2/2\sigma^2} \tag{2.29a}$$

where $\bar{x}$ and σ ($\sigma > 0$) are two real parameters (whose meaning is probably well known to the reader but will be shown later). The fact that the pdf (2.29a) satisfies the normalization condition (eq. (2.25b)) can be verified by writing

$$\frac{1}{\sigma\sqrt{2\pi}} \int_{-\infty}^{+\infty} \exp\left(-\frac{(t-\bar{x})^2}{2\sigma^2}\right) dt = \frac{1}{\sqrt{2\pi}} \int_{-\infty}^{+\infty} e^{-(y^2/2)}\, dy$$

where the second integral is obtained by the change of variable $y = (x-\bar{x})/\sigma$. Since from integrals tables we get $\int_{-\infty}^{+\infty} \exp(-ax^2)\, dx = \sqrt{\pi/a}$, eq. (2.25b) follows.

The PDF of eq. (2.29a) cannot be written in explicit analytical form but it is given by

$$F_X(x) = \frac{1}{\sigma\sqrt{2\pi}} \int_{-\infty}^{x} \exp\left(-\frac{(t-\bar{x})^2}{2\sigma^2}\right) dt = \frac{1}{\sqrt{2\pi}} \int_{-\infty}^{(x-\bar{x}/\sigma)} e^{-(y^2/2)}\, dy \tag{2.29b}$$

where, again, the second integral is obtained by the same change of variable as above and – due to its importance in statistics – can be easily found in numerical table form. However, if tables are not available, the approximation

$$F(z) \equiv \frac{1}{2\pi} \int_{-\infty}^{z} \exp(-y^2/2)\, dy \cong \frac{1}{1 + \exp\{-az(1+bz^2)\}} \tag{2.29c}$$

with $a = 1.5976$ and $b = 0.044715$ is sufficiently accurate for most applications (the maximum absolute error is $\leq 2 \times 10^{-4}$). Other examples of absolutely continuous probability laws will be given later.

As remarked above, in almost all practical cases the continuous singular part of the decomposition is generally absent. As a consequence, it is customary to speak of *mixed* random variables when neither the absolutely continuous part nor the discrete part of the PDF function are identically zero.

In this case there exist a number of points x_n for which eq. (2.17) holds; however eq. (2.18) is no longer true and we have

$$\sum_n p_n = \sum_n P\{X^{-1}(x_n)\} < 1 \tag{2.30}$$

This means that there exist at least a pair of neighboring points x_n and x_{n+1} such that $F(x_n) < F(x_{n+1}-)$. In other words, F can be written as the sum (called a 'convex' linear combination)

$$F(x) = \alpha F_{\mathrm{ac}}(x) + (1-\alpha)F_d(x) \tag{2.31}$$

where $0 \le \alpha \le 1$, F_{ac} is an absolutely continuous, monotonically increasing function and F_{d} is a jump function of the type (2.20). Obviously, $\alpha = 1$ corresponds to the absolutely continuous case and $\alpha = 0$ to the discrete case.

In both cases – and, clearly, also in the mixed case – we will see in the next section how the Lebesgue–Stieltjes integral is the appropriate tool used to calculate important quantities such as the mean value, the variance and, in general, many other parameters which describe in numerical form the behavior of a random variable.

Example 2.5 Suppose a r.v. has the following PDF

$$F(x) = \begin{cases} 0, & x < 0 \\ 1/2 - \mathrm{e}^{-x}/4, & x \in [0,1) \\ 1 - \mathrm{e}^{-x}/4, & x \ge 1 \end{cases}$$

(it is evident that this function satisfies the properties (D1)–(D3) of Section 2.3). Let us determine its decomposition according to eq. (2.28). First of all

$$F'(x) = \begin{cases} 0, & x < 0 \\ \mathrm{e}^{-x}/4, & x \ge 0, x \ne 1 \end{cases}$$

and therefore

$$\hat{F}_{\mathrm{ac}}(x) = \int_{-\infty}^{x} F'(t)\,\mathrm{d}t = \begin{cases} 0, & x < 0 \\ 1/4 - \mathrm{e}^{-x}/4, & x \ge 0 \end{cases}$$

Second, the function has two jumps at the points $x = 0$ and $x = 1$ of magnitude $p(0) = F(0) - F(0-) = 1/4$ and $p(1) = F(1) - F(1-) = 1/2$, respectively. Therefore

$$\hat{F}_{\rm d}(x) = \begin{cases} 0, & x < 0 \\ 1/4, & x \in [0, 1) \\ 3/4, & x \ge 1 \end{cases}$$

Then, by noting that $F = \hat{F}_{\rm ac} + \hat{F}_{\rm d}$ we get $F_{\rm sc} = F - \hat{F}_{\rm ac} - \hat{F}_{\rm d} = 0$, meaning that the singular continuous component of the decomposition is absent and eq. (2.28) reduces to eq. (2.31).

Now, when considered individually, the functions $\hat{F}_{\rm ac}$ and $\hat{F}_{\rm d}$ are not PDFs because – although being nondecreasing and right-continuous – they do not satisfy the limit condition at $+\infty$ of property (D2). As a consequence, if we want to write the decomposition of our PDF as in eq. (2.31) we have

$$F(x) = \frac{3}{4}\left(\frac{1}{3}H(x) + \frac{2}{3}H(x-1)\right) + \frac{1}{4}F_{\rm ac}(x)$$

where $H(x)$ is the Heaviside function (eq. (2.19)), the function within parenthesis is the discrete PDF and

$$F_{\rm ac}(x) = \begin{cases} 0, & x < 0 \\ 1 - {\rm e}^{-x}, & x \ge 0 \end{cases}$$

is the absolutely continuous PDF (which could also be written as $F_{\rm ac}(x) = (1 - {\rm e}^{-x})H(x)$).

2.3.2 *Numerical descriptors of random variables behaviour*

From the discussion of the preceding sections it is evident that the complete probabilistic description of a r.v. X is provided by its PDF F_X. Alternatively, we can use its mass distribution p_X if X is discrete or the pdf f_X if X is absolutely continuous. However, a certain degree of information – although incomplete in many cases – can be obtained by well-known numerical descriptors such as the mean value, the variance, etc. These quantities are special cases of a series of parameters called *moments* (of the r.v. X) whose general definition is given in terms of abstract Lebesgue integrals on (W, S, P), that is, the probability space on which X is defined. So, the mean (or expectation) of X, denoted by $E(X)$ or $E[X]$, is defined as

$$E(X) \equiv \int_W X \,{\rm d}P \tag{2.32a}$$

and, for its importance in statistics, is also often indicated by the symbol μ_X. Similarly, if k is any positive integer, the kth moment and the kth absolute moment of X are, respectively, the expectations of the random variables X^k and $|X|^k$, that is,

$$E(X^k) = \int_W X^k \, dP$$
$$E(|X|^k) = \int_W |X|^k \, dP \tag{2.32b}$$

Clearly $E(X^k) = E(|X|^k)$ if k is even. For odd values of k, on the other hand, we have the inequality $|E(X^k)| \leq E(|X|^k)$, which is a direct consequence of the properties of the integral.

The kth central moment – which makes sense only when $E(X)$ is finite – is defined as

$$E[(X - E(X))^k] = \int_W [X - E(X)]^k \, dP \tag{2.32c}$$

while $E[|X - E(X)|^k]$ is called the kth absolute central moment. Clearly, $E(X)$ is just the first moment of X and the first central moment (if it exists) is always zero because $\int_W dP = P(W) = 1$. Also, it is common to call *variance* the second central moment which is also frequently denoted by the special symbol σ_X^2 (or $\mathrm{Var}(X)$), that is, $\sigma_X^2 = E[(X - E(X))^2] = E[(X - \mu_X)^2]$. Its positive square root σ_X is called the *standard deviation* of X. The interpretation of μ_X and σ_X^2 – and of higher order moments in general – is probably known to the reader and will not be considered here because, in any case, it will become clearer as we proceed. Instead, we will turn to some of their basic properties which, as might be expected, are for the most part direct consequences of the properties of integrals (with respect to finite measures).

Proposition 2.12 *Liapunov inequality:* $E(|X|^k)^{1/k} \leq E(|X|^n)^{1/n}$ *for any two integers* k, n *such that* $k \leq n$. *More generally, if* $n > 1$ *and the nth moment of* X *is finite – that is,* $E(X^n) < \infty$ *– then both* $E(X^k)$ *and* $E(|X|^k)$ *are finite for* $1 \leq k \leq n$.

Mathematically this proposition can be expressed by saying that if $X \in L^n(W, S, P)$ then $X \in L^k(W, S, P)$ for $k \leq n$ and implies that $E(X^{m+1}), E(X^{m+2}), \ldots$ are not finite whenever $E(X^m)$ is not finite for some integer m.

Proposition 2.13 *(a) If* a *is a constant then* $E(a) = a$.

(b) Let a *be a constant and* A *an event, then – recalling that its indicator function* I_A *is a discrete random variable in its own right –* $E(aI_A) = aP(A)$. *In particular* $E(I_A) = P(A)$.

(c) *Linearity: Let a_j ($j = 1,2,\ldots,n$) be n constants and X_j n random variables, then* $E(\sum_{j=1}^{n} a_j X_j) = \sum_{j=1}^{n} a_j E(X_j)$.

(d) *Inequality preservation: If* $X_1 \le X_2$ *then* $E(X_1) \le E(X_2)$.

(e) $E(X)$ *exists if and only if* $E(|X|)$ *does. Moreover,* $|E(X)| \le E(|X|)$.

Proposition 2.14 *Let μ_X be the expectation of X. Then the central moments can be evaluated in terms of the ordinary moments by virtue of the binomial expansion theorem, that is,*

$$\begin{aligned} E[(X-\mu_X)^k] &= E\left[\sum_{j=0}^{k}(-1)^j \binom{k}{j} X^{k-j}\mu_X^j\right] \\ &= \sum_{j=0}^{k} \frac{(-1)^j k!}{j!(k-j)!}\mu_X^j E(X^{k-j}) \end{aligned} \tag{2.33}$$

A special case of eq. (2.33) is when $k = 2$; in this case we get the variance σ_X^2 as

$$\sigma_X^2 = E(X^2) - \mu_X^2 = E(X^2) - E^2(X) \tag{2.34}$$

Proposition 2.15 (a) *if b is a constant then* $\sigma_b^2 = 0$.

(b) *If X is a r.v. and b is a constant, then* $\sigma_{bX}^2 = b^2\sigma_X^2$.

(c) *Let X, Y be two random variables with finite mean and variance, then*

$$\begin{aligned} \sigma_{X\pm Y}^2 &= \sigma_X^2 + \sigma_Y^2 \pm 2(E(XY) - E(X)E(Y)) \\ &= \sigma_X^2 + \sigma_Y^2 \pm 2\mathrm{Cov}(X, Y) \end{aligned} \tag{2.35a}$$

and also, if a, b are two constants

$$\sigma_{aX\pm bY}^2 = a^2\sigma_X^2 + b^2\sigma_Y^2 \pm 2ab\mathrm{Cov}(X, Y) \tag{2.35b}$$

where we introduced the so-called covariance of two r.v.s defined as $Cov(X, Y) \equiv E[(X - E(X))(Y - E(Y))]$ *which, by the properties of expectation, is equal to* $E(XY) - E(X)E(Y)$.

It is not difficult to obtain, for example, eq. (2.35a). In fact, from eq. (2.34) we can write $\sigma_{X\pm Y}^2 = E[(X \pm Y)^2] - E^2(X \pm Y)$ and then note that the 1st term on the r.h.s. equals $E(X^2) + E(Y^2) \pm 2E(XY)$ while the second term equals $E^2(X) + E^2(Y) \pm 2E(X)E(Y)$; the difference of these two terms gives the desired result. Also, the reader is invited to show that the generalization of eq. (2.35a) to the case of the sum of n random variables $X_1, X_2, \ldots, X_n$ is

$$\operatorname{Var}\left(\sum_{i=1}^{n} X_i\right) = \sum_{i=1}^{n} \operatorname{Var}(X_i) + 2\sum_{i<j} \operatorname{Cov}(X_i X_j) \tag{2.35c}$$

which, again, is obtained by repeatedly exploiting the properties of expectation.

The next proposition states an important relation known as Chebyshev's (spelled variously in the literature as Tchebycheff or Tshebysheff) inequality which expresses an upper bound for the values assumed by any r.v. on subsets of the real line. Besides its importance in probability theory, the inequality can also be used as an estimating tool – although rather conservative – in many statistical applications.

Proposition 2.16 *Let b be any positive number, then Chebyshev's inequality can be written equivalently in the two forms*

$$\begin{aligned} &P(|X| \geq b) \leq \frac{E(|X|^k)}{b^k} \\ &P(|X| < b) \geq 1 - \frac{E(|X|^k)}{b^k} \end{aligned} \tag{2.36a}$$

Special cases of eq. (2.36a) are frequently found in other forms; for instance, if X has a finite variance, then

$$\begin{aligned} &P(|X - \mu_X| \geq b) \leq \frac{\sigma_X^2}{b^2} = \frac{\operatorname{Var}(X)}{b^2} \\ &P(|X - \mu_X| \geq r\sigma_X) \leq \frac{1}{r^2} \end{aligned} \tag{2.36b}$$

where in the second relation the constant b is expressed in standard deviation units, that is, $b = r\sigma_X$. Alternatively, one also finds

$$P(|X - \mu_X| < r\sigma_X) \geq 1 - \frac{1}{r^2} \tag{2.36c}$$

Other properties of the moments will be given and considered whenever needed in the course of the discussion.

Example 2.6 (Chebyshev's inequality as an estimating tool) Suppose the production of steel rods from a given industrial process is known to have a mean diameter of 20 mm and a standard deviation of 0.2 mm. Suppose further that these are the only available data about the process in question. For the future, the management decides that the production is considered satisfactory if at least 80% of the rods have diameters in the range 19.5–20.5 mm. Does the production process need to be changed?

Our r.v. X is the rod diameter and the question is whether $P(19.5 < X < 20.5) \geq 0.8$. In this case we have $r = 2.5$ and Chebychev's inequality in the form of eq. (2.36c) leads to

$$P(19.5 < X < 20.5) \geq 1 - \frac{1}{(2.5)^2} = 0.84$$

so that, according to the management's standards, the process can be considered satisfactory.

As stated above, Chebychev's inequality is rather conservative in the sense that the actual probability that X is in the range $\mu_X \pm r\sigma_X$ usually exceeds the lower bound $1 - 1/r^2$ by a significant amount. For example, if it was known that our r.v. follows a normal probability law (Example 2.4), then we would have $P(19.5 < X < 20.5) = 0.988$.

If now we consider the problem of actually calculating the moments of a random variable we note that it is not convenient to compute the abstract integrals on W given in Definitions (2.32) because in many cases the probability space (W, S, P) – even if it is known – is generally of little practical interest in applications. We then turn to the induced real probability space $(\mathbb{R}, \mathbb{B}, P_X)$ introduced in Section 2.3 and observe that, since P_X and P are strictly related and provide a complete probabilistic characterization of the random variable X, it should be possible to obtain its expectation, variance, etc. by computing Lebesgue–Stieltjes integrals (on $\mathbb{R}$) with respect to the probability measure P_X. In fact, we have the following result which can be proven within the framework of measure theory:

Proposition 2.17 *Let (W, S, P) be a probability space, X a r.v. on W and $g : \mathbb{R} \to \mathbb{R}$ a Borel function. Then the composite function $Z : W \to \mathbb{R}$ defined by $Z(w) \equiv g(X(w))$ is itself a r.v. (i.e. measurable) and*

$$E(Z) \equiv \int_W Z\,dP = \int_R g(x)\,dF_X \tag{2.37}$$

where the second integral is a Lebesgue–Stieltjes integral which, by definition, is an integral with respect to the measure P_X. Equation (2.37) is to be understood in the sense that if one of the two integrals exists, so does the other and they are equal (in other words, using a notation introduced in Appendix B, $Z \in L^1(W, S, P)$ if and only if $g \in L^1(\mathbb{R}, \mathbb{B}, P_X)$).

So, in particular, if $g(x) = x$ we get

$$E(X) = \int_{\mathbb{R}} x \, \mathrm{d}F_X \tag{2.38a}$$

and, clearly if $g(x) = x^k$

$$E(X^k) = \int_{\mathbb{R}} x^k \, \mathrm{d}F_X \tag{2.38b}$$

Also note that in the light of eq. (2.38a) – which apply to any r.v. – one can express the expectation of Z as $E(Z) = \int_{\mathbb{R}} z \, \mathrm{d}F_Z$ so that, by virtue of eq. (2.37), we are led to the equality

$$\int_{\mathbb{R}} z \, \mathrm{d}F_Z = \int_{\mathbb{R}} g(x) \, \mathrm{d}F_X \tag{2.39}$$

Which integral to use in order to calculate $E(Z)$ is merely a matter of convenience; in general the second integral is easier to use because it avoids having to determine the PDF F_Z, however, there may be cases in which the first integral is more efficient. Also, it is worth pointing out that these results imply that the expectation of a function of a r.v. depends only on its probability distribution: in other words, if the two r.v.s X, Y have the same PDF then $E[g(X)] = E[g(Y)]$ for all Borel functions $g(x)$.

So, considering the two cases of most practical importance in applications – namely the discrete and the absolutely continuous case – the moments of a r.v. X can be computed as

$$E(X^k) = \begin{cases} \sum_j x_j^k p_j \\ \int_{\mathbb{R}} x^k f_X(x) \, \mathrm{d}x \end{cases} \tag{2.40a}$$

respectively. Equations (2.40a) are, in fact, the explicit form taken on by the Lebesgue–Stieltjes integral of eq. (2.38b) in the two cases; the integration in $\mathrm{d}F_X$ becomes a sum if F_X is a jump function (clearly, the x_j are the values taken on by the discrete r.v. X) and a Lebesgue integral if F_X is absolutely continuous. In this latter case the pdf $f_X(x)$ is the Radon–Nikodym derivative $f_X = \mathrm{d}P_X/\mathrm{d}x$ of P_X with respect to the Lebesgue measure on $\mathbb{R}$ which, in turn, is denoted here by x instead of μ because in most practical cases the integral reduces to the ordinary Riemann integral of the function $x^k f_X(x)$. Clearly, these same conclusions apply to the more general case (i.e. when the function $g(x)$ is not equal to x^k) because eqs (2.40a) are just special cases of the relations

$$E[g(X)] = \begin{cases} \sum_j g(x_j) p_j \\ \int_{\mathbb{R}} g(x) f_X(x) \mathrm{d}x \end{cases} \tag{2.40b}$$

which represent the explicit form of eq. (2.39) in the discrete and absolutely continuous case, respectively.

Example 2.7 Suppose that X is an absolutely continuous r.v. with pdf $f_X(x) = \mathrm{e}^{-x} (x \geq 0)$. Suppose further that we want to obtain the first moment of the r.v. $Z = \sqrt{X}$. Equation (2.39) for the absolutely continuous case provides us with two options: we can either calculate

(a) $\int_0^\infty z f_Z(z)\, \mathrm{d}z$ or
(b) $\int_0^\infty \sqrt{x} f_X(x)\, \mathrm{d}x = \int_0^\infty \sqrt{x} \mathrm{e}^{-x}\, \mathrm{d}x$

We want to verify that these two integrals are, indeed, equal.

In case (a) we have to obtain the function $f_Z(z)$. Since it can be shown (Section 2.5, Example 2.10) that $f_Z(z) = 2z\mathrm{e}^{-z^2} (z \geq 0)$ the first integral becomes

$$\text{(a}'\text{)}\ 2\int_0^\infty z^2 \mathrm{e}^{-z^2}\, \mathrm{d}z$$

On the other hand, in order to compute the integral (b) we can make the change of variable $y = \sqrt{x}$ so that $x = y^2$, $\mathrm{d}x = 2y\mathrm{d}y$ and the integration limits remain unchanged. This way we get (b′) $2\int_0^\infty y^2 \mathrm{e}^{-y^2} \mathrm{d}y$, which, in fact, is the same as (a′).

Example 2.8 A discrete and an absolutely continuous case.

(a) An important case of discrete r.v. frequently encountered in applications is the so-called *binomial* r.v. Its mass distribution $p_X(x)$ is given by

$$p_X(x) = \binom{n}{x} p^x (1-p)^{n-x} = \frac{n!}{x!(n-x)!} p^x q^{n-x} \tag{2.41a}$$

where $x = 0, 1, 2, \ldots$ (we omit the index j) and $0 < p < 1$. This r.v. applies to all cases in which we perform a fixed (n) numbers of independent trials whose only possible outcomes are either a 'success' with probability p – which does not change from trial to trial – or a 'failure' with probability $q = 1 - p$ (this type of experiment is often described by saying that we are performing a series of n *'Bernoulli trials'*). The r.v. of interest X is the number of successes in n trials and obeys the probability law (2.41a) (the reader is invited to prove this by noting that p and q multiply because of the assumption of independent trials). By using eq. (2.40a) we get the mean $E(X) = np$. In fact,

$$\begin{aligned} E(X) &= \sum_{x=0}^{n} \frac{xn!}{x!(n-x)!} p^x q^{n-x} \\ &= np \sum_{x=1}^{n} \frac{(n-1)!}{(x-1)!(n-x)!} p^{x-1} q^{n-x} \\ &= np \sum_{x=1}^{n} \frac{(n-1)!}{(x-1)![(n-1)-(x-1)]!} p^{x-1} q^{(n-1)-(x-1)} \\ &= np \end{aligned} \tag{2.41b}$$

where the last equality holds because the sum is over all the ordinates of the distribution and must be unity for the normalization condition (2.18).

In order to obtain the variance of X we can use eq. (2.34) so that we only need the term $E(X^2)$. This is given by

$$\begin{aligned} E(X^2) &= \sum_{x=0}^{n} \frac{x^2 n!}{x!(n-x)!} p^x q^{n-x} = np \sum_{x=1}^{n} \frac{[(x-1)+1](n-1)!}{(x-1)!(n-x)!} p^{x-1} q^{n-x} \\ &= np \left((n-1)p \sum_{x=2}^{n} \frac{(n-2)!}{(x-2)![(n-2)-(x-2)]!} p^{x-2} q^{(n-2)-(x-2)} \right. \\ &\quad \left. + \sum_{x=1}^{n} \frac{(n-1)!}{(x-1)![(n-1)-(x-1)]!} p^{x-1} q^{(n-1)-(x-1)} \right) \\ &= np[(n-1)p + 1] \end{aligned}$$

Therefore

$$\sigma_X^2 = E(X^2) - n^2 p^2 = np(1-p) = npq \tag{2.41c}$$

(b) For the absolutely continuous case we use the eq. (2.40b) to determine, for instance, the mean and variance of a normal (Gaussian) r.v., that is, a r.v. whose pdf is given by eq. (2.29a). By performing (as in

Example 2.4) the change of variable $y = (x - \bar{x})/\sigma$ – so that $x = \sigma y + \bar{x}$ and $dx = \sigma\, dy$ – we get

$$E(X) = \frac{\sigma}{\sqrt{2\pi}} \int_{-\infty}^{\infty} y e^{-y^2/2}\, dy + \frac{\bar{x}}{\sqrt{2\pi}} \int_{-\infty}^{\infty} e^{-y^2/2}\, dy = \bar{x} \tag{2.42a}$$

because the first integral on the r.h.s. is zero ($y \exp(-y^2/2)$ is an odd function) and $\int_{-\infty}^{\infty} e^{-y^2/2}\, dy = \sqrt{2\pi}$.

By virtue of eq. (2.34) and with the same change of variable as above we obtain

$$\sigma_X^2 = \sigma^2 \tag{2.42b}$$

where we took into account that $\int_{-\infty}^{\infty} y^2 e^{-y^2/2}\, dy = \sqrt{2\pi}$. Equations (2.42a) and (2.42b) express the well-known result that the parameters $\bar{x}$ and σ appearing in the pdf of a normally distributed r.v. are its mean and standard deviation. In this light, it is customary to consider a normal r.v. with $\bar{x} = 0$ and $\sigma = 1$, which has a special name and is called a *standardized normal* r.v. Clearly, all the odd moments of a standardized normal r.v. are zero because its pdf is an even function (i.e. symmetrical about the ordinate axis). The moments of even order, on the other hand, are not zero and are given by

$$E(X^{2k}) = \frac{(2k)!}{2^k k!} \tag{2.42c}$$

for $k = 1, 2, \ldots$. Also, for a non-standard Gaussian r.v. it may be useful to know that the central higher order moments satisfy the recursion relation

$$\mu_k = (k-1)\sigma^2 \mu_{k-2} \tag{2.42d}$$

so that $\mu_2 = \sigma^2$, $\mu_4 = 3\sigma^4$, $\mu_6 = 5\sigma^2\mu_4 = 15\sigma^6$, etc. Clearly, all the odd central moments are zero because of the symmetry about the mean.

(c) In the case of a mixed r.v. (which, with applications in mind, we assume without continuous singular component) we know from the previous section that the measure P_X – and the PDF F_X – is expressed as the sum of an absolutely continuous part and a discrete part (eq. (2.31)). Then, by the properties of the Lebesgue–Stieltjes integral we have $\int x^k\, dF_X = \alpha \int x^k\, dF_{ac} + (1-\alpha) \int x^k\, dF_d$ and the same line of reasoning as above leads to

$$E(X^k) = \alpha \int_{-\infty}^{\infty} x^k f_X(x)\, dx + (1-\alpha) \sum_j x_j^k p_j \tag{2.43}$$

(the reader is invited to use eq. (2.43) to obtain the mean of the r.v. whose PDF is given in Example 2.5. The result is $E(X) = 0 + 1/2 + 1/4 = 3/4$).

2.4 Characteristic and moment-generating functions

In order to characterize completely the probabilistic behaviour of a random variable X, the PDF is not the only possibility at the analyst's disposal. In fact, let u be a real variable and consider the expectation of the function e^{iuX} (where i is the imaginary unit: $i = \sqrt{-1}$), that is,

$$E(e^{iuX}) = \int_{\mathbb{R}} e^{iux}\, dF_X \tag{2.44}$$

This is a (generally complex) function of u which is given the special name of *characteristic function* (CF) of the r.v. X and it is usually denoted by the symbol $\varphi_X(u)$. By virtue of eq. (2.40b) we have

$$\varphi_X(u) = \begin{cases} \sum_j e^{iux_j} p_j \\ \int_R e^{iux} f_X(x)\, dx \end{cases} \tag{2.45}$$

in the discrete and absolutely continuous case, respectively. Most readers will have probably noticed that the second of eq. (2.45) – besides the sign of the exponential which is generally written e^{-iux} in engineering and physics literature – is just the Fourier transform of f_X.

If, on the other hand, s is a real or complex variable the function $M_X(s)$ defined by the relation $M_X(s) \equiv E(e^{sX})$ is called the *moment-generating function* (MGF) of X. If s is in the form $s = iu$ then the MGF reduces to the CF; otherwise, if s is complex and the r.v. X is absolutely continuous, the MGF becomes the bilateral Laplace transform of the pdf f_X (which, again, in engineering literature is generally written with a minus sign in the exponential). In many works on probability, however, it is not unusual to consider s as a real variable.

For reasons of convergence, the CF is generally preferred. In fact, while $\varphi_X(u)$ exists for all values of u, the function $M_X(s)$ – where s is complex – may exist only for s in a particular region of the complex plane, the so-called region of convergence, which, in turn, is generally a vertical strip in the complex plane. In special cases this may be the whole plane, in other cases it is a proper strip (which, however, will always contains the imaginary axis because $\varphi_X(u) = M_X(s)$ there) but sometimes it is the degenerate strip consisting only of the imaginary axis. When the MGF exists, its properties are similar to the properties of the CF; therefore we will mainly consider characteristic functions with occasional remarks on MGFs.

The importance of the CF lies in the fact that there is a one-to-one correspondence between characteristic functions and probability distribution functions so that, as stated at the beginning of this section, knowledge of the CF of a random variable provides its complete probabilistic description. However, before examining why it is so, let us consider the main properties of characteristic functions.

Proposition 2.18 *Let $\varphi_X(u)$ be the CF of a r.v. X, then*

(a) $\varphi_X(0) = 1$ *and* $|\varphi_X(u)| \le 1$ *for all* u;
(b) $\varphi_X(-u) = \varphi_X^*(u)$ *where* $\varphi_X^*(u)$ *is the complex conjugate of* $\varphi_X(u)$;
(c) $\varphi_X(u)$ *is (uniformly) continuous on* $\mathbb{R}$;
(d) $\varphi_X(u)$ *is a non-negative definite function, meaning that for all n-tuples (where n is an arbitrarily chosen integer)* $u_1, u_2, \ldots, u_n \in \mathbb{R}$ *and all n-tuples of complex numbers* $a_1, a_2, \ldots, a_n$ *the quantity*

$$\sum_{j,k=1}^{n} a_j a_k^* \varphi_X(u_j - u_k)$$

is real and non-negative (as above the asterisk denotes complex conjugation).

Properties (a), (c) and (d) together characterize a CF in the sense that a function with these properties is necessarily the CF of some r.v. (this result is also known as Bochner's theorem). In this regard, however, given a function $\varphi(u)$, it may not be easy in practice to verify (d) in order to determine whether $\varphi(u)$ is a CF or not. This is why, in general, one usually checks properties (a), (b) and (c) and concludes that $\varphi(u)$ is not a CF if any one of them fails. If all of them are satisfied then, by necessity, property (d) must be considered.

The proofs of properties (a), (b) and (c) are not difficult. For property (d) we have, skipping some easy intermediate steps,

$$\sum_{j,k=1}^{n} a_j a_k^* \varphi_X(u_j - u_k) = \mathrm{E}\left(\sum_{j,k} a_j a_k^* \mathrm{e}^{\mathrm{i}u_j x} \mathrm{e}^{\mathrm{i}u_k x}\right)$$

$$= \mathrm{E}\left(\sum_j a_j \mathrm{e}^{\mathrm{i}u_j x} \sum_k a_k^* \mathrm{e}^{\mathrm{i}u_k x}\right) = \mathrm{E}\left(\left|\sum_j a_j \mathrm{e}^{\mathrm{i}u_j x}\right|^2\right) \ge 0$$

Also, a direct consequence of (b) is that the real part φ_{Re} of a CF φ is an even function – that is $\varphi_{\mathrm{Re}}(-u) = \varphi_{\mathrm{Re}}(u)$ – while its imaginary part φ_{Im} is odd, that is, $\varphi_{\mathrm{Im}}(-u) = -\varphi_{\mathrm{Im}}(u)$.

Proposition 2.19 *If $E(|X|^k) < \infty$ for some positive integer k then the kth derivative of $\varphi_X(u)$ exists for all u, is continuous and*

$$\frac{d^k \varphi_X(u)}{du^k} = \int_{\mathbb{R}} (ix)^k e^{iux}\, dF_X \tag{2.46}$$

Moreover, if for some even k the kth derivative of $\varphi_X(u)$ exists in $u = 0$, then $E(|X|^k) < \infty$ and eq. (2.46) holds.

Note that the first part of Proposition 2.19, in essence, states that the derivative can be taken under the integral sign or, in other words, that the kth derivative operator and the expectation operator commute, that is,

$$\frac{d^k}{du^k} E(e^{iux}) = E\left(\frac{d^k}{du^k} e^{iux}\right)$$

provided that the r.v. X has a finite kth absolute moment. We do not do it here but only point out that the proof of this result is based on the dominated convergence theorem (Appendix B). On the other hand, in the second part of the proposition (in whose proof one exploits Fatou's lemma) we cannot omit the condition of even k (in particular the existence of $d\varphi_X/du|_{u=0}$ does not imply $E(X) < \infty$).

In the light of these results let us now suppose that a r.v. has a finite kth order moment $E(X^k)$. Then by setting $u = 0$ in eq. (2.46) we get

$$\varphi_X^{(k)}(0) \equiv \left.\frac{d^k \varphi_X(u)}{du^k}\right|_{u=0} = i^k E(X^k) \tag{2.47a}$$

which means that the moments of a r.v. (when they exist) can be obtained by calculating the derivative of its CF in $u = 0$. In particular, we get

$$\begin{aligned} E(X) &= \varphi_X^{(1)}(0)/i = -i\varphi_X^{(1)}(0) \\ E(X^2) &= -\varphi_X^{(2)}(0) \end{aligned} \tag{2.47b}$$

As a side comment on the MGF of X it is not difficult to determine that the counterpart of eq. (2.47) is

$$M_X^{(k)}(0) = E(X^k) \tag{2.48}$$

with the important difference now that the condition $E(|X|^k) < \infty$ is no longer necessary because it can be shown that if $M_X(s)$ exists in some proper strip of the complex plane, then it is analytic there. This implies that its derivatives in $s = 0$ exist for all k and eq. (2.48) holds, thus justifying the name of MGF.

A direct consequence of the considerations above is that whenever X is such that $E(|X|^k) < \infty$ for all k then its CF can be expanded in Taylor series about the origin and

$$\varphi_X(u) = \sum_{k=0}^{\infty} \frac{(iu)^k}{k!} E(X^k) = 1 + \sum_{k=1}^{\infty} \frac{(iu)^k}{k!} E(X^k) \tag{2.49a}$$

within the interval of convergence of the series. For sake of completeness, it should be added that an expansion similar to eq. (2.49) is sometimes used for the function $\eta_X(u) \equiv \ln \varphi_X(u)$. This is written in the form

$$\eta_X(u) = \sum_{k=0}^{\infty} \frac{(iu)^k}{k!} C_k \tag{2.49b}$$

where C_k is called the kth order cumulant, or semi-invariant, of X. It is left to the reader to determine that the cumulants up to the third order are expressed in terms of moments and central moments by the relations

$$\begin{aligned} C_0 &= 0 \\ C_1 &= E(X) = \mu_X \\ C_2 &= E(X^2) - E^2(X) = \sigma_X^2 \\ C_3 &= 2E^3(X) - 3E(X)E(X^2) + E(X^3) = E[(X - \mu_X)^3] \end{aligned} \tag{2.49c}$$

Incidentally, we also note that the name 'semi-invariants' is due to the fact that the C_k ($k \geq 2$) are invariant under a translation of the random variable. In other words, if we change our r.v. from X to $\hat{X} = X + a$ then $\hat{C}_1 = C_1 + a$ and $\hat{C}_k = C_k$ for all $k \geq 2$, where $\hat{C}$ denotes the cumulants of $\hat{X}$.

Similarly to eqs (2.49a)–(2.49c), we can consider the MGF; within the proper strip of the complex plane where this function exists we can write the expansion

$$M_X(u) = \sum_{k=0}^{\infty} \frac{s^k}{k!} E(X^k) \tag{2.50}$$

Example 2.9(a) Consider a binomial random variable (i.e. a r.v. whose mass distribution is given by eq. (2.41a)). Using the first of eq. (2.45) we get

$$\varphi_X(u) = \sum_{x=0}^{n} \binom{n}{x} (pe^{iu})^x (1-p)^{n-x} = (1 - p + pe^{iu})^n \tag{2.51}$$

where in the last equality we used Newton's expansion theorem. Then, by calculating the derivatives of eq. (2.47) it is almost immediate to determine that $E(X) = np$ and $E(X^2) = np[(n-1)p+1]$ (in agreement with Example 2.8(a)).

(b) Consider now a Gaussian r.v. (its pdf is given in eq. (2.29a)). Its CF is given by

$$\varphi_X(u) = \frac{1}{\sigma\sqrt{2\pi}} \int_{-\infty}^{\infty} e^{iux} e^{-(x-\bar{x})^2/2\sigma^2} dx$$

In order to arrive to an explicit form we pass to the variable $y = (x-\bar{x})/\sigma$, so that skipping some easy intermediate steps we get

$$\begin{aligned}\varphi_X(u) &= \frac{e^{iu\bar{x}}}{\sqrt{2\pi}} \int_{-\infty}^{\infty} e^{iu\sigma y} e^{-y^2/2} dy = \frac{e^{iu\bar{x}}}{\sqrt{2\pi}} \int_{-\infty}^{\infty} e^{-(y-iu\sigma)^2/2} e^{-(\sigma u)^2/2} dy \\ &= \frac{e^{iu\bar{x}} e^{-(\sigma u)^2/2}}{\sqrt{2\pi}} \int_{-\infty}^{\infty} e^{-(y-iu\sigma)^2/2} dy = e^{iu\bar{x}-(1/2)\sigma^2 u^2}\end{aligned} \tag{2.52}$$

Then, using eq. (2.47) we obtain, as expected (Example 2.8(b)), $E(X) = \bar{x}$. Also, from the second derivative of $\varphi_X(u)$ we get $E(X^2) = \sigma^2 + \bar{x}^2$, which, taking eq. (2.34) into account, agrees with the result of eq. (2.42b). Clearly, the CF of a standardized normal r.v. can be obtained by setting $\bar{x} = 0$ and $\sigma = 1$ in eq. (2.52). Moreover, since the moments of this r.v. are given by eq. (2.42c) the series (2.49) converges for all u and we get

$$\varphi_X(u) = \sum_{k=0}^{\infty} i^{2k} \frac{u^{2k}}{(2k)!} \frac{(2k)!}{2^k k!} = \sum_{k=0}^{\infty} \frac{(-u^2/2)^k}{k!} = e^{-u^2/2}$$

because $i^{2k} = (-1)^k$ and in the last equality we used the well-known expansion $e^x = \sum_{k=0}^{\infty} x^k/k!$ of the exponential function.

The problem of finding the PDF of a r.v. when its CF is known is addressed by the so-called inversion formulas. In general, they may not be of easy use in practice but their importance lies in the fact that they justify and prove the statement that the CF provides a complete probabilistic description of the r.v. under study. We give without proof (which can be found, for example, in

Refs [1–3]) the following inversion formula:

Proposition 2.20 *Let $\varphi(u)$ and $F(x)$ (we omit here the subscript X) be the CF and PDF of a r.v. X, respectively. Then*

$$F(b) - F(a) = \lim_{c\to\infty} \frac{1}{2\pi} \int_{-c}^{c} \frac{e^{-iua} - e^{-iub}}{iu} \varphi(u)\, du \tag{2.53a}$$

for all points a, b $(a < b)$ where $F(x)$ is continuous.

This result implies that $F(x)$ is uniquely determined by $\varphi(u)$. In fact, if two PDFs have the same CF then – by Proposition 2.20 – they agree for every interval whose extremes are continuity points of $F(x)$. Therefore, by virtue of the characterizing properties of PDFs, the two PDFs must be identical.

For completeness, we give another inversion formula frequently found in literature. This is due to Feller and states that for all a, b $(a < b)$ where $F(x)$ is continuous

$$F(b) - F(a) = \lim_{c\to 0} \frac{1}{2p} \int_{\mathbb{R}} \frac{e^{-iua} - e^{-iub}}{iu} \varphi(u) e^{-c^2u^2}\, du \tag{2.53b}$$

Things are easier if $\varphi(u)$ is integrable on $\mathbb{R}$ because in this case we have

Proposition 2.21 *If the CF $\varphi(u)$ of a r.v. X is integrable on $\mathbb{R}$ then X is absolutely continuous and the inversion formula reads*

$$f(x) = \frac{1}{2\pi} \int_{-\infty}^{\infty} e^{-iux} \varphi(u)\, du \tag{2.54}$$

where $f(x)$ is the pdf of X (or, in other words, $F(x) = \int_{-\infty}^{x} f(t)\, dt$ where $F(x)$ is the PDF of the r.v. X).

It is not difficult to see that eq. (2.54) expresses the fact that $f(x)$ is just the inverse Fourier transform of $\varphi(u)$ and that the two functions $f(x)$ and $\varphi(u)$ – owing to the second of eq. (2.45) – are a Fourier transform pair.

Now, in the light of the 'moment-generating property' of CFs expressed by Proposition 2.19 and of eq. (2.49), the question could be asked if knowledge of all moments (when they exist) of a r.v. X determines uniquely its PDF, therefore allowing a complete probabilistic description of X. Similarly, one may ask: if two r.v.s X, Y (defined on the same probability space) have the same moments for all $k = 1, 2, \ldots$, can we conclude that they have the same PDF? This is, in general, a delicate problem which is beyond our scope; therefore we will limit ourselves to some general considerations. We noted above that the series expansion (2.49) determines the CF $\varphi_X(u)$ within the interval of convergence of the series on the r.h.s. However, since the knowledge of this function over a finite interval is not sufficient for a unique determination of $F(x)$, it follows that if the series converges only for $u = 0$ (to $E(X^0) = 1$) the answer to the question above is negative. In other cases, provided that some conditions are satisfied, the mathematical process of analytic continuation (frequently used in complex analysis) leads, in general, to a positive answer. In particular, we only state here the following results

Proposition 2.22 *Let $E(X^k)$, $k = 0, 1, 2 \ldots$, be the (finite) moments of a certain PDF $F(x)$. Suppose that the series $\sum_{k=0}^{\infty} u^k E(X^k)/k!$ is absolutely convergent for some $u > 0$. Then $F(x)$ is the only PDF that has the moments $E(X^0), E(X), E(X^2), \ldots$.*

Proposition 2.23 *(a) If a r.v. X has finite moments of all order then a sufficient condition in order that they identify its PDF is that there exists a number $c > 0$ such that*

$$\lim_{k\to\infty} \frac{c^{2k} E(X^{2k})}{(2k)!} = 0 \tag{2.55}$$

(b) If all moments of two r.v.s X, Y are finite, $E(X^k) = E(Y^k)$ for $k = 0, 1, 2 \ldots$ and eq. (2.55) holds, then $F_X(x) = F_Y(x)$ for all $x \in \mathbb{R}$.

Incidentally, it should be noted that this equality, in general, does not imply the equality $X = Y$. In fact, consider for example the two r.v.s $X = I_A$, $Y = I_{A^C}$ (i.e. the indicator functions of the sets A and A^C) on a probability space W such that $P(A) = P(A^C) = 1/2$. Owing to eq. (2.22) we have $F_X(x) = F_Y(x)$ for all x but $X(w) \neq Y(w)$ for all $w \in W$. On the other hand, it is obvious that the equality $X = Y$ necessarily implies $F_X(x) = F_Y(x)$.

A number of sufficient conditions other than (2.55) have been found for the sequence of moments to identify the probability distribution and the interested reader who desires to pursue this subject will find in literature the 'Riesz criterion', the 'Carlemen criterion' or the 'Ghizzetti criterion', just to name a few. We do not consider them here and turn our attention to the far-reaching concept of convergence in distribution.

Definition 2.5 Given the r.v.s $X_n (n = 1, 2, \ldots)$ and X with PDFs $F_n(x)$ and $F(x)$, respectively, we say that X_n *converges in distribution* to X if $\lim_{n\to\infty} F_n(x) = F(x)$ at all points x where $F(x)$ is continuous. In this case we will write $X_n \to X[D]$.

A few remarks are in order. First, it is important to note that the definition requires F_n to converge to a PDF. This is because there are cases in which a sequence F_n may converge to a function G which is not a PDF (typically,

G may fail to satisfy the conditions at infinity (D2) of Section 2.3). Second, we did not distinguish between discrete and continuous PDFs because a sequence of discrete PDFs may converge to a continuous PDF and conversely. Third, this type of convergence is also expressed by saying that the sequence F_n converges *weakly* to F and writing $F_n \to F[w]$. In the following, therefore, we will write indifferently $X_n \to X\ [D]$, $F_n \to F[w]$ or $X_n \to X\ [w]$.

Now, although the convergence of random variables will be considered in more detail in Chapter 4, Definition 2.5 has been given here in order to state an important result on characteristic functions. In fact, we have the following proposition:

Proposition 2.24 [Levy or Levy–Cramér theorem] *Let F_n be a sequence of PDFs and let φ_n be the sequence of their CFs. Then $F_n \to F[w]$ if and only if*

$$\lim_{n\to\infty} \varphi_n(u) = \varphi(u) \tag{2.56}$$

for all u, where F is a PDF whose CF is $\varphi(u)$.

Three points are worthy of mention:

(i) the convergence in eq. (2.56) is the ordinary pointwise convergence familiar from calculus;
(ii) Proposition 2.24 is a necessary and sufficient condition, that is, $F_n \to F[w]$ implies $\varphi_n(u) \to \varphi(u)$ and $\varphi_n(u) \to \varphi(u)$ implies $F_n \to F[w]$;
(iii) in addition to the statement of the proposition, it can also be shown that $F_n \to F[w]$ not only implies $\varphi_n(u) \to \varphi(u)$ but also that $\varphi_n(u) \to \varphi(u)$ uniformly on any bounded interval of $\mathbb{R}$.

As a final observation in this section one could ask if $X_n \to X\ [D]$ implies the convergence of moments. In general the answer is negative and some supplementary conditions are needed, as the following two propositions show.

Proposition 2.25(a) *Let $X_n \to X\ [D]$, then $E[g(X_n)] \to E[g(X)]$ for every bounded continuous function $g : \mathbb{R} \to \mathbb{R}$.*

As a matter of fact, the condition $\int_{\mathbb{R}} g(x)\, dF_n \to \int_{\mathbb{R}} g(x)\, dF$ – that is, $E[g(X_n)] \to E[g(X)]$ for any continuous and bounded $g(x)$ – is equivalent to Definition 2.5 (see, for example, Ref. [1]) and it is sometimes given as the definition of weak convergence while other authors [3] refer to Proposition 2.25(a) as the 'second generalized Helly's theorem'.

Proposition 2.25(b) *Let $X_n \to X[D]$. If there exist an integer $k > 0$ and a number C such that $E(|X_n|^k) \leq C$, then $E(|X|^k) < \infty$ and $E(X_n^j) \to E(X^j)$*

for $0 < j < k$ (where, clearly, this is the usual convergence of a sequence of real numbers).

As anticipated above, more about convergence of random variables is delayed to a later chapter.

2.5 Miscellaneous complements

In the light of the preceding developments it is our intention here to complement those ideas and concepts by discussing a few topics which – being worthy of attention in their own right – have only been considered briefly, if at all, in order not to interrupt the main line of reasoning.

2.5.1 Almost-sure and almost-impossible events

We start by going back to the probability axioms introduced in Section 2.2. Axiom (P1) states that $P(W) = 1$ for the sure event W; then, since $\emptyset = W^C$ this implies $P(\emptyset) = 0$ for the impossible event $\emptyset$. However, nothing is said about the possible existence of other events with probability one (or zero). Although this, at first sight, may seem in contrast with intuition, such events do exist.

So, if A ($A \neq W$) is an event with $P(A) = 1$ we say that A is an *almost-sure* event; consequently $P(A^C) = 0$ and we call A^C an *almost-impossible* event. The important point is that almost-sure and almost-impossible events behave, respectively, as W and $\emptyset$ as far as probabilities are concerned; in other words we can say that they are sure or impossible 'in probability'. In fact, we already know that given an (ordinary) event B then $B \cup \emptyset = B$ and $P(B \cup \emptyset) = P(B)$. If, on the other hand, A is an almost-impossible event the equality $B \cup A = B$ no longer holds but $P(B \cup A) = P(B)$ remains valid.

In this light, it is evident that the main properties of these events are as follows:

If A is an almost-impossible event, then

(a1) $P(B \cap A) = P(A) = 0$ and

(a2) $P(B \cup A) = P(B)$

for all $B \subset W$.
If C is an almost-sure event, then

(b1) $P(B \cap C) = P(B)$ and

(b2) $P(B \cup C) = P(C) = 1$

for all $B \subset W$.

The proof is easy and we only consider (a1) and (a2). Property (a1) is obtained by noting that $B \cap A \subset A$ which, by virtue of the monotonicity property, implies $P(B \cap A) \leq P(A) = 0$ and therefore $P(B \cap A) = 0$. For (a2) we have $B \cup A = B \cup (B^C \cap A)$ and since the two events on the r.h.s. are disjoint then $P[B \cup (B^C \cap A)] = P(B) + P(B^C \cap A) = P(B)$, where the last equality holds because of (a1).

A consequence of the definitions above is that the equalities $P(A) = 0$ or $P(A) = 1$ do not necessarily imply that $A = \emptyset$ or $A = W$, respectively. On the other hand, the equalities $A = \emptyset$ or $A = W$ do imply that $P(A) = 0$ or $P(A) = 1$, respectively (clearly, this remark is true for any two sets A, B; while $A = B$ implies $P(A) = P(B)$ the reverse statement, in general, is not true).

The considerations above may appear more familiar when considered from the measure viewpoint. In fact, for example, if $W = [0, 1]$ and S is the σ-algebra of its Borel subsets, the classical probability concept that any real number $0 \leq a \leq 1$ picked at random has the same probability of being chosen is obtained by assigning equal probabilities to all intervals of the same length. In fact, the condition of equal probabilities implies that any individual real number has a probability zero (i.e. is an almost-impossible event) and any countable union of such numbers has also probability zero by virtue of σ-additivity. However, σ-additivity holds at most for countable unions and an interval – an uncountable union of real numbers – may have a non-zero probability. Therefore, assigning equal probabilities to intervals with the same length determines uniquely a function P satisfying all the probability axioms. It is not difficult to see that in this case P is the Lebesgue measure μ on [0,1] which, as known from analysis, assigns zero measure to any individual point a (and to any finite or countable union of such points) and measure $\beta - \alpha$ to any subinterval (α, β) – open, closed or semiclosed – of [0, 1].

Incidentally, we note that the probability $P = \mu$ considered here defines the so-called *uniform probability distribution* on [0, 1]. More generally, by assigning equal probabilities to subintervals (of equal length) of a finite interval $[a, b]$ it is not difficult to see that we determine the probability measure $P = \mu/(b - a)$ which in turn, leads to the distribution function

$$F(x) = \begin{cases} 0, & x \leq a \\ \dfrac{x-a}{b-a}, & a < x \leq b \\ 1, & x > b \end{cases} \tag{2.57a}$$

called the *uniform* PDF on $[a, b]$. This function is absolutely continuous and its pdf is $f(x) = 1/(b - a)$ for $x \in (a, b)$ and zero elsewhere. Also, it is immediate to determine that a r.v. X whose PDF is given by (2.57a) has moments given by

$$E(X^k) = \int_a^b x^k f(x)\,dx = \frac{b^{k+1} - a^{k+1}}{(b-a)(k+1)} \tag{2.57b}$$

from which it follows $E(X) = (a+b)/2$ and, with a little algebra, $\mathrm{Var}(X) = (b^2 - a^2)/12$. The characteristic function is also obtained with little effort and we get

$$\varphi(u) = \frac{e^{iub} - e^{iua}}{iu(b-a)} \tag{2.57c}$$

which gives an indeterminate form 0/0 for $u = 0$; however $\varphi(u) \to 1$ as $u \to 0$. If $a = -b$ the function $F(x)$ is even, the CF is real and can be written as $\varphi(u) = \sin(bu)/bu$.

2.5.2 More on conditional probability

The second aspect we consider here deals with conditional probabilities. We introduced the concept informally in Section 1.2.1 and made some further comments in Section 2.2. There we pointed out that – given a probability space (W, S, P) and a conditioning event $G \in S$ with $P(G) > 0$ – the set function $P_G : W \to [0, 1]$ defined (for $A \in S$) by the relation $P_G(A) = P(A \cap G)/P(G)$ is a probability function in its own right which, often, is also denoted by $P(A|G)$. Also, if $X : W \to \mathbb{R}$ is a r.v. on (W, S, P) we observe that X is a r.v. (i.e. measurable) on the space (W, S, P_G) as well, because – we recall from Definition 2.3 – the measurability of functions is independent on the measure P.

We have now two probability measures on (W, S), that is, P and P_G, and the first thing to note is that P_G is absolutely continuous with respect to P because $P_G(A) = 0$ whenever $P(A) = 0$. Then, the Radon–Nikodym theorem states that there is an essentially unique function $H : W \to \mathbb{R}$ such that

$$P_G(A) = \int_A H\, dP$$

This function is called the Radon–Nikodym derivative of P_G with respect to P and it often symbolically denoted by dP_G/dP. We state now that

$$H = \frac{I_G}{P(G)} \tag{2.58}$$

where I_G is the indicator function of the set G. In fact, by the defining properties of abstract Lebesgue integral we have

$$P(A \cap G) = \int_{A\cap G} dP = \int_W I_{A\cap G}\, dP = \int_W I_A I_G\, dP = \int_A I_G\, dP$$

where the third equality holds because it is immediate to prove that $I_{A\cap G} = I_A I_G$. Substituting this result into the definition of P_G we get for every $A \in S$

$$P_G(A) = \frac{1}{P(G)} \int_A I_G \, dP \tag{2.59}$$

which, in the light of uniqueness of H, proves eq. (2.58).

Given a r.v. X on W, a more interesting – and useful in practice – case is when the conditioning set G is the counterimage through X of a Borel set $C \subset \mathbb{R}$, that is, when $G = X^{-1}(C)$. Then, calling P_C the probability measure defined by

$$P_C(A) = \frac{P[A \cap X^{-1}(C)]}{P(X^{-1}(C))} \tag{2.60}$$

(Incidentally, this notation may seem strange because P_C is a measure in (W, S) but C is a Borel set in the domain of X; rigorously one should write $P_{X^{-1}(C)}$ but then the notation would become too heavy) we can consider its image measure $P_{X|C}$ in $\mathbb{R}$ and note that it is absolutely continuous with respect to P_X (the image measure of P). By a similar argument as above, the Radon–Nikodym theorem applies. So – recalling the relation between the abstract Lebesgue integrals in dP and dP_X and noting that $X^{-1}(B) \cap X^{-1}(C) = X^{-1}(B \cap C)$ – from the chain of equalities

$$\begin{aligned} P_{X|C}(B) &= P_C[X^{-1}(B)] = \frac{P[X^{-1}(B \cap C]}{P(X^{-1}(C))} = \frac{1}{P(X^{-1}(C))} \int_{X^{-1}(B\cap C)} dP \\ &= \frac{1}{P_X(C)} \int_{B\cap C} dP_X = \frac{1}{P_X(C)} \int_B I_C \, dP_X \end{aligned} \tag{2.61}$$

it follows that the Radon–Nikodym derivative $dP_{X|C}/dP_X$ is

$$\frac{dP_{X|C}}{dP_X} = \frac{I_C}{P_X(C)} \tag{2.62}$$

If now we turn our attention to the conditional PDF defined as

$$F_{X|C}(x) = P_C[X^{-1}(J_x)] \tag{2.63}$$

where $J_x = (-\infty, x]$, we can use the basic result of eq. (2.61) to get

$$F_{X|C}(x) = \frac{1}{P_X(C)} \int_{J_x} I_C \, dP_X = \frac{1}{P_X(C)} \int_{J_x} I_C \, dF_X \tag{2.64}$$

where the second integral is a Lebesgue–Stieltjes integral. If, in addition, the function F_X is absolutely continuous on $\mathbb{R}$ then the pdf f_X exists and we can take the derivative of both sides to obtain the conditional pdf in terms of the unconditional one

$$f_{X|C}(x) = \frac{I_C f_X(x)}{P_X(C)} = \frac{I_C f_X(x)}{\int_C f_X(x)\, dx} \tag{2.65a}$$

On the other hand, if F_X is discrete we have

$$p_{X|C}(x_k) = \frac{I_C p_X(x_k)}{P_X(C)} = \frac{I_C p_X(x_k)}{\sum_{x_i \in C} p(x_i)} \tag{2.65b}$$

Equations (2.65a) and (2.65b) show that $f_{X|C}$ (or $p_{X|C}$) is zero outside the set C while in C, besides the multiplicative constant $1/P_X(C)$, coincides with the unconditioned pdf (or pmf). The factor $1/P_X(C)$ is necessary in order to satisfy the normalization condition.

Using the conditional characteristics we can define and calculate the conditional moments just as we did in the unconditioned case. So, we can define

$$E(X^k|C) = \int_W X^k \, dP_C = \int_{\mathbb{R}} x^k \, dP_{X|C} = \int_{\mathbb{R}} x^k \, dF_{X|C} \tag{2.66a}$$

where the second equality holds because of the relation between P_C and its image measure $P_{X|C}$ and the third equality holds because of the definition of Lebesgue–Stieltjes integral. Then, by virtue of eq. (2.62) we also have

$$E(X^k|C) = \frac{1}{P_X(C)} \int_C x^k \, dF_X \tag{2.66b}$$

where the integral on the r.h.s. is a sum or a Lebesgue integral on $\mathbb{R}$ (which in most practical cases coincides with an ordinary Riemann integral) depending on whether F_X is discrete or absolutely continuous. Similarly, we can define the conditional-CF as $\varphi_{X|C}(u) = E(e^{iuX}|C)$ and, more generally, the expectation of a (Borel) function $g(X)$. In the absolutely continuous case, for example, we get

$$E[g(X)|C] = \frac{\int_C g(x) f_X(x)\, dx}{\int_C f_X(x)\, dx} \tag{2.67}$$

Clearly, when all events $B \subset \mathbb{R}$ are independent of the conditioning event C then the conditioned characteristics coincide with the unconditioned ones; the simplest case of this situation is when $C = \mathbb{R}$ or, in the space (W, S), when $G = W$.

We close this section with a final result regarding conditional expectations which is somehow a counterpart of the total probability formula of eq. (1.12). This result is one form of the so-called total expectation theorem and is given in the following proposition:

Proposition 2.27 *(Total expectation theorem)* *Let the sets $G_i \in S$ be such that $W = \cup_i G_i$, $G_i \cap G_j = \emptyset$ for $i \neq j$ and $P(G_i) > 0$ for all $i = 1, 2, \ldots$. Then*

$$E(X) = \sum_i P(G_i) E(X|G_i) \tag{2.68}$$

In fact, as a consequence of the Radon–Nikodym theorem, we can write $E(X|G_i) = [P(G_i)]^{-1} \int_{G_i} X\, dP$ but then, owing to the properties of the abstract Lebesgue integral we get

$$E(X) = \int_W X\, dP = \int_{\cup_i G_i} X\, dP = \sum_i \int_{G_i} X\, dP$$

so that Proposition 2.26 follows from these two results.

For the moment, the considerations above suffice and we leave further developments on conditional probability to future sections. In particular, for continuous random variables we will show how one can condition on an event with zero probability, that is an event of the form $X = x_0$ where x_0 is a specified value and $P_X\{x_0\} = P(X^{-1}\{x_0\}) = 0$.

2.5.3 *Functions of random variables*

The third topic we consider here deals with random variables with a known functional dependence on another random variable. So, let X be a r.v. with known probability distribution F_X and let $g : \mathbb{R} \to \mathbb{R}$ be a well-behaved Borel function. Since we already know that the function $Y(w) \equiv g(X(w)) : W \to \mathbb{R}$ is itself a random variable, we may ask for its probability distribution.

This is not always simple and we will consider only some frequently encountered cases.

Suppose first that X is absolutely continuous with pdf f_X and g is monotonically increasing. Then, given a value y, we have that $Y \le y$ whenever $X \le x$ (where $y = g(x)$) and Y is also absolutely continuous. Moreover, in this case the inverse function g^{-1} exists is single-valued and $g^{-1}(y) = x$; therefore

$$F_Y(y) = P(Y \le y) = P[X \le g^{-1}(y)] = F_X(g^{-1}(y)) \tag{2.69a}$$

then, taking the derivative with respect to y we obtain the pdf of Y as

$$f_Y(y) = f_X(g^{-1}(y))\frac{\mathrm{d}g^{-1}(y)}{\mathrm{d}y} \tag{2.69b}$$

If, on the other hand, g is monotonically decreasing, then we have

$$\begin{aligned} F_Y(y) &= P(Y \le y) = P[X > g^{-1}(y)] \\ &= 1 - P[X \le g^{-1}(y)] = 1 - F_X(g^{-1}(y)) \end{aligned} \tag{2.70a}$$

and differentiating

$$f_Y(y) = -f_X(g^{-1}(y))\frac{\mathrm{d}g^{-1}(y)}{\mathrm{d}y} \tag{2.70b}$$

By noting that the derivative $\mathrm{d}g^{-1}/\mathrm{d}y$ is positive when g is monotonically increasing and negative when g is monotonically decreasing, eqs (2.69b) and (2.70b) can be combined into the single equation

$$f_Y(y) = f_X(g^{-1}(y))\left|\frac{\mathrm{d}g^{-1}(y)}{\mathrm{d}y}\right| \tag{2.71}$$

Example 2.10(a) Let X be a r.v. with pdf $f_X = \mathrm{e}^{-x}(x \ge 0)$ and let $Y = g(X) = \sqrt{X}$. Then $x = g^{-1}(y) = y^2$, $f_X(g^{-1}(y)) = \mathrm{e}^{-y^2}$ and $\mathrm{d}g^{-1}/\mathrm{d}y = 2y$, so that, by eq. (2.69b), we get $f_Y(y) = 2y\mathrm{e}^{-y^2}(y \ge 0)$. Clearly, if we note that $F_X(x) = 1 - \mathrm{e}^{-x}$ we can use eq. (2.69a) to get the PDF $F_Y(y) = 1 - \mathrm{e}^{-y^2}$, which, as expected, can also be obtained by computing the integral $F_Y(y) = \int_0^y f_Y(t)\mathrm{d}t$. As an easy exercise, the reader is invited to sketch a graph of f_X and f_Y and note that they are markedly different.

Example 2.10(b) Let us now consider the linear case $Y = g(X) = aX + b$. This is an increasing function for $a > 0$ and decreasing for $a < 0$. If $f_X(x)$

is the pdf of X, eq. (2.71) yields $f_Y(y) = |a|^{-1} f_X((y-b)/a)$. Also, $F_Y(y) = F_X((y-b)/a)$ if $a > 0$ and $F_Y(y) = 1 - F_X((y-b)/a)$ if $a < 0$.

So, for example, if the original r.v. X is normally distributed – that is, its pdf is given by eq. (2.29a) – and $a > 0$ we get

$$f_Y(y) = \frac{1}{a\sigma\sqrt{2\pi}} \exp\left(-\frac{(y - a\bar{x} - b)^2}{2a^2\sigma^2}\right)$$

meaning that Y is a Gaussian r.v. itself with mean $\bar{y} = a\bar{x} + b$ and variance $\sigma_Y^2 = a^2\sigma^2$.

Example 2.10(c) Starting again from the normal pdf of eq (2.29a) we can obtain the pdf of the standardized normal r.v. $Y = (X - \bar{x})/\sigma$. Noting that $g^{-1}(y) = \sigma y + x$ and $dg^{-1}/dy = \sigma$ we get

$$f_Y(y) = \frac{1}{\sqrt{2\pi}} \exp(-y^2/2)$$

which, as mentioned at the end of Example 2.8, is a normal r.v. with $\bar{x} = 0$ and $\sigma = 1$.

If g is not monotone, it can often be divided into monotone parts; the considerations above then apply to each part and in the end the sum of the various parts is taken. A simple example of this latter case is $Y = g(X) = X^2$ which is decreasing for $x < 0$ and increasing for $x > 0$. Since $g(x)$ is always positive for all x (or, stated differently, $g^{-1}(y) = \emptyset$ for $y < 0$) the r.v. Y cannot take on negative values and therefore $f_Y(y) = 0$ for $y < 0$. For $y > 0$ it is left to the reader to determine that the sum of the two parts leads to $f_Y(y) = (2\sqrt{y})^{-1}(f_X(-\sqrt{y}) + f_X(\sqrt{y}))$.

Example 2.11 In applications it is often of interest to have a probabilistic description of the maximum or minimum of a number n of r.v.s. As we will see in later chapters, an important case is when the r.v.s $X_1, X_2, \ldots, X_n$ are independent and have the same PDF $F(x)$. Now, first of all it can be shown that the function $Y = \max\{X_1, \ldots, X_n\}$ is itself a r.v. (as is the minimum). Then, since $F_Y(y) = P(\max\{X_1, \ldots, X_n\} \le y) = P(X_1 \le y, \ldots, X_n \le y)$ the assumption of independence leads to

$$F_Y(y) = P(X_1 \le y, \ldots, X_n \le y) = \prod_{i=1}^{n} P(X_1 \le y) = (F(y))^n \tag{2.72a}$$

and therefore, if F is absolutely continuous

$$f_Y(y) = n(F(y))^{n-1} f(y) \tag{2.72b}$$

where f is the derivative of F. We note here that the expression $P(X_1 \le y, \dots, X_n \le y)$ is written in the usual 'shorthand' notation of probability theory; in rigorous mathematical symbolism, however, this probability is written $P\left(\cap_{i=1}^{n} X_i^{-1}(-\infty, y]\right)$. The rigorous notation is useful when we consider the minimum of the r.v.s $X_1, X_2, \dots, X_n$. In fact, if $J_y = (-\infty, y]$ we have

$$\begin{aligned} F_Y(y) &= P(\min\{X_1, \dots, X_n\}) = P\left(\bigcup_i X_i^{-1}(J_y)\right) \\ &= P\left[\bigcap_i (X^{-1}(J_y))^C\right]^C = 1 - P\left[\bigcap_i (X^{-1}(J_y))^C\right] \end{aligned}$$

where in the third equality we used de Morgan's law. Then, by virtue of independence $P\left[\cap_i (X_i^{-1}(J_y))^C\right] = \prod_i P\left[X_i^{-1}(J_y)\right]^C = (1 - F(y))^n$, so that putting the pieces together we finally get

$$F_Y(y) = 1 - (1 - F(y))^n \tag{2.73a}$$

and if F is absolutely continuous ($F' = f$)

$$f_Y(y) = n(1 - F(y))^{n-1} f(y) \tag{2.73b}$$

So, for instance, if F is the uniform PDF (eq. (2.57a)) on the interval $[a, b] = [0, 1]$ then $F_Y(y) = n(1 - y)^{n-1}$ where $0 \le y \le 1$.

If X is a discrete r.v. whose range is the set $A_X = \{x_1, x_2, \dots\} = \{x_i\}$ then $Y = g(X)$ is also a discrete r.v. because its range is the set $A_Y = \{y_1, y_2, \dots\} = \{y_k\}$ (note that the elements of A_X and A_Y are labelled by different indexes because, in general, a given value y_k may be the image – through g – of more than one x_i). In this case, in general, it is not convenient to go through the PDF but it is better to determine the mass distribution p_Y by first identifying the values y_k and then using the relation

$$p_Y(y_k) = P(Y = y_k) = P(X = g^{-1}(y_k)) = \sum_{x_i} p_X(g^{-1}(y_k)) \tag{2.74}$$

where the sum is taken on all values x_i (when there is more than one) which are mapped in y_k. So, for instance, let X be such that $A_X = \{x_1 = -1,$

$x_2 = 0, x_3 = 1\}$, $p_X(-1) = p_X(0) = 0.25$ and $p_X(1) = 0.5$ and let $Y = X^2$. Then $A_Y = \{y_1, y_2\} = \{0, 1\}$ and $g^{-1}(y_2) = -1 \cup 1 = x_1 \cup x_3$ so that in calculating $p_Y(y_2)$ we must sum the probabilities $p_X(x_1)$ and $p_X(x_3)$. Therefore $p_Y(y_2) = 0.75$. By contrast, the sum is not needed in calculating $p_Y(y_1) = 0.25$.

2.6 Summary and comments

This chapter introduces the axiomatic approach to probability by giving a number of fundamental concepts and results which are at the basis of all further developments in both fields of probability theory and statistics. In essence, the axiomatic approach consists in calling 'probability' any set function that satisfies certain properties, together with the definition of what exactly is meant by the term 'event'. Clearly, in order to speak of probability this latter definition is a necessary prerequisite because probabilities can only be assigned to events. Both definitions, probability and events, are given in Section 2.2 by introducing the concepts of *elementary probability spaces* – which apply to all cases with a finite number of possible outcomes – and *probability spaces*, where the restriction of finiteness is relaxed. These notions are sufficiently general to include as special cases all the definitions of probability considered in Chapter 1.

In mathematical terms, a probability space is just a finite measure space and a probability P is a σ-additive measure defined on a σ-algebra of subsets (the events) of a 'universal' set W with the property $P(W) = 1$. The domain of P is taken to be a σ-algebra because measure theory – by virtue of the construction of the Lebesgue extension of measures – guarantees that a knowledge of the values taken on by P on a limited number of 'elementary' events (which, in general, form a semialgebra of subsets of W) is sufficient in order to determine uniquely P on a much broader class of events, this class being, in fact, a σ-algebra. The extension procedure is outlined in Section 2.2.1 and is summarized by the result known as *Caratheodory extension theorem*.

A fundamental aspect of probability which distinguishes it from measure theory is the notion of *independent events*. Due to its far-reaching consequences in both the theory and real-word applications of statistics, this concept is discussed in some detail in Section 2.2.2, where it is pointed out that the intuitive idea of independence as the absence of causal relation between two (or more) events is translated into mathematical language by a product rule between the probabilities of the events themselves.

At this point we consider the fact that in many applications the analyst is mainly interested in assigning probabilities to numerical quantities associated with the outcomes of an experiment rather than to the outcomes themselves. This task is accomplished by introducing the concept of r.v., that is, a real-valued function defined on W and satisfying a 'measurability' condition with respect to the σ-algebra of W and the σ-algebra of Borel sets of the real line $\mathbb{R}$. The condition is formulated by requiring that the inverse image of any Borel set (in the domain of the random variable) be an element of the σ-algebra of W, thus allowing the possibility of assigning probabilities to subsets of $\mathbb{R}$ (in the form of open, closed, semiclosed intervals or of individual real numbers, just to mention the most frequently encountered cases).

A r.v. X, in turn, induces a probability measure P_X on the real line and therefore a real probability space $(\mathbb{R}, \mathbb{B}, P_X)$. This space, in general, is all that is needed in applications because, through the measure P_X, the analyst can obtain a complete probabilistic description of X by defining the so-called PDF of X, usually denoted by $F_X(x)$. Clearly, P_X and F_X are strictly related; in fact, mathematically speaking, P_X is the Lebesgue–Stieltjes measure corresponding F_X and F_X is the generating function the measure P_X (this name comes from the fact that in analysis one usually defines a Lebesgue–Stieltjes measure by means of its generating function and not the other way around as it is done in probability).

With the concept of PDF at our disposal, Section 2.3.1 classifies the various types of random variables according to the continuity properties of their PDFs. A first classification distinguishes between discrete and continuous r.v.s by also introducing the concept of *probability mass distribution* for discrete r.v.s. Then, among continuous r.v.s a further classification distinguishes between absolutely continuous and singular continuous r.v.s, the distinction being due to the fact that for the former type – by far the most important in applications – it is possible to express their PDF by means of the ordinary Lebesgue integral of an appropriate pdf $f_X(x)$ which, in turn, is the derivative of $F_X(x)$. This possibility relies ultimately on an important result of analysis (given in Appendix B) known as Radon–Nikodym theorem.

The conclusion is that the PDF of the most general type of r.v. can be expressed as the sum of a discrete part, an absolutely continuous part and a continuous singular part which, however, is generally absent in most practical cases. Moreover, for the discrete and the absolutely continuous case, respectively, the mass distribution and the pdf provide a complete probabilistic description of the r.v. under study.

Proceeding in the discussion of fundamentals, Section 2.3.2 introduces the most common numerical descriptors of r.v.s – the so-called *moments* of a r.v. – which are defined by means of abstract Lebesgue integrals on the space W. Special cases of moments – the first and second moment, respectively – are the familiar quantities known as mean and variance. The properties of moments are then considered together with the important result of Chebychev's inequality. Subsequently, the problem of actually calculating moments is considered by first noting that it is generally not convenient to compute abstract integrals on W. In this regard, in fact, it is shown that moments can be obtained as Lebesgue–Stieltjes integrals on the real line and that these integrals, in the most common cases of discrete and absolutely continuous r.v.s respectively, reduce to a sum and to an ordinary Riemann integral. A few examples are then given in order to illustrate some frequently encountered cases.

Besides the PDF, another way of obtaining a complete probabilistic description of a r.v. X is its *characteristic function* $\varphi_X(u)$. The concept is introduced in Section 2.4 together with some comments on the 'parallel' notion of moment generating function (denoted $M_X(s)$). The main properties of CFs are given by also showing how moments (when they exist) can be easily computed by differentiating φ_X.

The fact that the CF provides a complete probabilistic description of a r.v. is due to the existence of a one-to-one relationship between PDFs and CFs. The problem of obtaining the CF from the PDF is given by the definition of CF itself while the reverse problem is addressed by the so-called inversion formulas which, in the general case, are important for their mere existence but are of little practical use. In the particular case of absolutely continuous r.v.s, however, things are easier because the correspondence reduces to the fact that the pdf f_X and the CF φ_X are a Fourier transform pair and the notion of Fourier transform is well known and widely used in Engineering and Physics literature. The section closes with a brief discussion of *convergence in distribution* (or weak convergence) of sequences of r.v.s for its strict relation with pointwise convergence of characteristic functions.

Finally, in Section 2.5 we consider a number of complementary ideas and concepts which are worthy of mention in their own right but have been delayed in order not to interrupt the main line of reasoning. Section 2.5.1 introduces the notion of almost-sure (and almost impossible) events by pointing out that there exist events with probability one (and zero) which are different from W (and $\emptyset$). This is not surprising in the light of measure theory when, for example, one considers the Lebesgue measure on a finite interval of the real line. Subsequently (Section 2.5.2) we extend the notion of conditional probability by showing how, in general – given an event with strictly positive probability – there is no difficulty in defining such concepts as the conditional PDF of a r.v., the conditional CF, etc. The chapter closes with Section 2.5.3 where it is shown with some examples how to obtain the PDF, pdf or mass distribution of a function $g(X)$ when the PDF (pdf or mass distribution) of the r.v. X is known.

References and further reading

[1] Ash, R.B., Doléans-Dade, C., '*Probability and Measure Theory*', Harcourt Academic Press, San Diego (2000).
[2] Cramer, H., '*Mathematical Methods of Statistics*', Princeton Landmarks in Mathematics, Princeton University Press,19th printing (1999).
[3] Gnedenko B.V., '*Teoria della Probabilità*', Editori Riuniti, Roma (1987).
[4] Haaser, N.B., Sullivan, J.A., '*Real Analysis*', Dover, New York (1991).
[5] Kolmogorov, A.N., '*Foundations of Probability*', AMS Chelsea Publishing, Providence, Rhode Island (2000).

[6] Kolmogorov, A.N., Fomin, S.V., *'Introductory Real Analysis'*, Dover, New York (1975).

[7] Kolmogorov, A.N., Fomin, S.V., *'Elementi di Teoria delle Funzioni e di Analisi Funzionale'*, Edizioni Mir, Mosca (1980).

[8] McDonald, J.N., Weiss, N.A., *'A Course in Real Analysis'*, Academic Press, San Diego (1999).

[9] Monti, C.M., Pierobon, G., *'Teoria della Probabilità'*, Decibel editrice, Padova (2000).

[10] Rudin, W., *'Principles of Mathematical Analysis'*, 3rd ed., McGraw-Hill, New York, (1976).

[11] Rudin, W., *'Real and Complex Analysis'*, McGraw-Hill, New York (1966).

[12] Taylor, J.C., *'An Introduction to Measure and Probability'*, Springer-Verlag, New York (1997).

3 The multivariate case: random vectors

3.1 Introduction

The scope of this chapter is to proceed along the line of reasoning of Chapter 2 by turning our attention to cases in which two or more random variables are considered together. With this in mind, we will introduce the new concept of 'random vector' by considering measurable vector-valued functions defined on probability spaces. The main mathematical aspects parallel closely the one-dimensional case but it is worth pointing out that now the notion of stochastic independence will play a major role. In fact, this concept is peculiar to probability theory and distinguishes it from being merely an application of analysis.

3.2 Random vectors and their distribution functions

The definition of random vector is a straightforward generalization of the concept of random variable; in fact

Definition 3.1 Given a probability space (W, S, P), an n-dimensional random vector is a function $\mathbf{X}$: $W \to \mathbb{R}^n$ such that $\mathbf{X}^{-1}(B) \in S$ for every Borel set $B \in \mathbb{B}(\mathbb{R}^n)$ where, as customary, we denote by $\mathbb{B}(\mathbb{R}^n)$ or $\mathbb{B}_n$ the σ-algebra of all Borel sets of $\mathbb{R}^n$. In this regard it is important to note that the σ-algebra $\mathbb{B}_n$ is the cartesian product of the n terms $\mathbb{B} \times \mathbb{B} \times \cdots \times \mathbb{B}$, meaning, in other words, that every n-dimensional Borel set $A \in \mathbb{B}_n$ is of the form $A = A_1 \times A_2 \times \cdots \times A_n$ where $A_1, \ldots, A_n$ are one-dimensional Borel sets.

So, in other words, a random vector is a measurable function from W to $\mathbb{R}^n$ just as a random variable is a measurable function from W to the real line $\mathbb{R}$. In the present case, however, the vector-valued function $\mathbf{X}$ has n components – that is, $\mathbf{X} = (X_1, X_2, \ldots, X_n)$ – and the question on the measurability of each individual function $X_i (i = 1, \ldots, n)$ arises. The main result is that $\mathbf{X}$ is measurable if and only if each function X_i is measurable, or, equivalently:

Proposition 3.1 *The vector-valued function* $\mathbf{X}$ *is a random vector if and only if each one of its components* X_i *is a random variable.*

As in the one-dimensional case, the original probability space (W, S, P) is of little importance in applications and one should not worry too much about measurability because the concept is sufficiently broad to cover almost all cases of practical interest. Therefore, given a random vector $\mathbf{X}$, the analyst's main concern is the (real) induced probability space $(\mathbb{R}^n, \mathbb{B}_n, P_{\mathbf{X}})$, where the probability measure $P_{\mathbf{X}}$ is defined by the relation

$$P_{\mathbf{X}}(B) \equiv P[\mathbf{X}^{-1}(B)] = P\{w \in W : \mathbf{X}(w) \in B\} \tag{3.1a}$$

for all $B \in \mathbb{B}_n$ (it is not difficult to show that $P_{\mathbf{X}}$ is, indeed, a probability measure). Again, we note that a common 'shorthand' notation is to write $P(\mathbf{X} \in B)$ to mean the probability defined by eq. (3.1a). Also, in the light of the fact that B can be expressed as the cartesian product of n one-dimensional Borel sets $B_1, \ldots, B_n$, the explicit form of eq. (3.1a) is

$$P_{\mathbf{X}}(B) \equiv P[\mathbf{X}^{-1}(B)] = P[\mathbf{X}^{-1}(B_1 \times \cdots \times B_n)] = P\left(\bigcap_{i=1}^{n} X_i^{-1}(B_i)\right) \tag{3.1b}$$

By means of $P_{\mathbf{X}}$ we can define the so-called *joint probability distribution function* (joint-PDF) $F_{\mathbf{X}} : \mathbb{R}^n \to [0,1]$ as

$$F_{\mathbf{X}}(\mathbf{x}) = P_{\mathbf{X}}\{w \in W : X_1(w) \le x_1, X_2(w) \le x_2, \ldots, X_n(w) \le x_n\} \tag{3.2a}$$

where $\mathbf{x} \in \mathbb{R}^n$ is the vector whose components are $x_1, x_2, \ldots, x_n$ and it is understood that all the inequalities on the r.h.s. of eq. (3.2a) must hold simultaneously. In rigorous (and rather cumbersome) notation it may be worth noting that $F_{\mathbf{X}}(\mathbf{x})$ can be expressed in terms of the original probability P as

$$F_{\mathbf{X}}(\mathbf{x}) = P\left(\bigcap_{i=1}^{n} \{X_i^{-1}(-\infty, x_i]\}\right) \tag{3.2b}$$

(and probably this is why, in agreement with the 'shorthand' notation above, one often finds the less intimidating $F_{\mathbf{X}}(\mathbf{x}) = P(X_1 \le x_1, X_2 \le x_2, \ldots, X_n \le x_n)$).

The main properties of the joint-PDF are the natural extensions of (D1)–(D3) given in Chapter 2 and can be summarized as follows:

(D1′) $F_{\mathbf{X}}(\mathbf{x}) = F_{\mathbf{X}}(x_1, x_2, \ldots, x_n)$ is non-decreasing and continuous to the right in each variable x_i $(i = 1, \ldots, n)$,

(D2′) $\lim_{x_i \to -\infty} F_{\mathbf{X}}(\mathbf{X}) = 0$ and $\lim_{\mathbf{X} \to \infty} F_{\mathbf{X}}(\mathbf{x}) = 1$, where it should be noted that the first limit holds for any particular x_i tending to $-\infty$ (with all other coordinates fixed) whereas the second property requires that all x_i tend to $+\infty$. So, in a different notation, the two properties can be expressed as

$$F_{\mathbf{X}}(-\infty, x_2, \ldots, x_n) = F_{\mathbf{X}}(x_1, -\infty, \ldots, x_n)$$
$$= \cdots = F_{\mathbf{X}}(x_1, x_2, \ldots, -\infty) = 0; \quad F_{\mathbf{X}}(+\infty, +\infty, \ldots, +\infty) = 1$$

respectively.

In mathematical terminology – as in the one-dimensional case (Section 2.3) – one refers to $P_{\mathbf{X}}$ as the Lebesgue–Stieltjes measure determined by $F_{\mathbf{X}}$ and, conversely, to $F_{\mathbf{X}}$ as the generating function of the (finite) measure $P_{\mathbf{X}}$.

If now we turn our attention to the property expressed by eq. (2.15), we find that its multi-dimensional generalization is a bit more involved. For simplicity, let us consider the two-dimensional case first. The two-dimensional counterpart of the half-open interval $(a, b]$ is a rectangle R whose points satisfy the inequalities $a_1 < x_1 \le b_1$ and $a_2 < x_2 \le b_2$; with this in mind it is not difficult to determine that

$$P(\mathbf{X} \in R) = F_{\mathbf{X}}(b_1, b_2) - F_{\mathbf{X}}(b_1, a_2) - F_{\mathbf{X}}(a_1, b_2) + F_{\mathbf{X}}(a_1, a_2) \tag{3.3a}$$

and going over to the more complicated n-dimensional case we get

$$P(\mathbf{X} \in R) = \sum (-1)^k F_{\mathbf{X}}(c_1, c_2, \ldots, c_n) \tag{3.3b}$$

where now (i) R is the n-dimensional parallelepiped $(a_1, b_1] \times (a_2, b_2] \times \cdots \times (a_n, b_n]$, (ii) the sum is extended to all the 2^n possible choices of the c_i's being equal to a_i or b_i – that is, the vertexes of the parallelepiped – and (iii) k represents the number of c_i's being equal to a_i. For instance, if $n = 3$ we get

$$\begin{aligned} P(\mathbf{X} \in R) = F_{\mathbf{X}}(b_1, b_2, b_3) &- F_{\mathbf{X}}(b_1, b_2, a_3) - F_{\mathbf{X}}(b_1, a_2, b_3) \\ &- F_{\mathbf{X}}(a_1, b_2, b_3) + F_{\mathbf{X}}(b_1, a_2, a_3) - F_{\mathbf{X}}(a_1, b_2, a_3) \\ &+ F_{\mathbf{X}}(a_1, a_2, b_3) - F_{\mathbf{X}}(a_1, a_2, a_3) \end{aligned}$$

So, to every random vector there corresponds a joint-PDF which satisfies the properties above. The reverse statement, however, is not true in general unless we add another property to (D1′) and (D2′): for a function F to be the joint-PDF of some random vector the sum on the r.h.s. of eq. (3.3b) must be non-negative for any a_i, b_i such that $a_i \le b_i$ $(i = 1, 2, \ldots, n)$. If a function F satisfies these three properties, then it is the joint-PDF of some random vector although, as in the one-dimensional case, this vector is not uniquely determined by F. This is only a minor inconvenience without significant consequences in most practical cases.

If all the components of a random vector are discrete random variables, we speak of discrete random vector. More specifically, a random vector $\mathbf{X} = (X_1, X_2, \ldots, X_n)$ is discrete if there is a finite or countable set $A_{\mathbf{X}} \subset \mathbb{R}^n$ such that $P[(X_1, X_2, \ldots, X_n) \in A_{\mathbf{X}}] = 1$; in this case – besides being understood that the set $A_{\mathbf{X}}$ is the range of $\mathbf{X}$ – the function $p_{\mathbf{X}} : \mathbb{R}^n \to [0,1]$ defined by

$$p_{\mathbf{X}}(x_1, x_2, \ldots, x_n) \equiv P(X_1 = x_1, X_2 = x_2, \ldots, X_n = x_n) \tag{3.4}$$

is called the joint probability mass function (joint-pmf) of $\mathbf{X}$ and satisfies the normalization condition

$$\sum_{\text{all } \mathbf{x}} p_{\mathbf{X}} = \sum_{\text{all } x_1} \cdots \sum_{\text{all } x_n} p_{\mathbf{X}}(x_1, \ldots, x_n) = 1 \tag{3.5}$$

The other type of random vector commonly encountered in applications is called jointly absolutely continuous. In this case there is a measurable non-negative function $f_{\mathbf{X}}$ on $\mathbb{R}^n$ such that for all $B \in \mathbb{B}_n$ we have

$$P_{\mathbf{X}}(B) = \int_B f_{\mathbf{X}} \, \mathrm{d}\mu_n \tag{3.6}$$

where μ_n denotes here the n-dimensional Lebesgue measure. The function $f_{\mathbf{X}}(x_1, x_2, \ldots, x_n)$ is called the joint probability density function (joint-pdf) of $\mathbf{X}$ and its main properties are the generalization of the one-dimensional case (eqs (2.23), (2.25a) and (2.25b)), that is

$$F_{\mathbf{X}}(x_1, x_2, \ldots, x_n) = \int_{-\infty}^{x_1} \int_{-\infty}^{x_2} \cdots \int_{-\infty}^{x_n} f_{\mathbf{X}}(t_1, t_2, \ldots, t_n) \, \mathrm{d}t_1 \, \mathrm{d}t_2 \cdots \mathrm{d}t_n \tag{3.7a}$$

$$f_{\mathbf{X}}(x_1, \ldots, x_n) = \frac{\partial F_{\mathbf{X}}(x_1, \ldots, x_n)}{\partial x_1 \cdots \partial x_n} \tag{3.7b}$$

$$\int_{-\infty}^{\infty} \int_{-\infty}^{\infty} \cdots \int_{-\infty}^{\infty} f_{\mathbf{X}}(x_1, x_2, \ldots, x_n) \, \mathrm{d}x_1 \, \mathrm{d}x_2 \cdots \mathrm{d}x_n = 1 \tag{3.7c}$$

As in the one-dimensional case, discrete and absolutely continuous random vectors, or combinations thereof, are not the only possibilities because Proposition 2.11 on the decomposition of measures still holds in $\mathbb{R}^n$ and the decomposition of a general PDF, in turn, reflects the decomposition of its probability measure. The cases shown above, however, are by far the most common in applications and there is generally no need – besides a specific theoretical interest – to introduce further complications which, if and whenever necessary, will be considered in future discussions. So, we close this section here and turn our attention, once again, to the important role of stochastic independence.

3.2.1 *Marginal distribution functions and independent random variables*

In the preceding section we pointed out that each individual component X_i of a random vector $\mathbf{X} = (X_1, \ldots, X_n)$ is a random variable itself. Consequently, it becomes important to examine the relation between the joint-PDF of $\mathbf{X}$ and the PDF of its components in order to answer the following two questions:

(i) given the joint-PDF of $\mathbf{X}$ is it possible to determine the PDF of each X_i or the joint-PDF of some of the X_i taken together and forming a random vector with $m(m < n)$ components?
(ii) given all the PDFs $F_i(x)$ of each X_i is it possible to obtain the joint-PDF $F_{\mathbf{X}}(x_1, \ldots, x_n)$ of the random vector $\mathbf{X}$?

Let us consider question (i) first. The answer is always yes because the joint-PDF of $\mathbf{X}$ implicitly contains the joint-PDF of any vector obtained by eliminating some of its components. This PDF can be determined from $F_{\mathbf{X}}$ by letting all the components to be eliminated tend to infinity; so, if we call $\mathbf{Y}$ the vector obtained by eliminating the kth component of $\mathbf{X}$ we have

$$F_{\mathbf{Y}}(x_1, \ldots, x_{k-1}, x_{k+1}, \ldots, x_n) = \lim_{x_k \to \infty} F_{\mathbf{X}}(x_1, \ldots, x_n) \tag{3.8}$$

Similarly, if we eliminate any $2, 3, \ldots, n-1$ components of $\mathbf{X}$ the PDF of the new vector will be a function of the remaining $n-2, n-3, \ldots, 1$ variables, respectively, and the r.h.s. of eq. (3.8) will be a multiple limit where all the variables to be eliminated tend to $+\infty$. All the possible 'sub-PDFs', so to speak, obtained like this are called *marginal-PDFs* of the original vector $\mathbf{X}$. Consider the two-dimensional case as an example; here we have a vector $\mathbf{X} = (X, Y)$ whose joint-PDF is the function $F_{\mathbf{X}}(x, y)$ (often also denoted by $F_{XY}(x, y)$) and the two marginal-PDFs are the one-dimensional PDFs of the random variables X and Y, respectively, that is,

$$\begin{aligned} F_X(x) &= F_{XY}(x, \infty) \equiv \lim_{y \to \infty} F_{XY}(x, y) \\ F_Y(y) &= F_{XY}(\infty, y) \equiv \lim_{x \to \infty} F_{XY}(x, y) \end{aligned} \tag{3.9}$$

where, clearly, $F_X(x)$ is associated to the probability measure P_X: $\mathbb{B} \to [0,1]$ and $F_Y(y)$ is associated to the probability measure P_Y: $\mathbb{B} \to [0,1]$ (we recall that $\mathbb{B}$ is the collection of all Borel sets of the real line). So, the first part of eq. (3.9) tells us that $F_X(x)$ is the probability that the r.v. X takes on a value less than or equal to x when all the possible values of Y have been taken into account and, similarly, the second part of eq. (3.9) states that $F_Y(y)$ is the probability that the r.v. Y takes on a value less than or equal to y when all the possible values of X have been taken into account. Therefore – continuing with the two-dimensional case for simplicity – it is not difficult to see that the marginal-pmfs of a discrete random vector are given by

$$p_X(x) = \sum_{\text{all } y} p_{XY}(x,y)$$
$$p_Y(y) = \sum_{\text{all } x} p_{XY}(x,y) \tag{3.10a}$$

where $p_{XY}(x,y)$ is the joint-pmf of the two r.v.s X, Y forming the vector $\mathbf{X}$. Similarly, if $\mathbf{X}$ is a two-dimensional absolutely continuous random vector with joint-pdf $f_{XY}(x,y) = \partial F_{XY}/\partial x \partial y$, the two marginal (one-dimensional) pdfs are

$$f_X(x) = \int_{-\infty}^{+\infty} f_{XY}(x,y)\,\mathrm{d}y$$
$$f_Y(y) = \int_{-\infty}^{+\infty} f_{XY}(x,y)\mathrm{d}x \tag{3.10b}$$

Example 3.1(a) Let $\mathbf{X}$ be a discrete two-dimensional random vector whose joint-pmf is

$$p_{XY}(x,y) = (1-q)^2 q^{x+y} \tag{3.11}$$

where q is a constant $0 \le q < 1$ and the variables x, y can only take on natural values (i.e. $0, 1, 2, \ldots$). Then eq. (3.10a) yields for the marginal-pmfs

$$p_X(x) = q^x(1-q)^2 \sum_{\text{all } y} q^y = (1-q)q^x$$
$$p_Y(y) = q^y(1-q)^2 \sum_{\text{all } x} q^x = (1-q)q^y \tag{3.12}$$

because the series $\sum_n q^n$ converges to $(1-q)^{-1}$ whenever $0 \le q < 1$. In addition, the reader is invited to verify that the joint-pmf (3.11) satisfies the normalization condition of eq. (3.5) which, in this case, is written

$$\sum_{\text{all } x}\sum_{\text{all } y}(1-q)^2 q^{x+y} = 1$$

Example 3.1(b) Let the two-dimensional absolutely continuous random vector $\mathbf{X}$ have the joint-pdf

$$f_{\mathbf{X}}(x,y) = \frac{1}{2\pi\sqrt{3}}\exp\left\{-\frac{1}{3}(x^2+xy+y^2)\right\} \tag{3.13a}$$

then, using the first of eqs (3.10b) we can obtain the marginal-pdf of the r.v. X by integrating (3.13a) in dy over the entire real line. In order to do so we start by rewriting $f_{\mathbf{X}}(x,y)$ as

$$f_{\mathbf{X}}(x,y) = \frac{1}{2\pi\sqrt{3}}\exp\left(-\frac{x^2}{4}\right)\exp\left(-\frac{1}{3}\left(y+\frac{x}{2}\right)^2\right) \tag{3.13b}$$

so that the first exponential can be factored out of the integral in dy. Then, by performing the change of variable $t = y + x/2$ we get

$$\begin{aligned} f_X(x) &= \frac{1}{2\pi\sqrt{3}}\exp(-x^2/4)\int_{-\infty}^{\infty}\exp(-t^2/3)\,\mathrm{d}t \\ &= \frac{1}{2\sqrt{\pi}}\exp(-x^2/4) \end{aligned} \tag{3.14a}$$

where in the last equality we used the result $\int_{-\infty}^{\infty}\exp(-ax^2)\,\mathrm{d}x = \sqrt{\pi/a}$ (which can easily be found in integral tables). Finally, by symmetry, it is immediate to obtain

$$f_Y(y) = \frac{1}{2\sqrt{\pi}}\exp(-y^2/4) \tag{3.14b}$$

If now we consider question (ii) posed at the beginning of this section it turns out that its answer, in the general case, is no. More specifically, one cannot determine the joint-PDF of the random vector $\mathbf{X} = (X_1,\ldots,X_n)$ from the PDFs of its components X_i unless they are independent. In mathematical terms the following proposition holds

Proposition 3.2(a) *Let* $X_1,\ldots,X_n$ *be random variables on the probability space* (W,S,P)*, let* $F_i(x_i)$ *be the PDF of* $X_i (i = 1,\ldots,n)$ *and* $F_{\mathbf{X}}(x_1,\ldots,x_n)$

be the joint-PDF of the vector $\mathbf{X} = (X_1, \ldots, X_n)$. *Then* $X_1, \ldots, X_n$ *are independent if and only if*

$$F_{\mathbf{X}}(x_1, \ldots, x_n) = F_1(x_1)F_2(x_2)\ldots F_n(x_n) \tag{3.15}$$

for all real $x_1, \ldots, x_n$.

It should be noted that Proposition 3.2(a) is an 'if and only if' statement; this means that if $X_1, \ldots, X_n$ are independent then their joint-PDF can be obtained by taking the product of the individual PDFs and, conversely, if the joint-PDF of a random vector $\mathbf{X}$ is the product of n one-dimensional PDFs, then the components $X_1, \ldots, X_n$ are independent random variables.

At this point, however, we must take a step back and return to the notion of stochastic independence introduced in Chapter 2. In Section 2.2.2, in fact, we discussed in some detail the notion of stochastic independence of events but nothing has been said on independent random variables; we do it now by giving the following definition

Definition 3.2 A collection of random variables $X_1, X_2, \ldots$ is called an independent collection if for any arbitrarily chosen class of Borel sets $B_1, B_2, \ldots$ the events $X_1^{-1}(B_1), X_2^{-1}(B_2), \ldots$ are collectively independent.

This definition means that the product rule (2.9) must apply. So, in particular, n random variables $X_1, \ldots, X_n$ are called independent if for any choice of Borel sets $B_1, \ldots, B_n$ we have

$$P\left(\bigcap_{k=1}^{n} X_k^{-1}(B_k)\right) = \prod_{k=1}^{n} P[X_k^{-1}(B_k)]$$

which, in turn, implies that $P_{\mathbf{X}}(B_1 \times B_2 \times \cdots \times B_n) = P_{X_1}(B_1)P_{X_2}(B_2)\cdots P_{X_n}(B_n)$. We have the following result:

Proposition 3.2(b) *Let* $X_1, \ldots, X_n$ *be random variables on the probability space* (W, S, P), *then they are independent if and only if the measure* $P_{\mathbf{X}}$ *is the product of the* n *individual* P_{X_i} $(i = 1, 2, \ldots, n)$.

In the light of the fact that the individual PDFs $F_i(x_i)$ are defined by the probabilities P_{X_i}, it is not surprising that Proposition 3.2(a), as a matter of fact, is a consequence of Proposition 3.2(b). Also – as it has been done for events – one can introduce the concept of 'collection of pairwise independent random variables' – that is, a set of r.v.s $X_1, X_2, \ldots$ where X_i is independent of X_j for each pair of distinct indexes i, j – and note that pairwise independence does not imply independence. The converse, however, is true and it is evident that these two statements parallel closely the remarks of Chapter 2 (Section 2.2.2).

For discrete and absolutely continuous random variables independence can be characterized in terms of pmfs and pdfs, respectively, because the product rule applies to these functions. More specifically, if $X_1, \ldots, X_n$ are a set of independent random variables on a probability space (W, S, P) then, with obvious meaning of the symbols,

$$p_{\mathbf{X}}(x_1, \ldots, x_n) = p_1(x_1)p_2(x_2)\cdots p_n(x_n) \tag{3.16a}$$

in the discrete case and

$$f_{\mathbf{X}}(x_1, \ldots, x_n) = f_1(x_1)f_2(x_2)\cdots f_n(x_n) \tag{3.16b}$$

in the absolutely continuous case. Conversely, if – as appropriate – eq. (3.16a) or (3.16b) applies then the random variables $X_1, \ldots, X_n$ are independent. In this regard, for example, it may be worth noting that the two random variables X, Y of Example 3.1(a) are independent because (see eqs (3.11) and (3.12)) $p_{XY}(x, y) = p_X(x)p_Y(y)$. On the other hand, the random variables X, Y of Example 3.1(b) are not independent; in fact, in this case the joint-pdf $f_{\mathbf{X}}(x, y)$ cannot be factored as required in eq. (3.16b) because of the cross-term xy in the exponential. One word of caution on the absolutely continuous case is in order: if the random vector $\mathbf{X}$ has a pdf $f_{\mathbf{X}}$ then each X_i has a pdf f_i and eq. (3.16b) holds if $X_1, \ldots, X_n$ are independent. However, from the fact that each X_i has a density it does not necessarily follow that $\mathbf{X} = (X_1, \ldots, X_n)$ has a density; it does if $X_1, \ldots, X_n$ are independent and this density $f_{\mathbf{X}}$ is given – as shown by eq. (3.16b) – by the product of the n pdfs f_i.

As a final remark for this section we point out an important property of independent random variables: measurable functions of independent r.v.s are independent r.v.s. More specifically, we can state the following result whose proof, using the definition of independence of the X_i, is almost immediate.

Proposition 3.3 *Let $X_1, \ldots, X_n$ be a set of independent random variables and $g_1, \ldots, g_n$ a set of Borel functions. Then the random variables $Z_1, \ldots, Z_n$ (we recall from Chapter 2 that Borel functions of r.v.s are r.v.s themselves) defined by the relations $Z_i \equiv g_i(X_i)$ $(i = 1, 2, \ldots, n)$ are independent.*

More generally, if $\mathbf{Y}_1, \ldots, \mathbf{Y}_m$ are sub-vectors of a vector $\mathbf{X}$ such that none of the components of $\mathbf{X}$ is a component of more that one of the $\mathbf{Y}_j$ and $g_1, \ldots, g_m$ are measurable functions, then $Z_i \equiv g_i(\mathbf{Y}_i)(i = 1, 2, \ldots, m)$ are independent. Also, one can proceed further. In fact, it is possible to extend Definition 3.1 in order to define the independence of n random vectors $\mathbf{X}_1, \ldots, \mathbf{X}_n$ and determine that also in this case the factorization property of the PDFs is a necessary and sufficient condition for independence. For the moment, however, the results given here will suffice and we delay further considerations on independence to later sections.

3.3 Moments and characteristic functions of random vectors

For simplicity, let us first consider a two-dimensional random vector $\mathbf{X} = (X, Y)$ defined on a probability space (W, S, P). This is a frequently encountered case in applications and it is worthy of consideration in its own right before generalizing to higher dimensional vectors.

As in the one-dimensional case (Section 2.3.2), the moments are defined as abstract Lebesgue integrals in $\mathrm{d}P$. So, if i, j are two non-negative integers the joint-moments of order $i+j$ – denoted by $E(X^i Y^j)$ or m_{ij} – are defined as

$$m_{ij} = E(X^i Y^j) = \int_W X^i Y^j \,\mathrm{d}P \tag{3.17a}$$

which, in the absolutely continuous case, becomes

$$m_{ij} = \int_{-\infty}^{\infty}\int_{-\infty}^{\infty} x^i y^j f(x, y)\,\mathrm{d}x\,\mathrm{d}y \tag{3.17b}$$

and the integrals are replaced by the appropriate sums in the discrete case.

In the light of eq. (3.10b) and their discrete counterparts (3.10a), it is then clear that the first-order moments m_{10} and m_{01}, respectively, are simply $\mu_X = E(X)$ and $\mu_Y = E(Y)$, that is, the mean values of the individual random variables X and Y and, similarly, m_{i0} and m_{0j} are the ith moment of X and the jth moment of Y.

The central (joint) moments of order $i+j$, in turn, are defined as (provided that $\mu_X, \mu_Y < \infty$)

$$\mu_{ij} = E[(X - \mu_X)^i (Y - \mu_Y)^j] \tag{3.18}$$

where, in the important case $i + j = 2$ (second-order central moments) we have $\mu_{20} = \sigma_X^2 = \mathrm{Var}(X)$ and $\mu_{02} = \sigma_Y^2 = \mathrm{Var}(Y)$. The moment μ_{11} is given a special name and is called the *covariance* of the two variables X, Y. For this reason μ_{11} is often denoted by $\mathrm{Cov}(X, Y)$ – although the symbols Γ_{XY}, σ_{XY} and K_{XY} are also frequently found in literature. Besides the immediate relations $\mathrm{Cov}(X, X) = \sigma_X^2, \mathrm{Cov}(Y, Y) = \sigma_Y^2$ and $\mathrm{Cov}(X, Y) = \mathrm{Cov}(Y, X)$ it should also be noted that the notion of covariance was mentioned in passing in Proposition 2.15 (Section 2.3.2) where, in addition, it was shown that

$$\operatorname{Cov}(X, Y) = E(XY) - E(X)E(Y) = m_{11} - \mu_X \mu_Y \tag{3.19a}$$

This equation is, broadly speaking, the 'mixed-variables' counterpart of eq. (2.34), which we rewrite here for the two individual r.v.s X, Y

$$\begin{aligned} \sigma_X^2 &= \mu_{20} = E(X^2) - E^2(X) = m_{20} - \mu_X^2 \\ \sigma_Y^2 &= \mu_{02} = E(Y^2) - E^2(Y) = m_{02} - \mu_Y^2 \end{aligned} \tag{3.19b}$$

Some of the main properties of the abstract integral can be immediately be re-expressed in terms of moments. We have already considered linearity for n random variables (Proposition 2.13(c)) – which, in our present case of two r.v.s reads $E(aX + bY) = aE(X) + bE(Y)$ where a, b are any two real or complex constants – but, in particular, we want to point out here the following inequalities

(a) Holder's inequality: let p, q be two numbers such that $p > 1, q > 1$ and $1/p + 1/q = 1$, then

$$E(|XY|) \leq [E(|X|^p)]^{1/p}[E(|Y|^q)]^{1/q} \tag{3.20}$$

(b) Cauchy–Schwarz inequality

$$E(|XY|) \leq \sqrt{E(X^2)}\sqrt{E(Y^2)} \tag{3.21}$$

Both relations are well known to the reader who is familiar with the theory of Lebesgue-integrable function spaces and, clearly, eq. (3.21) is a special case of (3.20) when $p = q = 2$. In particular – since the Cauchy–Schwarz inequality holds for any two r.v.s – there is no loss of generality in considering the two centered r.v.s $W = X - \mu_X, Z = Y - \mu_Y$ and rewriting (3.21) as $E(WZ) = \operatorname{Cov}(XY) \leq \sigma_W \sigma_Z = \sigma_X \sigma_Y$, where the last equality holds because the variance of a constant is zero. Therefore, if one defines the correlation coefficient ρ_{XY} as

$$\rho_{XY} = \frac{\operatorname{Cov}(XY)}{\sigma_X \sigma_Y} \tag{3.22}$$

it is immediate to determine that $-1 \leq \rho_{XY} \leq 1$. The fact that whenever $\rho_{XY} = -1$ or $\rho_{XY} = 1$ there is a perfect linear relationship between the two r.v.s X and Y (i.e. a relation of the type $Y = aX + b$, where a, b are two constants) is not so immediate and requires some explanation. In order to do this as an exercise, note that both equalities $\rho_{XY} = \pm 1$ imply

$E^2(WZ)/E(W^2)E(Z^2) = 1$, where W, Z are the two centered r.v.s defined above. This relation can be rewritten as

$$-\frac{E^2(WZ)}{E(W^2)} = -E(Z^2) \text{ or, equivalently}$$

$$\frac{E^2(WZ)}{E^2(W^2)}E(W^2) - 2\frac{E(WZ)}{E(W^2)}E(WZ) + E(Z^2) = 0$$

so that setting $E(WZ)/E(W^2) = a$ we get $a^2E(W^2) - 2aE(WZ) + E(Z^2) = 0$, that is $E[(aW - Z)^2] = 0$. This, in turn, means that $aW - Z = const$, that is, that W and Z are linearly related. Then, since (by definition) there is a linear relation between W and X and between Z and Y, it follows that also X and Y must be linearly related. On the other hand, in order to prove that $Y = aX + b$ implies $\rho_{XY} = 1$ or $\rho_{XY} = -1$ (depending on whether $a > 0$ or $a < 0$) it is sufficient to note that in this case $\text{Cov}(XY) = a\sigma_X$ and $\sigma_Y = |a|\sigma_X$; consequently, $\rho_{XY} = \pm 1$ follows from the definition of correlation coefficient.

The opposite extreme to maximum correlation occurs when $\rho_{XY} = 0$, that is, when $\text{Cov}(XY) = 0$ (if, as always implicitly assumed here, both σ_X, σ_Y are finite and different from zero). In this case we say that X and Y are uncorrelated and then, owing to eq. (3.19a), we get $E(XY) = E(X)E(Y)$. This form of 'multiplication rule' for expected values may suggest independence of the two random variables because the following proposition holds:

Proposition 3.4 *If X, Y are two stochastically independent r.v.s then*

$$E(XY) = E(X)E(Y) \tag{3.23}$$

and therefore $\text{Cov}(XY) = 0$.

This result can be proven by using the factorization properties given in eqs (3.15), (316a) and (3.16b), but the point here is that the reverse statement of Proposition 3.4 is not, in general, true (unless in special cases which will be considered in future sections). In fact, it turns out that uncorrelation – that is, $\text{Cov}(XY) = 0$ – is a necessary but not sufficient condition for stochastic independence. In other words, two uncorrelated r.v.s are not necessarily unrelated (a term which, broadly speaking, is a synonym of independent) because uncorrelation implies a lack of linear relation between them but not necessarily a lack of relation in general. The following example illustrates this situation.

Example 3.2 Consider a random vector (X, Y) which is uniformly distributed within a circle of radius r centered about the origin. This means that the vector is absolutely continuous with joint-pdf given by

$$f_{XY}(x,y) = \begin{cases} 1/\pi r^2, & x^2 + y^2 < r^2 \\ 0, & \text{otherwise} \end{cases}$$

so that, for instance, if $X = 0$ then Y can have any value between $-r$ and r but if $X = r$ then Y can only be zero. Therefore, since knowledge of X provides some information on Y, the two variables are not independent. On the other hand, they are uncorrelated because the symmetry of the problem leads to the result $\text{Cov}(XY) = 0$. In fact, denoting by C the domain where $f_{XY} \neq 0$, we can calculate the covariance as (see eq. (3.44))

$$\text{Cov}(XY) = \int_C xy\, f(x,y)\, dx\, dy = \frac{1}{\pi r^2} \int_C xy\, dx\, dy$$

and all the integrals in the four quadrants have the same absolute value. However, the function xy under the integral sign is positive in the first and third quadrant and negative in the second and fourth quadrant so that summing all the four contributions yields $\text{Cov}(XY) = 0$.

As a simpler example consider a r.v. X with the following characteristics: (a) its pdf (or pmf if it is discrete) is symmetrical about the ordinate axis and (b) it has a finite fourth moment. Then, if we define the r.v. $Y = X^2$ it is immediate to determine that X and Y are uncorrelated but not independent.

The definition of characteristic function (or, more precisely, joint-CF) for a two-dimensional random vector is a simple extension of the one-dimensional case of Section 2.4 and we have

$$\phi(u,v) = E[e^{i(uX+vY)}] \tag{3.24a}$$

which, in view of generalization to higher dimensions, can be expressed more synthetically with the aid of matrix algebra. We denote by **u** the vector whose components are the two real variables u, v and write

$$\varphi_{\mathbf{X}}(\mathbf{u}) = E[\exp(i\mathbf{u}^T\mathbf{X})] \tag{3.24b}$$

where, following the usual matrix notation, two-dimensional vectors are expressed as column matrices and their transpose (indicated by the upper T) are therefore row matrices. So, in eq. (3.24b) it is understood that the matrix multiplication in the exponential reads

$$\mathbf{u}^{\mathrm{T}}\mathbf{X} = (u, v)\begin{pmatrix} X \\ Y \end{pmatrix} = uX + vY$$

The properties of the CF of a random vector parallel closely the one-dimensional case; in particular $\varphi_{\mathbf{X}}(\mathbf{u})$ is uniformly continuous on $\mathbb{R}^2$ and, in addition

$$\begin{aligned} &\varphi_{\mathbf{X}}(\mathbf{0}) = 1 \\ &|\varphi_{\mathbf{X}}(\mathbf{u})| \leq 1 \quad \text{for all } \mathbf{u} \in \mathbb{R}^2 \\ &\varphi_{\mathbf{X}}(-\mathbf{u}) = \varphi_{\mathbf{X}}^{*}(\mathbf{u}) \end{aligned} \tag{3.25}$$

where $\mathbf{0} = (0, 0)$ is the zero vector and the asterisk denotes complex conjugation. Also, if we set $n = j + k$ (where j, k are two integers) and the vector $\mathbf{X}$ has finite moments of order n, then $\varphi_{\mathbf{X}}(u, v)$ is j times derivable with respect to u and k times derivable with respect to v and

$$i^n m_{jk} = i^n E(X^j Y^k) = \left. \frac{\partial^n \varphi_{\mathbf{X}}(u, v)}{\partial^j u \partial^k v} \right|_{u=v=0} \tag{3.26}$$

which is the two-dimensional counterpart of eq. (2.47a). Equation (3.26) shows that the moments of a random vector coincide – besides the multiplicative factor $1/i^n$ – with the coefficients of the MacLaurin expansion of $\varphi_{\mathbf{X}}$. This implies that the existence of all moments allows one to construct the MacLaurin series of the CF although, as in the one-dimensional case, in general it does not allow to reconstruct $\varphi_{\mathbf{X}}$ itself.

The marginal CFs of any 'sub-vector' of $\mathbf{X}$ can be obtained from $\varphi_{\mathbf{X}}$ by simply setting to zero all the arguments corresponding to the random variable(s) which do not belong to the sub-vector in question. This is an immediate consequence of the definition of CF and in the two-dimensional case under study we have

$$\begin{aligned} \varphi_X(u) &= \varphi_{\mathbf{X}}(u, 0) \\ \varphi_Y(v) &= \varphi_{\mathbf{X}}(0, v) \end{aligned} \tag{3.27}$$

As final remarks to this section, two results are worthy of notice. The first is somehow expected and states:

Proposition 3.5 *Two random variables X, Y forming a vector $\mathbf{X}$ are stochastically independent if and only if*

$$\varphi_{\mathbf{X}}(u, v) = \varphi_X(u)\varphi_Y(v) \tag{3.28}$$

meaning that the product rule (3.28) is a necessary and sufficient condition for independence.

A word of caution is in order here because Proposition 3.5 should not be confused with a different result (see also Example 3.3) which states that if X, Y are independent, then

$$\varphi_{X+Y}(u) = \varphi_X(u)\varphi_Y(u) \tag{3.29}$$

In fact, the independence of X and Y imply the independence of the r.v.s e^{iuX} and e^{iuY} and consequently $E(e^{iu(X+Y)}) = E(e^{iuX}e^{iuY}) = E(e^{iuX})E(e^{iuY})$ by virtue of Proposition 3.4. Then, by the definition of CF we get $E(e^{iuX})E(e^{iuY}) = \varphi_X(u)\varphi_Y(u)$. The converse of this result, however, is not true in general and the equality (3.29) does not imply the independence of X and Y. These same considerations, clearly, can be extended to the case of more than two r.v.s.

The second remark – here already given in the n-dimensional case – has to do with the important fact that the joint-CF $\varphi_{\mathbf{X}}$ provides a complete probabilistic description of the random vector $\mathbf{X} = (X_1, \dots, X_n)$ because the joint-PDF $F_{\mathbf{X}}(x_1, \dots, x_n)$ is uniquely determined by $\varphi_{\mathbf{X}}(u_1, \dots, u_n)$. The explicit result, which we state here for completeness, is in fact the n-dimensional counterpart of Proposition 2.20 and is expressed by the relation

$$\begin{aligned} P(a_k < X_k \le b_k) = \lim_{c \to \infty} \frac{1}{(2\pi)^n} \int_{-c}^{c} \cdots \int_{-c}^{c} \\ \times \prod_{k=1}^{n} \left(\frac{e^{iu_k a_k} - e^{iu_k b_k}}{iu_k} \right) \varphi_{\mathbf{X}}(u_1, \dots, u_n)\, du_1 \cdots du_n \end{aligned} \tag{3.30}$$

where the real numbers $a_k, b_k (k = 1, \dots, n)$ delimitate a bounded parallelepiped (i.e. an interval in $\mathbb{R}^n$) whose boundary has zero probability measure.

3.3.1 *Additional remarks: the multi-dimensional case and the practical calculation of moments*

3.3.1.1 *The multi-dimensional case*

In the preceding section we have been mainly concerned with two-dimensional random vectors but it is reasonable to expect that most of the considerations can be readily extended to the n-dimensional case. We only outline this extension here because it will not be difficult for the reader to fill in the missing details. It is implicitly assumed, however, that the reader has some familiarity with matrix notation and basic matrix properties (if not, one may refer, for example, to Chapter 11 of Ref. [3] at the end of this chapter, to the excellent booklet [16] or to the more advanced text [8]).

Given a n-dimensional random vector $\mathbf{X} = (X_1, \ldots, X_n)$ and a positive integer k, the kth order moments are defined as

$$m(k_1, k_2, \ldots, k_n) = E\left(X_1^{k_1} X_2^{k_2} \ldots X_n^{k_n}\right) \tag{3.31}$$

where $k_1, \ldots, k_n$ are n non-negative integers such that $k = k_1 + k_2 + \cdots + k_n$. This implies that we have now n first-order moments – which can be denoted $m_1, m_2, \ldots, m_n$ – and n^2 second-order ordinary and central moments. These latter quantities are often conveniently arranged in the so-called covariance matrix

$$\mathbf{K} = \begin{pmatrix} K_{11} & K_{12} & \ldots & K_{1n} \\ K_{21} & K_{22} & \ldots & K_{2n} \\ \vdots & \vdots & \vdots & \vdots \\ K_{n1} & K_{n2} & \ldots & K_{nn} \end{pmatrix} = E[(\mathbf{X} - \mathbf{m})(\mathbf{X} - \mathbf{m})^{\mathrm{T}}] \tag{3.32a}$$

where $K_{ij} = \mathrm{Cov}(X_i X_j)$ with $i, j = 1, \ldots, n$ and in the second expression $(\mathbf{X} - \mathbf{m})$ is the $n \times 1$ column matrix whose elements are $X_1 - m_1, \ldots, X_n - m_n$, that is, the difference of the two column matrices $\mathbf{X} = (X_1, \ldots, X_n)^{\mathrm{T}}$ and the first-order moments matrix $\mathbf{m} = (m_1, \ldots, m_n)^{\mathrm{T}}$. In this light, it is easy to notice that the covariance matrix can be written as

$$\mathbf{K} = E(\mathbf{X}\mathbf{X}^{\mathrm{T}}) - \mathbf{m}\mathbf{m}^{\mathrm{T}} \tag{3.32b}$$

The matrix $\mathbf{K}$ is obviously symmetric (i.e. $K_{ij} = K_{ji}$ or, in matrix symbolism, $\mathbf{K} = \mathbf{K}^{\mathrm{T}}$) so that there are only $n(n+1)/2$ distinct elements; also it is evident that the elements on the main diagonal are the variances of the individual r.v.s – that is, $K_{ii} = \mathrm{Var}(X_i)$. Similar considerations of symmetry and of number of distinct elements apply to the correlation matrix $\mathbf{R}$ defined as

$$\mathbf{R} = \begin{pmatrix} 1 & \rho_{12} & \ldots & \rho_{1n} \\ \rho_{21} & 1 & \ldots & \rho_{2n} \\ \vdots & \vdots & \vdots & \vdots \\ \rho_{n1} & \rho_{n2} & \ldots & 1 \end{pmatrix} \tag{3.32c}$$

where (eq. (3.22)) $\rho_{ij} = K_{ij}/\sigma_i\sigma_j$ and, for brevity, we denote by $\sigma_i = \sqrt{\mathrm{Var}(X_i)}$ the standard deviation of the r.v. X_i (assuming that σ_i is finite for each $i = 1, \ldots, n$). The relation between $\mathbf{K}$ and $\mathbf{R}$ – as it is immediately verified by using the rules of matrix multiplication – is

$$\mathbf{K} = \mathbf{S}\,\mathbf{R}\,\mathbf{S} \tag{3.33}$$

where we called $\mathbf{S} = \text{diag}(\sigma_1, \sigma_2, \ldots, \sigma_n)$ the matrix whose the only non-zero elements are $\sigma_1, \ldots, \sigma_n$ on the main diagonal. If the n r.v.s are pairwise uncorrelated – or, which is a stronger condition, pairwise independent – then both $\mathbf{K}$ and $\mathbf{R}$ are diagonal matrices and in particular $\mathbf{R} = \mathbf{I}$, where $\mathbf{I}$ is the identity, or unit, matrix (its only non-zero elements are ones on the main diagonal). Clearly, this holds true if the r.v.s X_i are mutually independent; in this case we can also generalize Proposition 3.4 on first-order moments to the multiplication rule

$$E\left(\prod_{i=1}^{n} X_i\right) = \prod_{i=1}^{n} E(X_i) \tag{3.34}$$

If we pass from the random vector $\mathbf{X}$ to a m-dimensional random vector $\mathbf{Y}$ by means of a linear transformation $\mathbf{Y} = \mathbf{AX}$ – where $\mathbf{A}$ is a $m \times n$ matrix of real numbers – we can use the second expression of (3.32a) to determine the covariance matrix $\mathbf{K_Y}$ of $\mathbf{Y}$ in terms of $\mathbf{K_X}$. In fact, calling for brevity $\widetilde{\mathbf{Y}}$ and $\widetilde{\mathbf{X}}$ the 'centered' matrices $\mathbf{Y} - \mathbf{m_Y}$ and $\mathbf{X} - \mathbf{m_X}$, respectively, we get

$$\begin{aligned} \mathbf{K_Y} = E(\widetilde{\mathbf{Y}}\widetilde{\mathbf{Y}}^{\mathrm{T}}) &= E[(\mathbf{A}\widetilde{\mathbf{X}})(\mathbf{A}\widetilde{\mathbf{X}})^{\mathrm{T}}] \\ &= E[\mathbf{A}\widetilde{\mathbf{X}}\widetilde{\mathbf{X}}^{\mathrm{T}}\mathbf{A}^{\mathrm{T}}] = \mathbf{A}\,E(\widetilde{\mathbf{X}}\widetilde{\mathbf{X}}^{\mathrm{T}})\mathbf{A}^{\mathrm{T}} = \mathbf{A}\,\mathbf{K_X}\mathbf{A}^{\mathrm{T}} \end{aligned} \tag{3.35}$$

where we used the well-known relation stating that the transpose of a product of matrices equals the product of the transposed matrices taken in reverse order – that is, in our case $(\mathbf{A}\widetilde{\mathbf{X}})^{\mathrm{T}} = \widetilde{\mathbf{X}}^{\mathrm{T}}\mathbf{A}^{\mathrm{T}}$. We mention here in passing a final property of the covariance and correlation matrices: both $\mathbf{K}$ and $\mathbf{R}$ are positive semi-definite. As it is known from matrix theory, this means that

$$\mathbf{z}^{\mathrm{T}}\mathbf{K}\mathbf{z} \geq 0 \tag{3.36}$$

where $\mathbf{z}$ is a column vector of n real or complex variables and $\mathbf{x}^{\mathrm{T}}\mathbf{K}\mathbf{x}$ is the so-called quadratic form of the (symmetric) matrix $\mathbf{K}$. Equation (3.36) implies that $\det(\mathbf{K})$ – that is, the determinant of $\mathbf{K}$, often also denoted by $|\mathbf{K}|$ – is non-negative. Clearly, the same considerations apply to $\mathbf{R}$.

The characteristic function of a n-dimensional random vector is the straightforward extension of eq. (3.24b) and the generalization of eq. (3.26) reads

$$i^k m(k_1, \ldots, k_n) = \left.\frac{\partial^k \varphi_{\mathbf{X}}(\mathbf{u})}{\partial u_1^{k_1} \cdots \partial u_n^{k_n}}\right|_{\mathbf{u}=0} \tag{3.37}$$

provided that the moment of order $k = k_1 + k_2 + \cdots + k_n$ exists.

The marginal CFs can be obtained from $\varphi_X(u_1, \ldots, u_n)$ as stated in Section 3.3. For example, $\varphi_X(u_1, u_2, \ldots, u_{n-1}, 0)$ is the joint-CF of the vector $(X_1, \ldots, X_{n-1})$, $\varphi_X(u_1, 0, \ldots, 0)$ is the one-dimensional CF of the r.v. X_1, etc., and the multiplication rule

$$\varphi_X(u_1, \ldots, u_n) = \prod_{i=1}^{n} \varphi_{X_i}(u_i) \tag{3.38}$$

is a necessary and sufficient condition for the mutual independence of the r.v.s $X_i (i = 1, \ldots, n)$. Similarly, all the other considerations apply. In addition, we can determine how a joint-CF changes under a linear transformation from $\mathbf{X}$ to a m-dimensional random vector $\mathbf{Y}$. As above, the transformation is expressed in matrix form as $\mathbf{Y} = \mathbf{AX}$ and we assume here that $\varphi_X(\mathbf{u})$ is known so that

$$\begin{aligned} \varphi_Y(\mathbf{v}) &= E[e^{i\mathbf{v}^T\mathbf{Y}}] = E[e^{i\mathbf{v}^T\mathbf{AX}}] \\ &= E[e^{i(\mathbf{A}^T\mathbf{v})^T\mathbf{X}}] = \varphi_X(\mathbf{A}^T\mathbf{v}) \end{aligned} \tag{3.39a}$$

In the more general case $\mathbf{Y} = \mathbf{AX} + \mathbf{b}$ – where $\mathbf{b} = (b_1, \ldots, b_m)^T$ is a column vector of constants – then it is immediate to determine

$$\varphi_Y(\mathbf{v}) = e^{i\mathbf{v}^T\mathbf{b}} \varphi_X(\mathbf{A}^T\mathbf{v}) \tag{3.39b}$$

In the preceding section nothing has been said about moment-generating functions (MGFs) but by now it should be clear that the definition is

$$M_X(s_1, \ldots, s_n) = E[\exp(\mathbf{s}^T\mathbf{X})] \tag{3.40}$$

where $s_1, \ldots, s_n$ is a set of n variables. Within the limitations on the existence of M_X outlined in Section 2.4 we have the n-dimensional version of eq. (2.48), that is,

$$m(k_1, \ldots, k_n) = \left. \frac{\partial^k M_X(\mathbf{s})}{\partial s_1^{k_1} \cdots \partial s_n^{k_n}} \right|_{\mathbf{s}=0} \tag{3.41}$$

3.3.1.2 *The practical calculation of expectations*

Many quantities introduced so far – moments in the first place but CFs and MGFs as well – are defined as expectations, which means, by definition, as abstract Lebesgue integral in dP. Therefore, the problem arises of how these integrals can be calculated in practice. With the additional slight complication of n-dimensionality, the general line of reasoning parallels closely all that has been said in Chapter 2. We will briefly repeat it here.

Given a random vector $\mathbf{X}$ on the probability space (W, S, P) – that is, a measurable function $\mathbf{X}: W \to \mathbb{R}^n$ – we can make probability statements regarding any Borel set $B \in \mathbb{B}_n$ by considering the probability measure $P_{\mathbf{X}}$ defined by eq. (3.1) and working in the induced real probability space $(\mathbb{R}^n, \mathbb{B}_n, P_{\mathbf{X}})$. This is the space of interest in practice, $P_{\mathbf{X}}$ being a Lebesgue–Stieltjes measure (on $\mathbb{R}^n$) which, by virtue of eq. (3.2), can be associated with the PDF $F_{\mathbf{X}}$. At this point we note that the n-dimensional version of Proposition 2.17 applies, with the consequence that our abstract Lebesgue integral on W can be calculated as a Lebesgue–Stieltjes integral on $\mathbb{R}^n$. Then, depending on the type of random vector under study – or, equivalently, on the continuity properties of $F_{\mathbf{X}}$ – this integral turns into a form amenable to actual calculations. As in the one-dimensional case, there are three possible cases: the discrete, the absolutely continuous and the singular continuous case, the first two (or a mixture thereof) being by far the most important in practice.

If $\mathbf{X}$ is a discrete random vector its range is a discrete subset $A_{\mathbf{X}} \subset \mathbb{R}^n$ and its complete probabilistic description can be given in terms of the mass distribution $p_{\mathbf{X}}(\mathbf{x})$ (see also eq. (3.4)) which, in essence, is a finite or countable set of real non-negative numbers $p_{i_1,i_2,\ldots,i_n}$ (the n indexes $i_1, \ldots, i_n$ mean that the ith r.v. X_i can take on the values $x_{i1}, x_{i2}, \ldots$) such that the normalization condition

$$\sum_{\mathbf{x} \in A_{\mathbf{X}}} p_{\mathbf{X}}(\mathbf{x}) = \sum_{i_1,\ldots,i_n} p_{i_1,\ldots,i_n} = 1 \tag{3.42}$$

holds. In this light, given a Borel measurable function $g(\mathbf{x})$ its expectation is given by the sum

$$E[g(\mathbf{x})] = \sum_{\mathbf{x} \in A_{\mathbf{X}}} g(\mathbf{x}) p_{\mathbf{X}}(\mathbf{x}) \tag{3.43}$$

Also, the marginal mass distribution of any group of any $m (m < n)$ random variables is obtained by summing the $p_{i_1,i_2,\ldots,i_n}$ over all the $n - m$ remaining variables; so, for example, the marginal mass distribution of the vector $(X_1, \ldots, X_{n-1})$ is given by $\sum_{i_n} p_{i_1,i_2,\ldots,i_n}$. This is just a straightforward generalization of eq. (3.10a).

If $\mathbf{X}$ is absolutely continuous there exists a density function $f_{\mathbf{X}}(\mathbf{x})$ such that eqs (3.7a–3.7c) hold. Then, the expectation of a measurable function $g(\mathbf{x})$ becomes a Lebesgue integral and reads

$$E[g(\mathbf{x})] = \int_{\mathbb{R}^n} g(\mathbf{x}) f_{\mathbf{X}}(\mathbf{x}) \, d\mathbf{x} \tag{3.44}$$

where $d\mathbf{x}$ is the Lebesgue measure on $\mathbb{R}^n$. As one might expect – since the Lebesgue integral is, broadly speaking, a generalization of the ordinary Riemann integral of basic calculus – the integral (3.44) coincides with Riemann's (when this integral exists).

In the multi-dimensional case, however, a result of fundamental importance is Fubini's theorem (Appendix B) which guarantees that a Lebesgue multiple integral can be calculated as an iterated integral under rather mild conditions on the integrand function. This theorem is the key to the practical evaluation of multi-dimensional integrals.

Two final comments are in order before closing this section. First, we recall eq. (3.10b) and note that their n-dimensional extension is immediate; in fact, the marginal pdfs of any 'subvector' of $\mathbf{X}$ of $m(m < n)$ components is obtained by integrating $f_{\mathbf{X}}$ with respect to the remaining $n - m$ variables. So, for instance, $f_{X_1}(x_1) = \int_{\mathbb{R}^{n-1}} f_{\mathbf{X}}(x_1, \ldots, x_n)\, dx_2 \cdots dx_n$ is the pdf of the r.v. X_1 and $f_{\mathbf{Y}}(x_1, \ldots, x_{n-1}) = \int_{-\infty}^{\infty} f_{\mathbf{X}}(x_1, \ldots, x_n)\, dx_n$ is the $(n-1)$-dimensional joint-pdf of the random vector $\mathbf{Y} = (X_1, X_2, \ldots, X_{n-1})$.

The second comment has to do with the CF of an absolutely continuous random vector $\mathbf{X}$. Owing to eq. (3.44), in fact, $\varphi_{\mathbf{X}}(\mathbf{u})$ and $f_{\mathbf{X}}(\mathbf{x})$ turn out to be a Fourier transform pair so that we have

$$\varphi_{\mathbf{X}}(\mathbf{u}) = \int_{\mathbb{R}^n} f_{\mathbf{X}}(\mathbf{x})\, e^{i\,\mathbf{u}^T\mathbf{x}}\, d\mathbf{x} \tag{3.45a}$$

with the inversion formula

$$f_{\mathbf{X}}(\mathbf{x}) = \frac{1}{(2\pi)^n} \int_{\mathbb{R}^n} \varphi_{\mathbf{X}}(\mathbf{u})\, e^{-i\,\mathbf{u}^T\mathbf{x}}\, d\mathbf{u} \tag{3.45b}$$

3.3.2 *Two important examples: the multinomial distribution and the multivariate Gaussian distribution*

A multinomial trial with parameters $p_1, p_2, \ldots p_n$ is a trial with n possible outcomes where the probability of the ith outcome is p_i $(i = 1, 2, \ldots, n)$ and, clearly, $p_1 + p_2 + \cdots + p_n = 1$. If we perform an experiment consisting of N independent and identical multinomial trials (so that the p_i do not change from trial to trial) we may call X_i the number of trials that result in outcome i so that each X_i is a r.v. which can take on any integer value between zero and N. Forming the vector $\mathbf{X} = (X_1, \ldots, X_n)$, its joint-pmf is called multinomial (it can be obtained with the aid of eq. (1.16)) and we have

$$p(x_1, \ldots, x_n) = \frac{N!}{x_1! x_2! \ldots x_n!} p_1^{x_1}, p_2^{x_2} \cdots p_n^{x_n} \tag{3.46a}$$

where, since each x_i represents the number of times in which i occurs

$$\sum_{i=1}^{n} x_i = N \tag{3.46b}$$

The name multinomial is due to fact that the expression on the r.h.s. of (3.46a) is the general term in the expansion of $(p_1 + \cdots + p_n)^N$. If $n = 2$ eq. (3.46a) reduces to the binomial pmf considered in Examples 2.8 and 2.9 (a word of caution on notation: in eq. (2.41a) n is the total number of trials while here n is the number of possible outcomes in each trial).

As an easy example we can consider three throws of a fair die. In this case $N = 3$ and $n = 6$, x_i is the number of times the face 1 shows up, x_2 is the number of times the face 2 shows up, etc. and $p_1 = \cdots = p_6 = 1/6$. As a second example we can think of a box with, say, 50 balls of which 10 are white, 22 yellow and 18 are red. The experiment may consist in extracting – with replacement – 5 balls from the box and then counting the extracted balls of each color. In this case $N = 5$, $n = 3$ and $p_1 = 0.20$, $p_2 = 0.44$, $p_3 = 0.36$. Note that after each extraction it is important to replace the ball in the box, otherwise the probabilities p_i would change from trial to trial and one of the basic assumptions leading to (3.46) would fail.

The joint-CF of the multinomial distribution is obtained from eq. (3.24b) by noting that in this discrete case the Lebesgue–Stieltjes integral defining the expectation becomes a sum on all the x_is. Therefore

$$\begin{aligned}\varphi_{\mathbf{X}}(u_1, \ldots, u_n) &= \sum \frac{N!}{x_1! \ldots x_n!} p_1^{x_1} \cdots p_n^{x_n} \mathrm{e}^{\mathrm{i}(u_1 x_1 + \cdots + u_n x_n)} \\ &= \sum \frac{N!}{x_1! \ldots x_n!} (p_1 \mathrm{e}^{\mathrm{i}u_1})^{x_1} \cdots (p_n \mathrm{e}^{\mathrm{i}u_n})^{x_n} \\ &= (p_1 \mathrm{e}^{\mathrm{i}u_1} + \cdots + p_n \mathrm{e}^{\mathrm{i}u_n})^N\end{aligned} \tag{3.47}$$

As a second step, let us obtain now the marginal CF of one of the r.v.s X_i, for example, X_1. In order to do this (recall Sections 3.2 and 3.3) we must set $u_2 = u_3 = \cdots = u_n = 0$ in eq. (3.47) thus obtaining

$$\varphi_{X_1}(u_1) = (p_1 \mathrm{e}^{\mathrm{i}u_1} + p_2 + \cdots + p_n)^N = (1 - p_1 + p_1 \mathrm{e}^{\mathrm{i}u_1})^N \tag{3.48}$$

which is the CF of a one-dimensional binomial r.v. (eq. (2.51)). Also, using the first of (2.47b) we can obtain the first moment of X_1, that is,

$$E(X_1) = Np_1 \tag{3.49}$$

in agreement with eq. (2.41b) and with the result we would get by using eq. (3.37a) and calculating the derivative $\partial \varphi_{\mathbf{X}}(\mathbf{u})/\partial u_1|_{\mathbf{u}=0}$ of the joint-CF

(3.47). Clearly, by substituting the appropriate index, both eqs (3.48) and (3.49) apply to each one of the X_i.

At this point one may ask about the calculation of $E(X_1)$ by directly using eq. (3.43) without going through the CF. This calculation is rather cumbersome but we outline it here for the interested reader. We have

$$\begin{aligned} E(X_1) &= \sum_{\text{all } x_i} \frac{N!\, x_1}{x_1! \ldots x_n!} p_1^{x_1} \cdots p_n^{x_n} = N(1-p_1)^{N-1} \sum_{x_1=0}^{N} \frac{x_1}{x_1!} p_1^{x_1} \\ &= Np_1(1-p_1)^{N-1} \sum_{x_1=1}^{N} \frac{x_1}{x_1!} p_1^{x_1-1} \end{aligned} \tag{3.50}$$

where we first isolated the sum over x_1, then used the multinomial theorem for the indexes $2, 3, \ldots, N$ and took into account that $p_2 + p_3 + \cdots + p_n = 1 - p_1$. Then, starting from the multinomial theorem

$$\sum_{\text{all } x_i} \frac{N!}{x_1! \cdots x_n!} p_1^{x_1} \cdots p_n^{x_n} = (p_1 + p_2 + \cdots + p_n)^N$$

we can differentiate both sides with respect to p_1 and then, on the l.h.s. of the resulting relation, isolate the sum on x_1. This procedure leads in the end to

$$(1-p_1)^{N-1} \sum_{x_1=1}^{N} \frac{x_1}{x_1!} p_1^{x_1-1} = 1$$

which, in turn, can be substituted in the last expression of (3.50) to give the desired result $E(X_1) = Np_1$.

After this, it is evident that the shortest way to determine the covariance between any two r.v.s X_k, X_m (where k, m are two integers $< n$ with $k \neq m$) is by using the CF. If we recall that $\text{Cov}(X_k X_m) = E(X_k X_m) - E(X_k)E(X_m)$ then we only need to calculate the first term on the r.h.s. because, owing to (3.49), the second term is $N^2 p_k p_m$. Performing the prescribed calculations we get

$$E(X_k X_m) = -\left.\frac{\partial^2 \varphi_{\mathbf{X}}(\mathbf{u})}{\partial u_k \partial u_m}\right|_{\mathbf{u}=0} = N(N-1)p_k p_m \tag{3.51}$$

and therefore the off-diagonal terms of the covariance matrix $\mathbf{K}$ are given by

$$\text{Cov}(X_k X_m) = -Np_k p_m \tag{3.52}$$

By the same token, for any index $1 \le k \le n$, it is not difficult to obtain $E(X_k^2) = Np_k[(N-1)p_k + 1]$ so that the elements on the main diagonal of **K** are

$$\mathrm{Var}(X_k) = Np_k(1 - p_k) \tag{3.53}$$

At first sight – since we spoke of independent trials – the fact that the variables X_i are correlated may seem a bit surprising. The correlation is due to the 'constraint' eq. (3.46b) and the covariances are negative (eq. (3.52)) because an increase of any one x_i tends to decrease the others. The fact that there exists one constraint equation on the x_i implies that **K** is singular (i.e. $\det(\mathbf{K}) = 0$) and has rank $n - 1$ so that, in essence, the n-dimensional vector **X** belongs to the $(n - 1)$-dimensional Euclidean space.

Let us consider now the multi-dimensional extension of the Gaussian (or normal) probability law considered in Examples 2.4, 2.8 and 2.9(b). For simplicity, we begin with the two-dimensional case. In the light of eq. (3.16b), the joint-pdf of two independent and individually normal r.v.s X, Y forming a vector **X** must be

$$f_{\mathbf{X}}(x, y) = \frac{1}{2\pi\sigma_1\sigma_2} \exp\left[-\frac{1}{2}\left(\frac{(x - m_1)^2}{\sigma_1^2} + \frac{(y - m_2)^2}{\sigma_2^2}\right)\right] \tag{3.54}$$

where $m_1 = E(X), \sigma_1^2 = \mathrm{Var}(X)$ and $m_2 = E(Y), \sigma_2^2 = \mathrm{Var}(Y)$. Also, using the result of eqs (2.52) and (3.28) of Proposition 3.5, the joint-CF of **X** is

$$\varphi_{\mathbf{X}}(u, v) = \exp\left[\mathrm{i}(um_1 + vm_2) - \frac{1}{2}(\sigma_1^2 u^2 + \sigma_2^2 v^2)\right] \tag{3.55}$$

which is easy to cast in matrix form as

$$\varphi_{\mathbf{X}}(\mathbf{u}) = \exp\left(\mathrm{i}\mathbf{u}^{\mathrm{T}}\mathbf{m} - \frac{1}{2}\mathbf{u}^{\mathrm{T}}\mathbf{K}\mathbf{u}\right) \tag{3.56}$$

where, in the present case, it should be noted that $\mathbf{K} = \mathrm{diag}(\sigma_1^2, \sigma_2^2)$ because of independence – and therefore uncorrelation – between X and Y. The matrix form of the pdf (3.54) is a bit more involved but only a small effort is required to show that we can write

$$f_{\mathbf{X}}(\mathbf{x}) = \frac{1}{2\pi\sqrt{\det(\mathbf{K})}} \exp\left(-\frac{1}{2}(\mathbf{x} - \mathbf{m})^{\mathrm{T}}\mathbf{K}^{-1}(\mathbf{x} - \mathbf{m})\right) \tag{3.57}$$

where $\sqrt{\det(\mathbf{K})} = \sigma_1\sigma_2$ and $\mathbf{K}^{-1} = \mathrm{diag}(1/\sigma_1^2, 1/\sigma_2^2)$.

From the vector **X** we can pass to the vector **Z** of standardized normal r.v.s by means of the linear transformation $\mathbf{Z} = \mathbf{S}^{-1}(\mathbf{X} - \mathbf{m})$, where **S** is the diagonal matrix introduced in eq. (3.33). By virtue of eq. (3.39b) the CF of **Z** is

$$\varphi_{\mathbf{Z}}(\mathbf{v}) = e^{-1/2\,\mathbf{v}^T\mathbf{R}\,\mathbf{v}} = \exp\left[-\frac{1}{2}(v_1^2 + v_2^2)\right] \tag{3.58}$$

where **R** is the correlation matrix which, in our case of independent r.v.s, equals the identity matrix $\mathbf{I} = \mathrm{diag}(1, 1)$. As expected, the CF (3.58) is the product of two standardized one-dimensional CFs and, clearly, the joint-pdf will also be in the form of product of two standardized one-dimensional pdfs, that is,

$$f_{\mathbf{Z}}(z_1, z_2) = f_{Z_1}(z_1) f_{Z_2}(z_2) = \frac{1}{2\pi}\exp\left[-\frac{1}{2}\left(z_1^2 + z_2^2\right)\right] \tag{3.59}$$

If, on the other hand, the two normal variables X, Y are correlated eqs (3.56) and (3.57) are still valid but it is understood that now **K** – and therefore $\mathbf{K}^{-1}$ – are no longer diagonal because $K_{12} = \mathrm{Cov}(X, Y) \neq 0$. So, the explicit expression of the joint-CF becomes

$$\varphi_{\mathbf{X}}(u, v) = \exp\left[i(u\,m_1 + v\,m_2) - \frac{1}{2}\left(\sigma_1^2 u^2 + \sigma_2^2 v^2 + 2K_{12}uv\right)\right] \tag{3.60}$$

from which – by setting $u = 0$ or $v = 0$, as appropriate – it is evident that both marginal distributions are one-dimensional CFs of Gaussian random variables (see eq. (2.52)). Using eq. (3.57), the explicit form of the joint-pdf is written

$$f_{\mathbf{X}}(x, y) = \frac{1}{2\pi\sigma_1\sigma_2\sqrt{(1-\rho^2)}}\, e^{-\gamma(x,y)/2} \tag{3.61a}$$

where $\rho = K_{12}/\sigma_1\sigma_2$ is the correlation coefficient between X and Y and the function $\gamma(x, y)$ in the exponential is

$$\gamma(x, y) = \frac{1}{1-\rho^2}\left[\frac{(x-m_1)^2}{\sigma_1^2} - 2\rho\frac{(x-m_1)(y-m_2)}{\sigma_1\sigma_2} + \frac{(y-m_2)^2}{\sigma_2^2}\right] \tag{3.61b}$$

From eqs (3.60) and/or (3.61) we note an important property of jointly Gaussian random variables: the condition $\mathrm{Cov}(X, Y) = 0$ – and therefore $\rho = 0$ if both σ_1, σ_2 are finite and different from zero – is necessary and sufficient for X and Y to be independent. In this case, in fact, the joint-CF (pdf) becomes the product of two one-dimensional Gaussian CFs (pdfs). The equivalence of uncorrelation and independence for Gaussian r.v.s is noteworthy because – we recall from Section 3.3 – uncorrelation does not, in general, imply independence.

We consider now another important result for jointly Gaussian r.v.s. Preliminarily, we notice that – referring to a three-dimensional system of coordinate axes x, y, z – the graph of the pdf (3.61) is a bell-shaped surface with maximum of height $z = (2\pi\sigma_1\sigma_2\sqrt{1-\rho^2})^{-1}$ above the point $x = m_1, y = m_2$. If we cut the surface with a horizontal plane (i.e. parallel to the x, y-plane), we obtain an ellipse whose projection on the x, y-plane has equation

$$\frac{(x-m_1)^2}{\sigma_1^2} - 2\rho\frac{(x-m_1)(y-m_2)}{\sigma_1\sigma_2} + \frac{(y-m_2)^2}{\sigma_2^2} = \text{const.} \tag{3.62}$$

which, in particular, is a circle whenever $\rho = 0$ and $\sigma_1 = \sigma_2$. On the other hand, as ρ approaches $+1$ or -1, the ellipse becomes thinner and thinner and more and more needle-shaped until $\rho = 1$ or $\rho = -1$, when the ellipse degenerates into a straight line. In these limiting cases $\det(\mathbf{K}) = 0$, $\mathbf{K}^{-1}$ does not exist and one variable depends linearly on the other. In other words, we are no longer dealing with a two-dimensional random vector but with a single random variable and this is why one speaks of degenerate or singular Gaussian distribution.

Returning to our main discussion, the important result is the following: when the principal axes of the ellipse are parallel to the coordinate axes, then $\rho = 0$ and the two r.v.s are uncorrelated – and therefore independent. In other words, by means of a rotation of the coordinate axes x, y it is always possible to pass from a pair of dependent Gaussian variables – whose pdf is in the form (3.61) – to a pair of independent Gaussian variables. This property can be extended to n dimensions and is frequently used in statistical applications (see the following chapters).

Let us examine this property more closely. Given the ellipse (3.62), it is known from analytic geometry that the relation

$$\tan 2\alpha = \frac{2\rho\sigma_1\sigma_2}{\sigma_1^2 - \sigma_2^2} \tag{3.63}$$

determines the angles α between the x-axis and the principal axes of the ellipse (eq. (3.63) leads to two values α, namely α_1, α_2 where α_1, α_2 differ by $\pi/2$). If we rotate the x, y-plane through an angle α, the new coordinate axes are parallel to the ellipse principal axes and the cross-product term in (3.62) vanishes. If, in addition to this rotation, we perform now a rigid translation of the coordinate system which brings the origin to the point (m_1, m_2), the ellipse will also be centered in the origin. At this point, the original pdf (3.61) has transformed into

$$f(p, q) = \frac{1}{2\pi\sigma_p\sigma_q}\exp\left(-\frac{p^2}{2\sigma_p^2} - \frac{q^2}{2\sigma_q^2}\right) \tag{3.64a}$$

where we call p, q the final coordinate axes obtained by first rotating and then rigidly translating the original axes x, y. Moreover, it can be shown that the new variances σ_p^2, σ_q^2 are expressed in terms of the original variances by the relations

$$\begin{aligned} \sigma_p^2 &= \sigma_1^2 \cos^2\alpha + \rho\sigma_1\sigma_2 \sin 2\alpha + \sigma_2^2 \sin^2\alpha \\ \sigma_q^2 &= \sigma_1^2 \sin^2\alpha - \rho\sigma_1\sigma_2 \sin 2\alpha + \sigma_2^2 \cos^2\alpha \end{aligned} \tag{3.64b}$$

from which it follows $\sigma_p\sigma_q = \sigma_1\sigma_2\sqrt{1-\rho^2}$. Equation (3.64b) is obtained by noticing that the (linear, since α is fixed) relation between the x, y and the p, q axes – and therefore between the original random vector $\mathbf{X} = (X, Y)^T$ and the new random vector $\mathbf{P} = (P, Q)^T$ – is

$$\begin{pmatrix} p \\ q \end{pmatrix} = \begin{pmatrix} \cos\alpha & \sin\alpha \\ -\sin\alpha & \cos\alpha \end{pmatrix} \begin{pmatrix} x \\ y \end{pmatrix} - \begin{pmatrix} m_1 \\ m_2 \end{pmatrix} \tag{3.65}$$

so that eq. (3.64b) follows by virtue of eq. (3.35). As a side comment, eqs (3.64b) represent the diagonal terms of the covariance matrix of the two-dimensional vector $\mathbf{P}$. By using eqs (3.35) and (3.63) the reader is invited to determine that, as expected, $\mathrm{Cov}(P, Q) = 0$, that is, that the off-diagonal terms of the 'new' covariance matrix are zero.

From (3.64a), if needed, it is then possible to take a further step and pass to the standardized Gaussian random vector $\mathbf{Z}$ whose pdf and CF are given by eqs (3.59) and (3.58), respectively.

At this point we can consider a frequently encountered problem and determine the probability P_k that a point falls within the ellipse whose principal axes are k times the standard deviations σ_p, σ_q of the two variables. Calling E_k this ellipse (centered in the origin), the probability we are looking for is

$$P_k = P[(P, Q) \in E_k] = \int_{E_k} f(p, q)\, \mathrm{d}p\, \mathrm{d}q \tag{3.66a}$$

where $f(p, q)$ is given by eq. (3.64a). Passing to the standardized variables $z_1 = p/\sigma_p$ and $z_2 = q/\sigma_q$ the ellipse E_k becomes a circle C_k of radius k and

$$P_k = P\left[Z_1^2 + Z_2^2 \le k^2\right] = \frac{1}{2\pi} \int_{C_k} \mathrm{d}z_1\, \mathrm{d}z_2 \exp\left(-\frac{z_1^2}{2} - \frac{z_2^2}{2}\right) \tag{3.66b}$$

The integral on the r.h.s. can now be calculated by turning to the polar coordinates $z_1 = r\cos\theta, z_2 = r\sin\theta$ (recall from analysis that the Jacobian determinant of this transformation is $|J| = r$) and we finally get

$$P_k = \frac{1}{2\pi}\int_0^{2\pi}\int_0^k r\,\mathrm{e}^{-(r^2/2)}\mathrm{d}r\,\mathrm{d}\theta = \int_0^k r\,\mathrm{e}^{-(r^2/2)}\mathrm{d}r = 1 - \exp(-k^2/2) \tag{3.66c}$$

so that, for instance, we have the probabilities $P_1 = 0.393$, $P_2 = 0.865$ and $P_3 = 0.989$ for $k = 1$, $k = 2$ and $k = 3$, respectively. This result is the two-dimensional counterpart of the well-known fact that for a one-dimensional Gaussian variable X the probability of obtaining a value within k standard deviations is

$$P[|X - \mu_X| \le k\sigma_X] = \begin{cases} 0.683 & \text{for } k = 1 \\ 0.954 & \text{for } k = 2 \\ 0.997 & \text{for } k = 3 \end{cases} \tag{3.67}$$

Equation (3.67), in addition, can also be used to calculate the two-dimensional probability to obtain a value of the vector (P, Q) within the rectangle R_k of sides $2k\sigma_p, 2k\sigma_q$ centered in the origin. In fact, since P and Q are independent, the two-dimensional probability $P[(P, Q) \in R_k]$ is given by the product of the one-dimensional probabilities (3.67); consequently $P[(P, Q) \in R_1] = (0.683)^2 = 0.466$, $P[(P, Q) \in R_2] = (0.954)^2 = 0.910$, etc. and it should be expected that $P[(P, Q) \in R_k] > P[(P, Q) \in E_k]$ because the ellipse E_k is inscribed in the rectangle R_k.

We close this rather lengthy section with a few general comments on Gaussian random vectors in any number of dimensions:

(i) The property of being Gaussian is conserved under linear transformations (as the discussion above has shown more than once in the two-dimensional case).

(ii) The marginal distributions of a jointly-Gaussian are individually Gaussian. However, the reverse may not be true and examples can be given of individually Gaussian r.v.s which, taken together, do not form a Gaussian vector.

(iii) The CF and pdf of a jointly-Gaussian n-dimensional vector are written in matrix form as in eqs (3.56) and (3.57); in this latter equation, however, the factor 2π at the denominator becomes $(2\pi)^{n/2}$. In other words, at the denominator of (3.57) there must be a factor $\sqrt{2\pi}$ for each dimension.

(iv) Let us examine in the general case the possibility of passing from a Gaussian vector of correlated random variables (i.e. with a non-diagonal covariance matrix) to a Gaussian vector of independent – or even standardized – random variables (that is with a diagonal covariance matrix). In two-dimensional this was accomplished by first rotating the coordinate axes and then translating the origin to the point (m_1, m_2) but it is evident that we can first translate the axes and then rotate them without changing the final result. So, starting from a n-dimensional Gaussian vector $\mathbf{X}$ of correlated r.v.s $X_1, \ldots, X_n$ whose pdf is

$$f_{\mathbf{X}}(\mathbf{x}) = \frac{1}{(2\pi)^{n/2}\sqrt{\det(\mathbf{K})}} \exp\left(-\frac{1}{2}(\mathbf{x}-\mathbf{m})^{\mathrm{T}}\mathbf{K}^{-1}(\mathbf{x}-\mathbf{m})\right) \tag{3.68a}$$

we can translate the axes and consider the new 'centered' vector $\widehat{\mathbf{X}} = \mathbf{X} - \mathbf{m}$ with pdf

$$f_{\widehat{\mathbf{X}}}(\hat{\mathbf{x}}) = \frac{1}{(2\pi)^{n/2}\sqrt{\det(\mathbf{K})}} \exp\left(-\frac{1}{2}\hat{\mathbf{x}}^{\mathrm{T}}\mathbf{K}^{-1}\hat{\mathbf{x}}\right) \tag{3.68b}$$

Now, since $\mathbf{K}$ is symmetric and positive definite (i.e. the Gaussian vector is assumed to be non-degenerate), a theorem of matrix algebra states that there exists a non-singular matrix $\mathbf{H}$ such that $\mathbf{H}\mathbf{H}^{\mathrm{T}} = \mathbf{K}$. From this relation we get $\mathbf{H}\mathbf{H}^{\mathrm{T}}\mathbf{K}^{-1} = \mathbf{I}$ and then $\mathbf{H}\mathbf{H}^{\mathrm{T}}\mathbf{K}^{-1}\mathbf{H} = \mathbf{H}$ which, in turn, implies $\mathbf{H}^{\mathrm{T}}\mathbf{K}^{-1}\mathbf{H} = \mathbf{I}$. Using this same matrix $\mathbf{H}$ let us now pass to the new random vector $\mathbf{Z} = (Z_1, \ldots, Z_n)^{\mathrm{T}}$ defined by the relation $\widehat{\mathbf{X}} = \mathbf{HZ}$. The term at the exponential of (3.68b) becomes

$$\hat{\mathbf{x}}^{\mathrm{T}}\mathbf{K}^{-1}\hat{\mathbf{x}} = (\mathbf{Hz})^{\mathrm{T}}\mathbf{K}^{-1}\mathbf{Hz} = \mathbf{z}^{\mathrm{T}}\mathbf{H}^{\mathrm{T}}\mathbf{K}^{-1}\mathbf{Hz} = \mathbf{z}^{\mathrm{T}}\mathbf{Iz} = \mathbf{z}^{\mathrm{T}}\mathbf{z}$$

which is the sum of squares $z_1^2 + z_2^2 + \cdots + z_n^2$. Moreover, the Jacobian determinant of the transformation to $\mathbf{Z}$ is $\det(\mathbf{H})$ so that the multiplying factor before the exponential becomes $\det(\mathbf{H})/\sqrt{(2\pi)^n \det(\mathbf{K})}$; however, from $\mathbf{H}^{\mathrm{T}}\mathbf{K}^{-1}\mathbf{H} = \mathbf{I}$ we get $(\det \mathbf{H})^2 \det(\mathbf{K}^{-1}) = 1$ and since $\det(\mathbf{K}^{-1}) = [\det(\mathbf{K})]^{-1}$, then $\det(\mathbf{H}) = \sqrt{\det(\mathbf{K})}$. Consequently, our final result is

$$f_{\mathbf{Z}}(\mathbf{z}) = \frac{1}{(2\pi)^{n/2}} \exp\left(-\frac{1}{2}\mathbf{z}^{\mathrm{T}}\mathbf{z}\right) \tag{3.69}$$

which, as expected, is the pdf of a standardized Gaussian vector whose covariance matrix is $\mathbf{I}$. Also, it is now clear that the matrix $\mathbf{H}$ represents a n-dimensional rotation of the coordinate axes.

3.4 More on conditioned random variables

The subject of conditioning has been discussed in both Chapters 1 and 2 (see, in particular, Section 2.5.2). Here we return on the subject for two main reasons: first, because some more remarks are worthy of mention in their own right and, second, because a number of new aspects are due the developments of the preceding sections of this chapter.

Let us start with some additional results to what has been said in Section 2.5.2 by working mostly in the real probability space $(\mathbb{R}, \mathbb{B}, P_X)$. We do so because, as stated before, it is $(\mathbb{R}, \mathbb{B}, P_X)$ which is considered in practice while the original space (W, S, P) is an entity in the background with only occasional interest in applications.

Consider an absolutely continuous r.v. X defined on (W, S, P). This, we recall, implies that the measure P_X is absolutely continuous with respect to the Lebesgue measure on the real line and there exists a function f_X (the pdf of X) such that

$$P_X(B) = P(X^{-1}(B)) = \int_B f_X(x)\,\mathrm{d}x$$

Then, although the set $C = \{x\}$ is indeed a Borel set of $\mathbb{R}$, we cannot – at least in the usual way – define a conditional probability with respect to this event because $P_X(x) = 0$. However, for $h > 0$ consider the Borel set B_h defined as $B_h = (x - h, x + h]$. Then $P_X(B_h) > 0$ and we can condition on this event by defining, for every $A \in S$, the measure P_{B_h} exactly as we did in Section 2.5.2 (eq. (2.60); again with a slight misuse of notation because B_h is not a set of S. Rigorously, we should write $P_{X^{-1}(B_h)}$).

Now, with a slight change of notation let us call $P_X(\cdot \mid B_h)$ its image measure in $\mathbb{R}$ instead of $P_{X \mid B_h}$. Then, given $B \in \mathbb{B}$ we have

$$P_X(B \mid B_h) = \frac{P_X(B \cap B_h)}{P_X(B_h)} \tag{3.70}$$

and at this point we can try to define $P_X(B \mid x)$ as the limit of $P_X(B \mid B_h)$ as $h \to 0$. By virtue of Bayes' theorem (eq. (1.14)) together with the total probability formula of eq. (1.12) we can write $P_X(B \mid B_h)$ as

$$\begin{aligned} P_X(B \mid B_h) &= P_X(B)\frac{P_X(B_h \mid B)}{P_X(B_h)} \\ &= P_X(B)\frac{F_{X \mid B}(x+h) - F_{X \mid B}(x-h)}{F_X(x+h) - F_X(x-h)} \end{aligned}$$

(the second equality is due to the basic properties of the PDFs $F_{X \mid B}$ and F_B), then, dividing both the numerator and denominator by $2h$ and passing to the limit we get the desired result

$$P_X(B \mid x) = P_X(B)\frac{f_{X \mid B}(x)}{f_X(x)} \tag{3.71}$$

provided that the conditional density $f_{X \mid B}$ exists. The fact that $P_X(B \mid x)$ is not defined whenever $f_X(x) = 0$ is not a serious limitation because the set

$N = \{x : f_X(x) = 0\}$ has probability zero, meaning that, in practice, it is unimportant as far as probability statements are concerned. In fact

$$P_X(N) = \int_N f_X(x)\,dx = 0$$

If now in eq. (3.71) we move the factor $f_X(x)$ to the left-hand side and integrate both sides over the real line we get (since $f_{X|B}$ is normalized to unity)

$$P_X(B) = \int_{-\infty}^{\infty} f_X(x)\,P_X(B\,|\,x)\,dx \tag{3.72}$$

which, because of its analogy with eq. (1.12), is the continuous version of the total probability formula (see also eq. (3.80b)); the sum becomes now an integral and the probabilities of the conditioning events A_j are now the infinitesimal probabilities $f_X(x)\,dx$.

In the case of discrete r.v.s the complications above do not exist. If A_X is the (discrete) set of values taken on by the r.v. X and $x_i \in A_X$ is such that $P_X\{x_i\} = p_X(x_i) > 0$, then the counterpart of eq. (3.71) is

$$P_X(B\,|\,x_i) = P_X(B)\frac{p_{X|B}(x_i)}{p_X(x_i)} \tag{3.73}$$

and it is not defined if $p(x_i) = 0$. On the other hand, the total probability formula reads

$$P_X(B) = \sum_{x_i \in A_X} p_X(x_i) P_X(B\,|\,x_i) \tag{3.74}$$

We turn now to some new aspects of conditional probability brought about by the discussion of the previous sections. Consider a two-dimensional absolutely continuous random vector $\mathbf{X} = (X, Y)$ with joint-PDF $F_{XY}(x, y)$ and joint-pdf $f_{XY}(x, y)$; we want to determine the statistical description of, say, Y conditioned on a value taken on by the other variable, say $X = x$.

We have now three image measures, P_X, P_Y in $\mathbb{R}$ and the joint measure P_{XY} (or $P_{\mathbf{X}}$) in $\mathbb{R}^2$; all of them, however, originate from P in W. Therefore, if we look for a probabilistic description of an event relative to Y conditioned on an event relative to X it is reasonable to consider – in (W, S) – the conditional probability (where $J_y = (-\infty, y]$ and B_h is as above)

$$P[Y^{-1}(J_y)\,|\,X^{-1}(B_h)] = \frac{P[Y^{-1}(J_y) \cap X^{-1}(B_h)]}{P(X^{-1}(B_h))} = \frac{P_{XY}(J_y \cap B_h)}{P_X(B_h)}$$

and define the conditional PDF $F_{Y|X}(y\,|\,x)$ as the limit of this probability as $h \to 0$. As before, we divide both the numerator and denominator by $2h$ and pass to the limit to get

$$F_{Y|X}(y\,|\,x) = \left(\frac{1}{f_X(x)}\right)\frac{\partial F_{XY}(x,y)}{\partial x} \tag{3.75a}$$

Then, taking the derivative of both sides with respect to y we obtain the conditional pdf

$$f_{Y|X}(y\,|\,x) = \left(\frac{1}{f_X(x)}\right)\frac{\partial^2 F_{XY}(x,y)}{\partial y \partial x} = \frac{f_{XY}(x,y)}{f_X(x)} \tag{3.75b}$$

A few remarks are in order:

(a) the function $f_{Y|X}$ is not defined at the points x where $f_X(x) = 0$; however, as noticed above, this is not a serious limitation;
(b) $f_{Y|X}$ is a function of y alone and not a function of the two variables x, y. In this case x plays the role of a parameter: for a given value, say x_1, we have a function $f_{Y|X}(y\,|\,x_1)$ and we have a different function $f_{Y|X}(y\,|\,x_2)$ for $x_2 \neq x_1$;
(c) being a pdf in its own right, $f_{Y|X}$ is normalized to unity. In fact, recalling eq. (3.10b) we get

$$\int_{-\infty}^{\infty} f_{Y|X}(y\,|\,x)\,dy = \frac{1}{f_X(x)}\int_{-\infty}^{\infty} f_X(x,y)\,dy = 1$$

For the same reason it is clear that the usual relation between pdf and PDF holds, that is

$$F_{Y|X}(y\,|\,x) = \int_{-\infty}^{y} f_{Y|X}(t\,|\,x)\,dt$$

(d) the symmetry between the two variables leads immediately to the conditional-pdf $f_{X|Y}(x\,|\,y)$ of X given $Y = y$, that is,

$$f_{X|Y}(x\,|\,y) = \frac{f_X(x,y)}{f_Y(y)} \tag{3.76}$$

(e) if $\mathbf{X}$ is discrete and $A_X = \{x_1, x_2, \ldots\}$, $A_Y = \{y_1, y_2, \ldots\}$ are the ranges of X and Y, respectively, then the joint-pmf takes on values in $A_X \times A_Y$

and the counterpart of (3.76) can be written as

$$p_{X|Y}(x_i | y_k) = P(X = x_i | Y = y_k) = \frac{p_{\mathbf{X}}(x_i, y_k)}{p_Y(y_k)} \tag{3.77}$$

where it is assumed that $p_Y(y_k) \neq 0$.

In the light of the considerations above, we can now obtain some relations which are often useful in practical cases. We do so for the absolutely continuous case leaving the discrete case to the reader. First, by virtue of eqs (3.75b) and (3.76) we note that it is possible to express the joint-pdf of $\mathbf{X}$ in the two forms

$$\begin{aligned} f_{XY}(x,y) &= f_{Y|X}(y | x) f_X(x) \\ f_{XY}(x,y) &= f_{X|Y}(x | y) f_Y(y) \end{aligned} \tag{3.78}$$

Then, combining these two results we get

$$f_{X|Y}(x | y) = f_{Y|X}(y | x) \frac{f_X(x)}{f_Y(y)} \tag{3.79}$$

and a similar equation for $f_{Y|X}$. Next, in order to obtain the counterpart of the total probability expression of eq. (3.72), we can go back to the probability P by letting B be the event $Y^{-1}(J_y)$; then $P(Y^{-1}(J_y)) = F_Y(y)$ and we obtain the marginal-PDF of Y in terms of the conditional-PDF $F_{Y|X}$ and of the pdf of the conditioning variable, that is,

$$F_Y(y) = \int_{-\infty}^{\infty} F_{Y|X}(y | x) f_X(x) \, dx \tag{3.80a}$$

Differentiating with respect to y on both sides leads to the total probability formula

$$f_Y(y) = \int_{-\infty}^{\infty} f_{Y|X}(y | x) f_X(x) \, dx \tag{3.80b}$$

(which could also be obtained from the second of (3.10b) using (3.79)). By symmetry, it is then evident that

$$F_X(x) = \int_{-\infty}^{\infty} F_{X\,|\,Y}(x\,|\,y) f_Y(y)\, dy$$
$$f_X(x) = \int_{-\infty}^{\infty} f_{X\,|\,Y}(x\,|\,y) f_Y(y)\, dy \qquad (3.81)$$

If the expression (3.80b) for f_Y is inserted at the denominator of eq. (3.79) we note a formal analogy with Bayes' theorem of eq. (1.14); for this reason eq. (3.79) – and its counterpart for $f_{Y\,|\,X}$ – is also called Bayes' formula (in the continuous case).

In Section 3.2.1 we pointed out that two variables X, Y are independent if and only if $f_{XY}(x,y) = f_X(x) f_Y(y)$. Therefore, by virtue of eq. (3.78), independence implies

$$f_{Y\,|\,X}(y\,|\,x) = f_Y(y)$$
$$f_{X\,|\,Y}(x\,|\,y) = f_X(x) \qquad (3.82)$$

as it might be expected considering that knowledge of a specific outcome, say $X = x$, gives no information on Y.

At this point, the extension to more than two r.v.s is immediate and we only mention it briefly here, leaving the rest to the reader. If we call $\overline{\mathbf{X}}$ a multi-dimensional vector of components $X_1, \ldots, X_m, Y_1, \ldots, Y_n$ with joint-pdf $f_{\overline{\mathbf{X}}}$ then, for example,

$$f_{\mathbf{X}\,|\,\mathbf{Y}}(x_1, \ldots, x_m\,|\,y_1, \ldots, y_n) = \frac{f_{\overline{\mathbf{X}}}(x_1, \ldots, x_m, y_1, \ldots, y_n)}{f_{\mathbf{Y}}(y_1, \ldots, y_n)} \qquad (3.83)$$

where we denoted by $f_{\mathbf{Y}}$ the marginal-pdf relative to the n Y-type variables. Similarly, the generalization of eq. (3.80b) becomes

$$f_{\mathbf{Y}}(\mathbf{y}) = \int_{R^m} f_{\mathbf{Y}\,|\,\mathbf{X}}(\mathbf{y}\,|\,\mathbf{x}) f_{\mathbf{X}}(\mathbf{x})\, d\mathbf{x} \qquad (3.84)$$

where $f_{\mathbf{X}}$ the marginal-pdf of the X variables and $d\mathbf{x} = dx_1 \cdots dx_m$.

As an exercise to close this section, we also invite the reader to examine the case of a two-dimensional vector $\mathbf{X} = (X, Y)$ where X is absolutely continuous with pdf $f_X(x)$ and Y is discrete with pmf defined by the values $p_Y(y_i)$.

3.4.1 *Conditional expectation*

As noted in Section 2.5.2, the theory defines conditional expectations as abstract Lebesgue integrals in W with respect to an appropriate conditional measure which, in turn, depends on the conditioning event and is ultimately expressed in terms of the original measure P. In practice, however, owing to the relation between measures in W and their image measures (through a random variable or a random vector), expectations become in the end Lebesgue–Stieltjes integrals on $\mathbb{R}$, $\mathbb{R}^2$ or $\mathbb{R}^n$, whichever is the case. These integrals, in turn, are sums or ordinary Lebesgue integrals (i.e. Riemann integrals in most applications) depending on the type of distribution function.

Owing to the developments of the preceding section, it should be expected that the conditional expectation of X given the event $Y = y$ is expressed as

$$E(X\,|\,y) = \int_{\mathbb{R}} x\,\mathrm{d}F_{X\,|\,Y} \tag{3.85}$$

which, in the absolutely continuous case becomes

$$E(X\,|\,y) = \int_{-\infty}^{\infty} x f_{X\,|\,Y}(x\,|\,y)\,\mathrm{d}x = \frac{1}{f_Y(y)}\int_{-\infty}^{\infty} x f_X(x,y)\,\mathrm{d}x \tag{3.86}$$

where in the second equality we took eq. (3.76) into account. It is understood that analogous relations hold for $E(Y\,|\,x)$. On the other hand, in the discrete case (conditioning on the event $Y = y_k$) we have

$$E(X\,|\,y_k) = \sum_{\text{all } i} x_i p_{X\,|\,Y}(x_i\,|\,y_k) = \frac{1}{p_Y(y_k)}\sum_i x_i p_X(x_i, y_k) \tag{3.87}$$

More generally, if g is a measurable function of both X and Y we have the fundamental relations (their discrete counterparts are left to the reader)

$$\begin{aligned} E[g(X,Y)\,|\,y] &= \int_{-\infty}^{\infty} g(x,y) f_{X\,|\,Y}(x\,|\,y)\,\mathrm{d}x \\ E[g(X,Y)\,|\,x] &= \int_{-\infty}^{\infty} g(x,y) f_{Y\,|\,X}(y\,|\,x)\,\mathrm{d}y \end{aligned} \tag{3.88}$$

It is evident that eq. (3.86) coincides with the first of (3.88) when $g(x,y) = x$ and also that the expressions for all conditional moments can be obtained as special cases of eq. (3.88), depending on which one of the two variables is the conditioning one.

Being based on the properties of the integral, conditional expectations satisfy all the properties of expectation given in Chapter 2. In particular we mention, for example

(i) the conditional expectation of a constant is the constant itself;
(ii) if a, b are two constants and X, Y_1, Y_2 are random variables then linearity holds, that is, $E(aY_1 + bY_2 \mid x) = aE(Y_1 \mid x) + bE(Y_2 \mid x)$;
(iii) if $Y_1 \le Y_2$ then $E(Y_1 \mid x) \le E(Y_2 \mid x)$.

Now, so far we have spoken of conditional expectations of, say, X given $Y = y$ by tacitly assuming that y is a given, well-specified value. If we adopt a more general point of view we can look at expectations as functions of the values taken on by the random variable Y. In other words since, in general, we have a value of $E(X \mid y)$ for every given y we may introduce the real-valued function $g(Y) \equiv E(X \mid Y)$ defined on the range of Y. This function – which, clearly, takes on the value $E(X \mid y)$ when $Y = y$ – can be shown to be measurable and therefore it is a random variable itself. In this light it is legitimate to ask about its expectation $E[g(Y)] = E[E(X \mid Y)]$. The interesting result is that we get

$$E[E(X \mid Y)] = E(X) \tag{3.89a}$$

and, by symmetric arguments

$$E[E(Y \mid X)] = E(Y) \tag{3.89b}$$

In fact, in the absolutely continuous case, for example,

$$\begin{aligned} E[E(X \mid Y)] &= \int E(X \mid Y) f_Y(y)\,\mathrm{d}y = \int \left(\int x f_{X \mid Y}(x \mid y)\,\mathrm{d}x \right) f_Y(y)\,\mathrm{d}y \\ &= \int \int x f_X(x, y)\,\mathrm{d}x\,\mathrm{d}y = \int x \left(\int f_X(x, y)\,\mathrm{d}y \right) \mathrm{d}x \\ &= \int x f_X(x)\,\mathrm{d}x = E(X) \end{aligned}$$

(all integrals are from $-\infty$ to $+\infty$ and eqs (3.76) and (3.10b) have been taken into account).

Equations (3.89) – which are sometimes useful in practice – may appear confusing at first sight but they state a reasonable fact: for instance, eq. (3.89a) shows that $E(X)$ can be calculated by taking a weighted average on all the expected values of X given $Y = y$, each term being weighted by the probability of that particular conditioning event $Y = y$.

Equation (3.89a) can be generalized to

$$E[g(X)] = E[E(g(X) \mid Y)] \tag{3.90}$$

where $g(X)$ is a (measurable) function of X. With the appropriate modifications, the same obviously applies to (3.89b). By similar arguments, the reader is invited to prove that

$$\begin{aligned} \text{Var}(X) &= E[\text{Var}(X \mid Y)] + \text{Var}[E(X \mid Y)] \\ \text{Var}(Y) &= E[\text{Var}(Y \mid X)] + \text{Var}[E(Y \mid X)] \end{aligned} \tag{3.91}$$

(Hint: to prove the first of (3.91) start from $\text{Var}(X \mid Y) = E(X^2 \mid Y) - E^2(X \mid Y)$ and use eq. (3.90).

For our purposes, the discussion above suffices. However, for the interested reader we close this section with some additional remarks of theoretical nature on the function $E(X \mid Y)$. We simply outline the general ideas and more details can be found in the references at the end of the chapter.

Consider an event $G \in S$. As a consequence of eq. (2.58), the expectation of a r.v. X conditioned on G can be written as

$$E(X \mid G) = \int_W X\,\mathrm{d}P_G = \frac{1}{P(G)} \int_W I_G X\,\mathrm{d}P = \frac{1}{P(G)} \int_G X\mathrm{d}P \tag{3.92a}$$

which leads to

$$P(G)E(X \mid G) = \int_G X\,\mathrm{d}P \tag{3.92b}$$

This last expression makes no reference to the conditional measure P_G and can be assumed to be the defining relation of $E(X \mid G)$. Clearly, in the same way one can define $E(X \mid G^C)$. Then, noting that $\tilde{G} = \{\varnothing, G, G^C, W\}$ is a σ-algebra $\tilde{G} \subset S$ (the σ-algebra generated by G) one can define a function $E(X \mid \tilde{G})$ on $\tilde{G}$ as

$$E(X \mid \tilde{G}) = E(X \mid G)I_G + E(X \mid G^C)I_{G^C} \tag{3.93}$$

$E(X \mid \tilde{G})$ is a simple function (see the definition of simple function in Appendix B) which is measurable – and therefore a random variable – with respect to both S and $\tilde{G}$ and it is such that, for every set $A \in \tilde{G}$

$$\int_A E(X \mid \tilde{G})\,\mathrm{d}P = \int_A X\,\mathrm{d}P \tag{3.94}$$

(in this case A can only be one of the four sets $\varnothing, G, G^C, W$; using the definition of integral for simple functions, the reader is invited to verify eq. (3.94)). If, in particular $X = I_F$ (where $F \in S$) then the r.h.s. of (3.94) equals $P(F \cap A)$. Setting $P_F(A) = P(F \cap A)$ for every $A \in \widetilde{G}$ then by virtue of the Radon–Nikodym theorem we can define the conditional probability as a special case of conditional expectation, that is,

$$P(F \mid \widetilde{G}) = E(I_F \mid \widetilde{G}) \tag{3.95}$$

which agrees with the fact that the measure of a set is the expectation of its indicator function.

Now, besides this illustrative example, it can be shown that this same line of reasoning extends to any σ-algebra $\widetilde{S} \subset S$ and the resulting function $E(X \mid \widetilde{S})$ is called the conditional expectation of X given $\widetilde{S}$. In particular, if $\widetilde{S}$ is the σ-algebra generated by a collection of sets $G_1, G_2, \ldots, G_n \in S$ such that $W = \cup_{i=1}^{n} G_i$, then $E(X \mid \widetilde{S})$ is a r.v. on $(W, \widetilde{S}, P)$ which takes on the value $E(X \mid G_i)$ on G_i and satisfies eq. (3.94) for every $A \in \widetilde{S}$. Also, one can define $P(F \mid \widetilde{S})$ as above. However, it is not necessary for $\widetilde{S}$ to be determined by a finite collection of sets. Therefore, if Y is another r.v. defined on the space (W, S, P), for every Borel set $B \subset \mathbb{R}$ one can consider the σ-algebra $\widetilde{Y}$ generated by the inverse images $Y^{-1}(B)$ and introduce the function $E(X \mid \widetilde{Y})$ which satisfies the counterpart of (3.94), that is,

$$\int_{Y^{-1}(B)} E(X \mid \widetilde{Y})\, \mathrm{d}P = \int_{Y^{-1}(B)} X\, \mathrm{d}P \tag{3.96}$$

Since it can be shown that $E(X \mid \widetilde{Y})$ is constant on every set of the form $Y^{-1}(y)$ (where y is a fixed value in $\mathbb{R}$), then it follows that $E(X \mid \widetilde{Y})$ is a function of Y which takes on the value $E(X \mid y)$ for all the elements $w \in W$ such that $w \in Y^{-1}(y)$. Equation (3.96) then shows that $E[E(X \mid \widetilde{Y})] = E(X)$ which, on more theoretical grounds, justifies eq. (3.89a).

3.4.2 *Some examples and further remarks*

In order to illustrate with an example the considerations of the preceding two sections we start with the bivariate Gaussian distribution. If the two variables X, Y are correlated their joint-pdf is given by eqs (3.61a) and (3.61b). We could obtain the marginal-pdfs by using eq. (3.10b) but it is quicker to consider the joint-CF of eq. (3.60) and note that the marginal CFs are both one-dimensional Gaussian. It follows that $f_X(x)$ and $f_Y(y)$ are Gaussian pdfs with parameters $E(X) = m_1$, $\mathrm{Var}(X) = \sigma_1^2$ and $E(Y) = m_2$, $\mathrm{Var}(Y) = \sigma_2^2$, respectively. For the conditional pdfs we can use eq. (3.78) so that, say, $f_{Y|X}(y \mid x)$ is given by $f_{Y|X}(y \mid x) = f_{XY}(x, y)/f_X(x)$. Explicitly, after some manipulations we get

$$f_{Y|X}(y|x) = \frac{1}{\sigma_2\sqrt{2\pi(1-\rho^2)}}\exp[-h(x,y)] \tag{3.97a}$$

where the function in the exponential is

$$h(x,y) = \frac{1}{2(1-\rho^2)}\left(\frac{y-m_2}{\sigma_2} - \rho\frac{x-m_1}{\sigma_1}\right)^2 \tag{3.97b}$$

and can be rewritten in the form

$$h(x,y) = \frac{1}{2\sigma_2^2(1-\rho^2)}\left(y - m_2 - \rho\frac{\sigma_2}{\sigma_1}(x-m_1)\right)^2 \tag{3.97c}$$

from which it is evident that the conditional expectation and variance are

$$\begin{aligned} m_{Y|X} &= E(Y\,|\,X) = m_2 + \rho\frac{\sigma_2}{\sigma_1}(x-m_1) \\ \sigma_{Y|X}^2 &= \text{Var}(Y\,|\,X) = \sigma_2^2(1-\rho^2) \end{aligned} \tag{3.98}$$

Equation (3.98) show that (i) as a function of x, the conditional expectation of Y given x is a straight line (which is called the regression line of Y on X) and (ii) the conditional variance does not depend on x.

With the obvious modifications, relations similar to (3.97) and (3.98) hold for $f_{X|Y}(x\,|\,y), E(X\,|\,Y)$ and $\text{Var}(X\,|\,Y)$. If the two variables are independent – which, we recall, is equivalent to uncorrelated for the Gaussian case – then the conditional-pdfs coincide with the marginal pdfs and the conditional parameters coincide with the unconditioned ones.

Equations (3.97) and their counterparts for $f_{X|Y}(x\,|\,y)$, in addition, show that the conditional-pdfs of jointly Gaussian r.v.s are Gaussian themselves. We do not prove it here but it can be shown that this is an important property which extends to the n-dimensional case: all the conditional pdfs that can be obtained from a jointly Gaussian vector are Gaussian.

If now, as another example, we consider the joint-pdf of Example 3.1(b) (eq. (3.13a)), the reader is invited to determine that

$$\begin{aligned} f_{X|Y}(x\,|\,y) &= \frac{1}{\sqrt{3\pi}}\exp\left(-\frac{1}{3}(x+y/2)^2\right) \\ E(X\,|\,Y) &= -y/2 \end{aligned} \tag{3.99}$$

and also that $E(XY) = -1$. So, if we note from the marginal-pdfs (3.14a) and (3.14b) that $E(X) = E(Y) = 0$ and $\text{Var}(X) = \text{Var}(Y) = 2$, it follows from eqs (3.19a) and (3.22) that $\text{Cov}(X,Y) = -1$ and $\rho_{XY} = -1/2$.

In the bivariate Gaussian case above we spoke of regression line of Y on X. In the general case of a non-Gaussian pdf $E(Y \mid X)$ – as a function of x – may not be a straight line and then one speaks of regression curve of Y on X. For some non-Gaussian pdfs, however, it may turn out that $E(Y \mid X)$ is a straight line, that is, that we have

$$E(Y \mid X) = \int y f_{Y \mid X}(y \mid x)\, dy = a + bx \tag{3.100}$$

where a and b are two constants. Now, since $E[E(Y \mid X)] = E(Y)$ we can take eq. (3.100) into account to get

$$\begin{aligned} E(Y) &= \int\int y f_{XY}(x, y)\, dx\, dy = \int f_X(x) \left(\int y f_{Y \mid X}(y \mid x)\, dy \right) dx \\ &= \int f_X(x)(a + bx)\, dx = a + bE(X) \end{aligned} \tag{3.101}$$

showing that the regression line passes through the point $(E(X), E(Y))$.

By similar arguments, we also obtain (the easy calculations are left to the reader)

$$E(XY) = aE(X) + bE(X^2) \tag{3.102}$$

so that, in the end, the slope and intercept of the straight line are given by

$$\begin{aligned} b &= \frac{E(XY) - E(X)E(X)}{E(X^2) - E^2(X)} = \frac{\mathrm{Cov}(X, Y)}{\mathrm{Var}(X)} \\ a &= E(Y) - bE(X) \end{aligned} \tag{3.103a}$$

where eqs (3.19a) and (3.19b) have been taken into account in the second equality for b. By substituting eqs (3.103a) in (3.100) and recalling eq. (3.22) we see that in all cases where $E(Y \mid X)$ is a linear function of x the first of eq. (3.98) holds. On the other hand, now the second of (3.98) may no longer hold and

$$\sigma_{Y \mid X}^2 = \int \{y - E(Y \mid X)\}^2 f(y \mid x)\, dy$$

is, in general, a function of x. Nonetheless, the quantity $\sigma_2^2(1 - \rho^2)$ still has a meaning: it represents a measure of the average variability of Y around the regression line on X. In fact, it is left to the reader to show that by defining $\sigma_{Y(\mathrm{avg})}^2$ as the weighted (with the probability density of the x-values) average of $\sigma^2_{Y|X}$, then

$$\sigma^2_{Y(\text{avg})} \equiv \int \sigma^2_{Y|X} f_X(x)\,dx = \sigma_2^2(1-\rho^2)$$

(Hint: use (3.100) and the second of (3.103a) to determine that $[y - E(Y|X)]^2 = (y - E(Y))^2 + b^2(x - E(X))^2 - 2b(y - E(Y))(x - E(X))$, insert in the expression of $\sigma^2_{Y|X}$ and then use eq. (3.75b) and the first of (3.103a).) From the expression of $\sigma^2_{Y(\text{avg})}$ we note, however, that the second of eq. (3.98) holds whenever $\sigma^2_{Y|X}$ does not depend on x. In all these particular cases we have $\sigma^2_{Y(\text{avg})} = \sigma^2_{Y|X} = \sigma_2^2(1-\rho^2)$. The bivariate Gaussian – as we have seen – is one of these cases.

If now, in addition to (3.100), we also assume that $E(X|Y) = c + dy$ then

$$\begin{aligned} d &= \frac{\text{Cov}(X,Y)}{\text{Var}(Y)} \\ a &= E(X) - dE(Y) \end{aligned} \tag{3.103b}$$

and the geometric mean of b and d is the correlation coefficient, that is,

$$\sqrt{bd} = \frac{\text{Cov}(X,Y)}{\sigma_X \sigma_Y} = \rho_{XY} \tag{3.104}$$

which, as noted in Section 3.3, is a measure of the extent of the linear relationship between the two variables. As a final remark we point out that the fact that $E(Y|X)$ is a linear function of x does not necessarily imply, in general, that $E(X|Y)$ is a linear function of y – and conversely; the bivariate Gaussian distribution is, in this respect, an exception. More on linear regression in statistical applications is delayed to Chapter 7.

3.5 Functions of random vectors

It often happens that we have some information on one or more random variables but our interest – rather than in the variables themselves – lies in a function of these variables. In Section 2.5.3, we already touched this subject by considering mainly the one-dimensional case; we now move on from there extending the discussion to random vectors.

Let $\mathbf{X} = (X_1, \ldots, X_n)$ be a n-dimensional random vector and let $\mathbf{Z} = (Z_1, \ldots, Z_n)$ be such that $\mathbf{Z} = \mathbf{g}(\mathbf{X})$, where this symbol means that the function $\mathbf{g} : \mathbb{R}^n \to \mathbb{R}^n$ has components $g_1, \ldots, g_n$ and

$$\begin{aligned} Z_1 &= g_1(X_1, \ldots, X_n) \\ Z_2 &= g_2(X_1, \ldots, X_n) \\ &\vdots \\ Z_n &= g_n(X_1, \ldots, X_n) \end{aligned} \tag{3.105}$$

First of all we note that $\mathbf{Z}$ is a random vector if all the $g_k (k = 1, \ldots, n)$ are Borel functions, a condition which is generally true in most practical cases. If we suppose further that we are dealing with absolutely continuous vectors and that the joint-pdf $f_{\mathbf{X}}(\mathbf{x})$ is known, we can obtain $f_{\mathbf{Z}}(\mathbf{z})$ by using a well-known change-of-variables theorem of analysis. This result leads to an equation formally similar to (2.71), that is,

$$f_{\mathbf{Z}}(\mathbf{z}) = f_{\mathbf{X}}(\mathbf{g}^{-1}(\mathbf{z}))|\det(\mathbf{J})| \tag{3.106a}$$

where $\mathbf{J}$ is the Jacobian matrix

$$\mathbf{J} = \begin{bmatrix} \partial g_1^{-1}/\partial z_1 & \partial g_1^{-1}/\partial z_2 & \ldots & \partial g_1^{-1}/\partial z_n \\ \partial g_2^{-1}/\partial z_1 & \partial g_2^{-1}/\partial z_2 & \ldots & \partial g_2^{-1}/\partial z_n \\ \vdots & \vdots & \vdots & \vdots \\ \partial g_n^{-1}/\partial z_1 & \partial g_n^{-1}/\partial z_2 & \ldots & \partial g_n^{-1}/\partial z_n \end{bmatrix} \tag{3.106b}$$

The assumptions of the theorem require that

(a) $\mathbf{g}$ is one-to-one (so that $X_1 = g_1^{-1}(Z_1, \ldots, Z_n)$; $X_2 = g_2^{-1}(Z_1, \ldots, Z_n)$, etc.);
(b) all the derivatives are continuous;
(c) $\det(\mathbf{J}) \neq 0$.

If one (or more) of the g_k $(k = 1, \ldots, n)$ is not invertible (i.e. not one-to-one), one needs to divide the domains of $\mathbf{X}$ and $\mathbf{Z}$ in a sufficient number – say p – of mutually disjoint subdomains in such a way that – in these subdomains – there exists a one-to-one mapping between the two variables. Then eq. (3.106a) holds in each subdomain and the final result $f_{\mathbf{Z}}(\mathbf{z})$ is obtained by summing the p contributions.

All the results above can also be used if $\mathbf{Z}$ is m-dimensional, with $m < n$. In this case one introduces n–m auxiliary variables and proceeds as stated by the theorem. Provided that the requirements of the theorem are satisfied, the choice of the auxiliary variables is arbitrary; therefore it is understood that one should choose them in a way that keeps the calculations as simple as possible. So, for instance, if $\mathbf{X} = (X_1, X_2)$ is a two-dimensional vector and $Z = g(X_1, X_2)$ is one-dimensional, we can introduce the auxiliary variable

$Z_2 = X_2$; then, the transformation (3.105) is

$$\begin{aligned} Z_1 &= g(X_1, X_2) \\ Z_2 &= X_2 \end{aligned} \tag{3.107a}$$

and

$$\det(\mathbf{J}) = \det\begin{bmatrix} \partial g^{-1}/\partial z_1 & \partial g^{-1}/\partial z_2 \\ 0 & 1 \end{bmatrix} = \frac{\partial g^{-1}}{\partial z_1} \tag{3.107b}$$

Consequently, eq. (3.106) reads

$$f_{\mathbf{Z}}(z_1, z_2) = f_{\mathbf{X}}(g^{-1}(z_1), x_2)\left|\frac{\partial g^{-1}}{\partial z_1}\right| \tag{3.107c}$$

and the desired result – that is, the marginal pdf of Z_1 – is given by

$$f_{Z_1}(z_1) = \int_{-\infty}^{\infty} f_{\mathbf{Z}}(z_1, z_2)\,dz_2 \tag{3.107d}$$

Example 3.3 Let $Z = X_1 + X_2$. As above, we introduce the auxiliary variable $Z_2 = X_2$. Then

$$\begin{aligned} X_1 &= Z_1 - Z_2 \\ X_2 &= Z_2 \end{aligned}$$

and $\det(\mathbf{J}) = 1$. Therefore $f_{\mathbf{Z}}(z_1, z_2) = f_{\mathbf{X}}(z_1 - z_2, z_2)$ and

$$f_{Z_1}(z_1) = \int_{-\infty}^{\infty} f_{\mathbf{X}}(z_1 - z_2, z_2)\,dz_2 = \int_{-\infty}^{\infty} f_{\mathbf{X}}(z_1 - x_2, x_2)\,dx_2 \tag{3.108}$$

If the two variables X_1, X_2 are independent with pdfs $f_1(x_1), f_2(x_2)$, respectively, then $f_{\mathbf{X}}(x_1, x_2) = f_1(x_1)f_2(x_2)$ and (3.108) becomes

$$f_{Z_1}(z_1) = \int_{-\infty}^{\infty} f_1(z_1 - x_2)f_2(x_2)\,dx_2 \tag{3.109}$$

which is called the convolution integral of f_1 and f_2; this is a frequently encountered type of integral in applications of Physics and Engineering and is often denoted by the symbol $f_1 * f_2$. (Incidentally, we note that eq. (3.109) is in agreement with eq. (3.29) on CFs; in fact the Fourier transform of a convolution integral is given by the product of the individual Fourier transforms of the functions appearing in the convolution.) So, for instance, if X_1, X_2 are independent and are both uniformly distributed in such a way that

$$f_1(x) = f_2(x) = \begin{cases} 1/(b-a), & a < x \le b \\ 0, & \text{otherwise} \end{cases}$$

eq. (3.109) gives $f_{Z_1}(z_1) = (b-a)^{-1} \int_a^b f_1(z_1 - x_2)\,dx_2$ where z_1 ranges from a minimum value of $2a$ to a maximum of $2b$ and is zero otherwise. The integral can be divided into two parts considering that (i) if $z_1 - x_2 > a$ then $x_2 < z_1 - a$ and (ii) if $z_1 - x_2 < b$ then $x_2 > z_1 - b$. In the first case we have

$$f_{Z_1}(z_1) = \frac{1}{(b-a)^2} \int_a^{z_1 - a} dx_2 = \frac{z_1 - 2a}{(b-a)^2} \tag{3.110a}$$

which holds for $2a < z_1 \le a + b$ (the second inequality is due to the fact that we must have $z_1 - a \le b$, therefore $z_1 \le a + b$). In the second case

$$f_{Z_1}(z_1) = \frac{1}{(b-a)^2} \int_{z_1 - b}^{b} dx_2 = \frac{2b - z_1}{(b-a)^2} \tag{3.110b}$$

which holds for $a + b < z_1 \le 2b$ ($z_1 - b > a$ implies $z_1 > a + b$). The distribution given by eqs (3.110a) and (3.110b) is called Simpson's distribution.

If, turning to another case, X_1, X_2 are jointly-Gaussian and not independent (eqs (3.61a) and (3.61b)) then it can be shown (Refs [3, 4, 6, 17]) that the pdf of the r.v. $Z = X_1 + X_2$ is

$$f_Z(z) = \frac{1}{\sqrt{2\pi(\sigma_1^2 + 2\rho\sigma_1\sigma_2 + \sigma_2^2)}} \exp\left(-\frac{(z - m_1 - m_2)^2}{2(\sigma_1^2 + 2\rho\sigma_1\sigma_2 + \sigma_2^2)}\right) \tag{3.111}$$

which is also Gaussian. The reverse statement, in general, is not true and the fact that $Z = X_1 + X_2$ is Gaussian does not necessarily imply that X_1, X_2 are individually Gaussian. It does, however, if X_1, X_2 are independent (Cramer's theorem). In this case, $f_Z(z)$ is obtained by simply setting $\rho = 0$ in eq. (3.111). All these considerations on jointly-Gaussian vectors extend to n dimensions and the sum $Z = \sum_{i=1}^{n} X_i$ of n Gaussian r.v.s is itself Gaussian with $m_Z = \sum_i m_i$ and $\mathrm{Var}(Z) = \sigma_Z^2 = \sum_i \sigma_i^2$ if the X_i are independent and $m_Z = \sum_i m_i$ and $\sigma_Z^2 = \sum_i \sigma_i^2 + 2\sum_{i<j} \rho_{ij}\sigma_i\sigma_j$ if they are not independent (ρ_{ij} is the correlation coefficient between X_i and X_j).

Example 3.4(a) Consider the random variable $Z_1 = X_1X_2$. If we define $Z_2 = X_2$ then $X_1 = Z_1/X_2$ and $|\det(\mathbf{J})| = 1/|x_2|$. Therefore

$$f_{Z_1}(z_1) = \int_{-\infty}^{\infty} \frac{1}{|x_2|} f_{\mathbf{X}}(z_1/x_2, x_2)\, \mathrm{d}x_2 \tag{3.112}$$

Example 3.4(b) If, on the other hand, we consider the ratio $Z_1 = X_1/X_2$ – and, as above, we define $Z_2 = X_2$ – then $X_1 = Z_1X_2$ and $|\det(\mathbf{J})| = |x_2|$. Therefore

$$f_{Z_1}(z_1) = \int_{-\infty}^{\infty} |x_2| f_{\mathbf{X}}(z_1x_2, x_2)\, \mathrm{d}x_2 \tag{3.113a}$$

In addition, if the two original r.v.s are independent with pdfs $f_1(x_1), f_2(x_2)$

$$f_{Z_1}(z_1) = \int_{0}^{\infty} x_2 f_1(z_1x_2) f_2(x_2)\, \mathrm{d}x_2 - \int_{-\infty}^{0} x_2 f_1(z_1x_2) f_2(x_2)\, \mathrm{d}x_2 \tag{3.113b}$$

So, for instance, if X_1, X_2 are independent Gaussian r.v. with $m_1 = m_2 = 0$ and $\mathrm{Var}(X_1) = \sigma_1^2, \mathrm{Var}(X_2) = \sigma_2^2$ then the term at the exponentials in both integrals of eq. (3.113b) can be written as

$$-\frac{x_2^2}{2}\left(\frac{z_1^2\sigma_2^2 + \sigma_1^2}{\sigma_1^2\sigma_2^2}\right) = -x_2^2\frac{a}{b}$$

where we defined $a = z_1^2\sigma_2^2 + \sigma_1^2$ and $b = 2\sigma_1^2\sigma_2^2$. Eq. (3.113b) then becomes

$$f_{Z_1}(z_1) = \frac{1}{2\pi\sigma_1\sigma_2}\left\{\int_{0}^{\infty} x_2 \exp(-ax_2^2/b)\, \mathrm{d}x_2 - \int_{-\infty}^{0} x_2 \exp(-ax_2^2/b)\, \mathrm{d}x_2\right\}$$

and performing the change of variable $t = ax_2^2/b$ so that $(b/2a)\,\mathrm{d}t = x_2\,\mathrm{d}x_2$ we get

$$f_{Z_1}(z_1) = \frac{2}{2\pi\sigma_1\sigma_2}\left(\frac{b}{2a}\right)\int_{0}^{\infty} \mathrm{e}^{-t}\, \mathrm{d}t = \frac{\sigma_1\sigma_2}{\pi(z_1^2\sigma_2^2 + \sigma_1^2)} \tag{3.114}$$

where we took into account that the two integrals within braces are equal to twice the integral in $\mathrm{d}t$ from 0 to ∞ and we substituted the explicit expressions for a and b to obtain the final term on the r.h.s. of (3.114). The pdf of eq. (3.114) is a form of the so-called Cauchy distribution. In particular, if X_1, X_2 are (independent) standardized r.v.s, then $\sigma_1 = \sigma_2 = 1$ and

$$f_{Z_1}(z_1) = \frac{1}{\pi(z_1^2 + 1)} \tag{3.115}$$

which is the form of the Cauchy distribution commonly found in the literature.

3.5.1 *Numerical descriptors of functions of random variables*

In the preceding section we determined how to obtain the probability distribution of a random variable (vector) which is a function of another random variable (vector) when we know the distribution of the original r.v. Depending on the functional relation between the two variables (vectors), this may not always be an easy task. It often happens, however, that the analyst's interest lies in the numerical descriptors of $\mathbf{Z} = \mathbf{g}(\mathbf{X})$ rather than in a complete probabilistic description of $\mathbf{Z}$ (i.e. $f_\mathbf{Z}$ or $F_\mathbf{Z}$). Moreover, in most cases one is mainly interested in the first and second order moments of $\mathbf{Z}$. These quantities can be obtained – or, more generally, approximated – without going through the determination of $f_\mathbf{Z}$ or $F_\mathbf{Z}$.

Starting from the case in which Z is one-dimensional we have already considered (Propositions 2.13 and 2.15; see also eq. (2.35c)) the situation when Z is a linear function of X, that is, $Z = aX + b$ where a, b are two constants. Then $E(Z) = aE(X) + b$ and $\mathrm{Var}(Z) = a^2\mathrm{Var}(X)$, which, in turn, are special cases of the more general relations

$$\begin{aligned} E(Z) &= \sum_{i=1}^{n} a_i E(X_i) + b \\ \mathrm{Var}(Z) &= \sum_{i=1}^{n} a_i^2 \mathrm{Var}(X_i) + 2\sum_{i<j} a_i a_j \mathrm{Cov}(X_i, X_j) \\ &= \sum_{i=1}^{n} a_i^2 \mathrm{Var}(X_i) + \sum_{ij(i\neq j)} a_i a_j \mathrm{Cov}(X_i, X_j) \end{aligned} \tag{3.116}$$

which occur whenever Z is a linear function of more than one r.v., that is, when $Z = \sum_{i=1}^{n} a_i X_i + b$. If, in addition, the variables $X_1, \ldots, X_n$ are pairwise uncorrelated (or, more strictly, independent), the second of (3.116) becomes $\mathrm{Var}(Z) = \sum_i a_i^2 \mathrm{Var}(X_i)$.

Before turning to the general discussion, consider for instance the frequently encountered non-linear case $Z = XY$. Then, by the properties of covariance (Proposition 2.15 or eq. (3.19a)) we have

$$m_Z \equiv E(Z) = E(XY) = E(X)E(Y) + \text{Cov}(X, Y) \tag{3.117}$$

which becomes $E(Z) = E(XY) = E(X)E(Y)$ whenever X, Y are uncorrelated or independent. Moreover if X, Y are independent it is left to the reader to show that the variance of Z is given by

$$\begin{aligned} \text{Var}(Z) &= \text{Var}(X)\text{Var}(Y) + E^2(X)\text{Var}(Y) + E^2(Y)\text{Var}(X) \\ &= \sigma_X^2\sigma_Y^2 + m_X^2\sigma_Y^2 + m_Y^2\sigma_X^2 \end{aligned} \tag{3.118}$$

(Hint: start from the definition $\text{Var}(Z) = E[(Z - m_Z)^2]$ and then take into account that X^2, Y^2 are also independent r.v.s.)

Let us now tackle the general problem. We will do so in three steps: in the order (a) a one-dimensional variable function of another one-dimensional variable, (b) a one-dimensional variable function of a random vector and (c) a random vector function of another random vector.

Let now $Z = g(X)$ where both X and Z are assumed to be absolutely continuous. If the function g is invertible then we have

$$E(Z) = \int z f_Z(z)\, dz = \int g(x) f_X(x)\, dx \tag{3.119}$$

because $z = g(x)$, $dz = g'(x)\, dx$ (the prime indicates the derivative) and, from eq. (2.71), $f_Z(z) = f_X(x)/g'(x)$ since $dg^{-1}(z)/dz = 1/g'(x)$. However, we can expand $g(x)$ in a Taylor series around m_X as

$$z = g(x) = g(m_X) + (x - m_X)g'(m_X) + \frac{1}{2}(x - m_X)^2 g''(m_X) + \cdots \tag{3.120}$$

and insert this expression in (3.119) to get the approximate relation

$$E(Z) \cong g(m_X) + \frac{1}{2}g''(m_X)\,\text{Var}(X) \tag{3.121}$$

because it is easily verified that the term with the first derivative yields zero in the integration. The calculation of the variance is a bit more involved. Similarly to eq. (3.119) we can write

$$\text{Var}(Z) = \int (z - m_Z)^2 f_Z(z)\, dz = \int [g(x) - m_Z]^2 f_X(x)\, dx \tag{3.122}$$

and use (i) the Taylor expansion (3.120) to approximate $g(x)$ and (ii) eq. (3.121) to approximate $m_Z = E(Z)$. After a few passages we arrive at

$$\mathrm{Var}(Z) \cong [g'(m_X)]^2 \mathrm{Var}(X) + \frac{1}{4}[g''(m_X)]^2\{M_4 - \mathrm{Var}^2(X)\} + g'(m_X)g''(m_X)M_3 \tag{3.123a}$$

where we denoted by M_3 and M_4 the third and fourth-order central moments of X, respectively (i.e. $M_3 = E[(X - m_X)^3]$ and $M_4 = E[(X - m_X)^4]$; also, using this notation note that $\mathrm{Var}(X) = M_2$).

If the pdf $f_X(x)$ is symmetric about the mean, then $M_3 = 0$ and if, in addition, it is Gaussian then (eq. (2.42d)) $M_4 = 3\,M_2^2 = 3\,\mathrm{Var}^2(X)$; therefore

$$\mathrm{Var}(Z) \cong [g'(m_X)]^2 \mathrm{Var}(X) + \frac{1}{2}[g''(m_X)]^2\,\mathrm{Var}^2(X) \tag{3.123b}$$

For approximation purposes, one may sometimes use $m_Z = g(m_X)$ – which is equivalent to interchanging the expectation operator with the functional dependence, that is, $E[g(X)] = g[E(X)]$ – for the mean and $\sigma_Z^2 = [g'(m_X)]^2\sigma_X^2$ for the standard deviation; however, it should be kept in mind that these relations are exact only in case of a linear relation between X and Z.

Let now Z be a function of n random variables $X_1, \ldots, X_n$, that is, $Z = g(X_1, \ldots, X_n)$. In this case the linear approximation is frequently used; in other words one assumes that (i) the mean of the function equals the function of the X-means $m_1, \ldots, m_m$ and (ii) the variance of the function depends only on the first derivatives of g and on the variances $\sigma_1^2, \ldots, \sigma_n^2$ of $X_1, \ldots, X_n$. Although this may seem a rather crude approximation, it generally leads to acceptable result and consequently – besides specific applications where a higher accuracy is required – linearization is the main technique to deal with the case $Z = g(X_1, \ldots, X_n)$. So, linearizing the function g in a neighbourhood of $m_1, \ldots, m_m$ we have

$$g(\mathbf{x}) = g(\mathbf{m}) + \sum_{i=1}^{n} \left.\frac{\partial g}{\partial x_i}\right|_{\mathbf{x}=\mathbf{m}} (x_i - m_i) + \cdots \tag{3.124}$$

so that inserting this expression in

$$E(Z) = \int g(\mathbf{x})\, f_{\mathbf{X}}(\mathbf{x})\, d\mathbf{x} \tag{3.125}$$

all the terms with the first derivatives go to zero in the integration and

$$m_Z = E(Z) \cong g(\mathbf{m}) = g(m_1, m_2, \ldots, m_n) \tag{3.126}$$

Equation (3.126), in turn, can be used together with (3.124) in the expression

$$\operatorname{Var}(Z) = \int [g(\mathbf{x}) - m_Z]^2 f_X(\mathbf{x})\, d\mathbf{x} \tag{3.127}$$

to arrive at the (approximate) result

$$\sigma_Z^2 = \operatorname{Var}(Z) \cong \sum_{i=1}^{n} [D_i g(\mathbf{m})]^2 \sigma_i^2 + \sum_{i,j;\, i \neq j} [D_i g(\mathbf{m})][D_j g(\mathbf{m})]\, K_{ij} \tag{3.128a}$$

where, for short, we denoted $D_i g(\mathbf{m}) = \partial g/\partial x_i|_{\mathbf{x}=\mathbf{m}}$.

If the variables $X_1, \ldots, X_n$ are uncorrelated then

$$\operatorname{Var}(Z) \cong \sum_{i=1}^{n} [D_i g(\mathbf{m})]^2 \sigma_i^2 \tag{3.128b}$$

which is, nonetheless, an approximation due to the fact that we retained only the first-order terms in the Taylor expansion. Introducing the column matrix $\mathbf{D}$ whose elements are the first-order derivatives of g calculated at $\mathbf{x} = \mathbf{m}$, that is,

$$\mathbf{D} = \begin{bmatrix} D_1 g(\mathbf{m}) \\ D_2 g(\mathbf{m}) \\ \vdots \\ D_n g(\mathbf{m}) \end{bmatrix}$$

then eq. (3.128a) can be concisely written in matrix form as

$$\operatorname{Var}(Z) \cong \mathbf{D}^{\mathrm{T}} \mathbf{K} \mathbf{D} \tag{3.128c}$$

where $\mathbf{K}$ is the covariance matrix introduced in eq. (3.32a). If, in addition, the variables $X_1, \ldots, X_n$ are uncorrelated then $\mathbf{K} = \operatorname{diag}(\sigma_1^2, \ldots, \sigma_n^2)$ and eq. (3.128c) reduces to the sum of squares of eq. (3.128b).

A better approximation to $E(Z)$ and $\operatorname{Var}(Z)$ than eqs (3.126) and (3.128), respectively, can be obtained by retaining the next term in the Taylor expansion (3.124). This term contains the second-order derivatives of g and can be written as

$$\frac{1}{2} \sum_{i=1}^{n} D_i^2 g(\mathbf{m})(x_i - m_i)^2 + \frac{1}{2} \sum_{i,j;\, i \neq j} D_{ij}^2 g(\mathbf{m})(x_i - m_i)(x_j - m_j)$$

where $D_i^2 g(\mathbf{m}) = \partial^2 g/\partial x_i^2|_{\mathbf{x}=\mathbf{m}}$ and $D_{ij}^2 g(\mathbf{m}) = \partial^2 g/\partial x_i \partial x_j|_{\mathbf{x}=\mathbf{m}}$. In this approximation we are led to

$$m_Z \cong g(\mathbf{m}) + \frac{1}{2}\sum_i D_i^2 g(\mathbf{m})\,\sigma_i^2 + \frac{1}{2}\sum_{i,j;\,i\neq j} D_{ij}^2 g(\mathbf{m})\,K_{ij} \tag{3.129a}$$

or, if the variables are uncorrelated

$$m_Z \cong g(\mathbf{m}) + \frac{1}{2}\sum_i D_i^2 g(\mathbf{m})\,\sigma_i^2 \tag{3.129b}$$

For the variance we can limit the calculations to the independent (or uncorrelated) case – although eq. (3.128b) will, in general, suffice in this circumstance – and arrive at the rather lengthy relation

$$\begin{aligned}\sigma_Z^2 \cong{}& \sum_i [D_i g(\mathbf{m})]^2 \sigma_i^2 + \frac{1}{4}\sum_i \left[D_i^2 g(\mathbf{m})\right]^2 \{M_4(X_i) - \mathrm{Var}^2(X_i)\} \\ &+ \sum_i [D_i g(\mathbf{m})]\left[D_i^2 g(\mathbf{m})\right] M_3(X_i) + \sum_{i\neq j}\left[D_{ij}^2 g(\mathbf{m})\right]\sigma_i^2\sigma_j^2\end{aligned} \tag{3.130}$$

where we denoted by $M_3(X_i), M_4(X_i)$ the third and fourth-order central moments of the variable X_i, respectively. As an example, we can return to the case $Z = XY$ (X and Y independent) considered above. The reader can check that the approximation (3.128b) does not lead to the correct result (3.118) while, on the other hand, eq. (3.130) does.

Finally, we examine now the most general case of m r.v.s $Z_1, \ldots, Z_m$ which are functions of n r.v.s $X_1, \ldots, X_n$. The situation is as follows

$$\begin{aligned} Z_1 &= g_1(X_1, \ldots, X_n) \\ Z_2 &= g_2(X_1, \ldots, X_n) \\ &\vdots \\ Z_m &= g_m(X_1, \ldots, X_n) \end{aligned} \tag{3.131}$$

Denoting by $\overline{m}_1, \ldots, \overline{m}_m$ the means of $Z_1, \ldots, Z_m$, the linear approximation immediately yields

$$\overline{m}_k \cong g_k(m_1, \ldots, m_n), \quad k = 1, 2, \ldots, m \tag{3.132}$$

while the covariance matrix $\overline{\mathbf{K}}$ of the Z-variables is given by

$$\overline{\mathbf{K}} \cong \mathbf{D}^{\mathrm{T}}\mathbf{K}\mathbf{D} \tag{3.133a}$$

where $\mathbf{K}$ is the covariance matrix of the X-variables and we denoted by $\mathbf{D}$ the $n \times m$ matrix of derivatives

$$\mathbf{D} = \begin{bmatrix} \partial g_1/\partial x_1 & \partial g_2/\partial x_1 & \dots & \partial g_m/\partial x_1 \\ \partial g_1/\partial x_2 & \partial g_2/\partial x_2 & \dots & \partial g_m/\partial x_2 \\ \vdots & \vdots & \vdots & \vdots \\ \partial g_1/\partial x_n & \partial g_2/\partial x_n & \dots & \partial g_m/\partial x_n \end{bmatrix} = \begin{bmatrix} D_{11} & D_{21} & \dots & D_{m1} \\ D_{12} & D_{22} & \dots & D_{m2} \\ \vdots & \vdots & \vdots & \vdots \\ D_{1n} & D_{2n} & \dots & D_{mn} \end{bmatrix}$$

and it is understood that all derivatives are calculated at the point $\mathbf{x} = \mathbf{m}$. So, the (i, j)th element of the matrix is

$$\overline{K}_{ij} = \mathrm{Cov}(Z_i, Z_j) \cong \sum_{k,l} D_{ik} K_{kl} D_{jl} \tag{3.133b}$$

and $\overline{\mathbf{K}}$, being a covariance matrix, is clearly symmetric, that is, $\overline{K}_{ij} = \overline{K}_{ji}$. Clearly, eq. (3.133b) could also be directly obtained from the definition of covariance. In fact, for example, if $Z_1 = g_1(X_1, X_2)$ and $Z_2 = g_2(X_1, X_2)$ we have

$$\overline{K}_{12} = \mathrm{Cov}(Z_1, Z_2) = \int (z_1 - \overline{m}_1)(z_2 - \overline{m}_2) f_{\mathbf{Z}}(\mathbf{z})\, d\mathbf{z}$$

and by a similar line of reasoning as above we can expand both g_1, g_2 in a neighborhood of $\mathbf{m}$ and use this expansion together with eq. (3.132) to get

$$\overline{K}_{12} \cong \sum_{k,l} \frac{\partial g_1}{\partial x_k} \frac{\partial g_2}{\partial x_l} K_{kl}$$

which, as expected, is the same as eq. (3.133b).

As the next example will show, a final point worthy of notice is that independence of the X-variables does not, in general, imply independence of the Z-variables.

Example 3.5 Let X_1, X_2 be two uncorrelated r.v.s with variances $K_{11} = \sigma_1^2$, $K_{22} = \sigma_2^2$. Also let $Z_1 = 2X_1 + X_2$ and $Z_2 = 5X_1 + 3X_2$. Then

$$\mathbf{D}^{\mathrm{T}}\mathbf{K}\mathbf{D} = \begin{bmatrix} 2 & 1 \\ 5 & 3 \end{bmatrix} \begin{bmatrix} \sigma_1^2 & 0 \\ 0 & \sigma_2^2 \end{bmatrix} \begin{bmatrix} 2 & 5 \\ 1 & 3 \end{bmatrix} = \begin{bmatrix} 4\sigma_1^2 + \sigma_2^2 & 10\sigma_1^2 + 3\sigma_2^2 \\ 10\sigma_1^2 + 3\sigma_2^2 & 25\sigma_1^2 + 9\sigma_2^2 \end{bmatrix}$$

showing that Z_1, Z_2 are, as a matter of fact, correlated.

Example 3.6 Suppose that the coordinates x, y in a plane can be measured with uncertainties $\sigma_1 = 0.2$ cm for the x-coordinate and $\sigma_2 = 0.4$ cm for the y-coordinate. Assume further that the measured x, y values of a point in the plane are uncorrelated and they are considered the mean coordinates for that point. Our measurement yields $(x, y) = (1, 1)$; what are the uncertainties in polar coordinates? Now, the functional relations for the problem are

$$r = \sqrt{x^2 + y^2}$$
$$\theta = \arctan(y/x)$$

and the derivative matrix is

$$\mathrm{D} = \begin{bmatrix} x/r & -y/r^2 \\ y/r & x/r^2 \end{bmatrix}_{(x,y)=(1,1)} = \begin{bmatrix} 1/\sqrt{2} & -1/2 \\ 1/\sqrt{2} & 1/2 \end{bmatrix}$$

while $\mathbf{K} = \mathrm{diag}(\sigma_1^2, \sigma_2^2) = \mathrm{diag}(0.04, 0.16)$. Therefore

$$\overline{\mathbf{K}} \cong \mathbf{D}^{\mathrm{T}}\mathbf{K}\mathbf{D} = \begin{bmatrix} 0.100 & 0.042 \\ 0.042 & 0.050 \end{bmatrix}$$

and the uncertainties we are looking for are $\sigma_r = \sqrt{0.1} = 0.32$ cm and $\sigma_\theta = \sqrt{0.05} = 0.22$ radians. Consequently, we will express our measurement as $x = 1.0 \pm 0.2$; $y = 1.0 \pm 0.4$ cm in rectangular coordinates and $r = 1.41 \pm 0.32$ cm; $\theta = \pi/4 \pm 0.22$ radians in polar coordinates. Note that the transformation from rectangular to polar coordinates has introduced a positive correlation between r and θ.

3.6 Summary and comments

This chapter continues along the line of Chapter 2 by extending the discussion to the so-called multivariate case, that is, the case in which two, three, ..., n random variable are considered simultaneously. In this light it is useful to introduce the concept of random vector and – whenever convenient – exploit the brevity and compactness of vector and matrix notation.

A n-dimensional random vector $\mathbf{X}$ is, in essence, a measurable function from an abstract probability space (W, S, P) to $\mathbf{R}^n$ and this implies that each one of its components must be a random variable. In this light, Section 3.2 shows that the familiar concepts of induced probability measure, PDF and pdf (when it exists) can be readily extended to these vector-values functions. A new aspect, which has no counterpart in the one-dimensional case, is considered in Section 3.2.1 where the notion of marginal distribution functions is introduced. These functions have to do with the 'subvectors' of a given vector $\mathbf{X}$ and it is shown that the joint probability description of $\mathbf{X}$ contains implicitly the probabilistic description of each one of its possible 'subvectors'. In general, however, the reverse statement is not true unless its components are independent. In this case, in fact, a number of important 'product rules' hold and one can obtain the joint-PDF (or pdf) of the vector from the PDFs (pdfs) of its components.

Similarly to the one-dimensional case, the moments of a random vector are defined as abstract Lebesgue integrals in the probability space (W, S, P). The most important moments in applications are the first- and second-order moments which are given special names. So, in addition to the concepts of mean values and variances of $\mathbf{X}$, the notion of covariance is defined in Section 3.3 and some properties of these numerical descriptors are given. Particularly important in both theory and applications is the notion of uncorrelation of random variables which, broadly speaking, is a weak form of (pairwise) independence. Stochastic independence, in fact, implies uncorrelation but the reverse, in general, is not true. Besides this, Section 3.3 introduces the concept of joint-characteristic function by generalizing the one-dimensional case of Chapter 2; in particular, it is shown that independence implies the validity of a 'product rule' also for characteristic functions. Then, in Section 3.3.1 the discussion continues by noting the usefulness of matrix notation and by considering the actual calculations of moments and expectations in practice. In fact, mathematical analysis provides all the necessary results to show that the abstract Lebesgue integrals with respect to the measure P are evaluated as Lebesgue–Sieltjes integrals in $\mathbb{R}^n$; these, in turn, in most practical cases become either sums or ordinary Lebesgue integrals depending on the type of PDF – that is, $F_{\mathbf{X}}$ – induced by the random vector $\mathbf{X}$. Moreover, when the pdf exists the Lebesgue integrals coincide with the familiar Riemann integrals (it should be remembered, however, that Lebesgue integrals have a number of desirable properties which are not satisfied by Riemann integrals).

Next, Section 3.3.2 is more application-oriented and gives two important examples of multivariate distributions: a discrete one, the so-called multinomial distribution, and a continuous one, the multivariate Gaussian (or normal) distribution. This is done in order to show how the developments considered so far are translated into practice.

For its importance in both theory and practice, Sections 3.4 and 3.4.1 return on the subject of conditional probability. Here we extend the notion of conditioning to random variables by also considering, in the continuous case, the possibility of conditioning on events of zero probability. Then, since a conditional probability is a probability measure in its own right, the concepts of conditional PDF and pdf are introduced in the multivariate case and their relation to the joint and marginal functions is also shown. As one might expect, conditional expectations satisfy all the main properties of expectations. However, some additional properties are worthy of mention and these are given in Section 3.4.1 together with further theoretical remarks and examples.

Finally, the last two Sections 3.5 and 3.5.1, deal with the probabilistic description of functions of a given random vector $\mathbf{X}$, assuming that some information on $\mathbf{X}$ is available. More specifically – limiting for the most part the discussion to the continuous case – Section 3.5 considers the general problem of obtaining the joint-pdf of a vector $\mathbf{Z} = \mathbf{g}(\mathbf{X})$; then, in order to show practical cases, some examples are given. On the other hand, Section 3.5.1 addresses the problem of obtaining some information on **Z** without necessarily trying to describe it completely. The task is accomplished by calculating the lowest-order moments – typically means, variances and covariances – of **Z** only on the basis of the available information on **X**. In most cases one only arrives at approximate relations because linearization of the function **g** is often necessary. Nonetheless, this partial information – obtained, in addition, by means of approximate equations – is sufficient and sufficiently accurate in a large number of practical situations.

References and further reading

[1] Ash, R.B., Doléans-Dade, C., *'Probability and Measure Theory'*, Harcourt Academic Press, San Diego (2000).
[2] Brémaud, P., *'An Introduction to Probabilistic Modeling'*, Springer-Verlag, New York (1988).
[3] Cramer, H., *'Mathematical Methods of Statistics'*, Princeton Landmarks in Mathematics, Princeton University Press, 19th printing (1999).
[4] Dall'Aglio, G., *'Calcolo delle Probabilità'*, Zanichelli, Bologna (2000).
[5] Friedman, A., *'Foundations of Modern Analysis'*, Dover Publications, New York (1982).
[6] Gnedenko, B.V., *'Teoria della Probabilità'*, Editori Riuniti, Roma (1987).
[7] Heathcote, C.R., *'Probability, Elements of the Mathematical Theory'*, Dover Publications, New York (2000).
[8] Horn, R.A., Johnson, C.R., *'Matrix Analysis'*, Cambridge University Press (1985).
[9] Kolmogorov, A.N., Fomin, S.V., *'Introductory Real Analysis'*, Dover, New York (1975).
[10] McDonald, J.N., Weiss, N.A., *'A Course in Real Analysis'*, Academic Press, San Diego (1999).
[11] Monti, C.M., Pierobon, G., *'Teoria della Probabilità'*, Decibel editrice, Padova (2000).
[12] Pfeiffer, P.E., *'Concepts of Probability Theory'*, 2nd edn., Dover Publications, New York (1978).
[13] Rotondi, A., Pedroni, P., Pievatolo, A., *'Probabilità, Statistica e Simulazione'*, Springer-Verlag, Italia, Milano (2001).
[14] Biswas, S., *'Topics in Statistical Methodology'*, Wiley Eastern Limited, New Delhi (1991).
[15] Taylor, J.C., *'An Introduction to Measure and Probability'*, Springer-Verlag, New York (1997).
[16] Thompson, R.S.H.G., *'Matrices: Their Meaning and Manipulation'*, The English Univerities Press Ltd., London (1969).
[17] Ventsel, E.S., *'Teoria delle Probabilità'*, Mir Publisher, Moscow (1983).

4 Convergences, limit theorems and the law of large numbers

4.1 Introduction

In most issues where chance plays a part, things seem to behave rather erratically if one looks only at a few instances. On the other hand, this type of behaviour seems to 'smooth out' in the long run. In other words, as the number of observed instances – or trials or experiments – increases, a more and more orderly pattern seems to ensue and certain regularities become clearer and clearer. This is what happens, for example, when we toss a coin; after 10 tosses we would not be surprised to have, say, eight heads and two tails but we would surely be if we got 800 heads and 200 tails after 1000 tosses. In fact, in this case we would seriously suspect that the coin is biased. This state of affair would be intriguing but not particularly interesting if it applied only to coins and dice. As a matter of fact, however, a large number of experiences in many fields of human activities – from birth and death rates to accidents, from measurements in science and technology to the occurrence of hurricanes or earthquakes, just to name a few – behave in a similar manner when measured, tabulated and/or assigned numerical values. The appearance of long-term regularities as the number of trials increases has been known for centuries and goes under the name of 'law of large numbers'. The great achievement of probability theory is in having established the general conditions under which these regularities can and do occur.

We open here a short parenthesis. Returning to the coin example for a moment, it is worth pointing out that the law of large numbers does not justify certain mistaken beliefs such as, say: I tossed a fair coin 15 times and I got 14 heads, the next toss is very likely to result in a head. This is wrong because the process has no memory and the probability of a head is 0.50 for each toss. In other words, the coin has no responsibility whatsoever to 'make up' for a past run of many heads in a row. This misinterpretation (unfortunately, a rather common misinterpretation; consider, for example, the habit of betting on 'late' numbers in lotteries) of the law of large numbers is due to the fact that one fails to distinguish between a regularity 'in the ratio sense' and a regularity in an 'absolute sense'. The former concept refers to the number of heads (or tails) divided by the total number of tosses while the latter refers to the number of heads (or tails) in excess over tails (heads); as the number N of tosses increases, the above ratio tends to stabilize by getting closer and closer to 0.50 while the difference between heads and tails can become rather large (in fact, it generally increases).

So, returning to our main discussion, this chapter is intended to provide the mathematical rationale behind the general term 'law of large numbers' and since the concept implies a tendency towards something, it is easily guessed that its mathematical formalization entails some kind of limit. The first step, therefore, is to consider which kind of limits are involved in the long-term behaviour of experiments governed by chance.

4.2 Weak convergence

In the final part of Section 2.4 (Definition 2.5) we introduced the notion of weak convergence of random variables. This type of convergence is also known in probability theory as 'convergence in distribution' or 'convergence in law' to mean that the probability law (i.e. the PDF) of X_n converges to a function which is itself a probability law. We recall here some important points:

(a) $F_n \to F[w]$ – or equivalently $X_n \to X[D]$ – means that $\lim_{n\to\infty} F_n(x) = F(x)$ at all points where $F(x)$ is continuous (there is no ambiguity because $F(x)$, being a PDF, is right-continuous). Also, it is not difficult to see that Definition 2.5 of weak convergence is equivalent to stating that $\lim_{n\to\infty} P(X_n \le x) = P(X \le x)$ whenever $P(X = x) = 0$;

(b) since weak convergence does not refer directly to the r.v.s X_n and neither it involves directly the probability space on which they are defined (weak convergence is a property of the PDFs and not of the X_n themselves), the concept makes sense even if the X_n are defined on different probability spaces;

(c) sequences of discrete r.v.s may converge (weakly) to a continuous r.v.s and conversely. Moreover, the fact that a sequence X_n of absolutely continuous r.v.s with pdfs $f_n = F_n'$ converges in distribution to an absolutely continuous r.v. X whose pdf is $f = F'$ does not imply, in general, that the sequence f_n converges to f. It is worth noting, however, that if $f_n \to f$ pointwise (or even almost everywhere, see Section 4.3), then $X_n \to X[D]$.

The extension to random vectors is rather straightforward: if $(X_n^{(1)}, X_n^{(2)}, \dots, X_n^{(m)})$ converges weakly to the vector $(X^{(1)}, X^{(2)}, \dots, X^{(m)})$ then $X_n^{(i)} \to X^{(i)}[D]$ for every $i = 1, 2, \dots, m$. The reverse in general is not true and weak convergence of every individual component does not imply the vector weak convergence. This result should be hardly surprising; in fact, given $F^{(1)}$ and $F^{(2)}$ – we are considering the two-dimensional case for simplicity – there are infinite joint-PDFs for which $F^{(1)}, F^{(2)}$ are the marginal PDFs and therefore $F_n^{(i)} \to F^{(i)}[w]$ for $i = 1, 2$ gives no information on the convergence of $F_{\mathbf{X}_n}$. In addition, $F_{\mathbf{X}_n}$ may not even converge at all. Also – in the light of the definition of weak convergence – it should be noted that $X_n \to X[D]$ does not imply $X_n - X \to 0[D]$, as it is customary for ordinary convergence of real variables.

A fundamental result on D-convergence is given by Levy's theorem of Proposition 2.24 which brings into play pointwise convergence of characteristic functions and is often used in probability theory. We use it, for instance, to prove a first limit theorem:

Proposition 4.1 *Let X_n be a sequence of binomial r.v.s with parameters n and $p = \lambda/n$, where λ is a positive real number. Then, as $n \to \infty$, X_n converges in distribution to a Poisson r.v. of parameter λ.*

Before proving this proposition, some preliminary comments on the Poisson distribution are in order. As it is probably known to the reader, we call Poisson r.v. with parameter λ a discrete r.v. X whose pmf is given by

$$p_X(x) = \mathrm{e}^{-\lambda}\frac{\lambda^x}{x!} \qquad (x = 0, 1, 2, \ldots) \tag{4.1}$$

and it can be shown that $E(X) = \mathrm{Var}(X) = \lambda$. In fact, for example,

$$E(X) = \sum_{x=0}^{\infty} x\mathrm{e}^{-\lambda}\frac{\lambda^x}{x!} = \lambda \sum_{x=1}^{\infty}\frac{\lambda^{x-1}}{(x-1)!}\mathrm{e}^{-\lambda} = \lambda \tag{4.2a}$$

because on the r.h.s. we sum on all the ordinates of the distribution and therefore the sum equals 1. In addition, the CF of the Poisson distribution is easily obtained as

$$\begin{aligned} \varphi(u) = E(\mathrm{e}^{\mathrm{i}uX}) &= \mathrm{e}^{-\lambda}\sum_x \frac{(\lambda \mathrm{e}^{\mathrm{i}u})^x}{x!} \\ &= \mathrm{e}^{-\lambda}\exp(\lambda \mathrm{e}^{\mathrm{i}u}) = \exp[\lambda(\mathrm{e}^{\mathrm{i}u} - 1)] \end{aligned} \tag{4.2b}$$

from which, using eqs (2.47b) and (2.34), it is almost immediate to determine that $E(X^2) = \lambda + \lambda^2$ and $\mathrm{Var}(X) = \lambda$. For higher-order moments it may be more convenient to use the recursion relation

$$E(X^k) = \lambda\left(\frac{\mathrm{d}}{\mathrm{d}\lambda} + 1\right)E(X^{k-1}) \tag{4.2c}$$

with the starting assumption $E(X^0) = 1$. Therefore $E(X) = \lambda$, $E(X^2) = \lambda + \lambda^2$, $E(X^3) = \lambda + 3\lambda^2 + \lambda^3$, $E(X^4) = \lambda + 7\lambda^2 + 6\lambda^3 + \lambda^4$, etc.

A final remark on the Poisson distribution is as follows: let X, Y be two independent Poisson r.v.s with parameters λ_1, λ_2, respectively. Independence implies (eq. (3.29)) that the CF of the r.v. $X + Y$ is

$$\begin{aligned}\varphi_{X+Y}(u) &= \{\exp[\lambda_1(e^{iu} - 1)]\}\{\exp[\lambda_2(e^{iu} - 1)]\} \\ &= \exp[(\lambda_1 + \lambda_2)(e^{iu} - 1)]\end{aligned} \tag{4.3}$$

which is the CF of a Poisson r.v. with parameter $\lambda_1 + \lambda_2$. This property of reproducing itself by addition of independent variables – possessed also by the Gaussian distribution – is noteworthy and often useful in practice. Moreover, a result by Rajkov shows that the reverse is also true: if the sum of two independent r.v. has a Poisson distribution then each individual r.v. is Poisson distributed. This, we recall (remark in Example 3.3) is true also for Gaussian r.v.s.

Now, returning to our main discussion, we know from eq. (2.51) that the CF of the binomial r.v. X_n is given by $\varphi_n(u) = (1 - \lambda/n + \lambda e^{iu}/n)^n$. Passing to the limit as $n \to \infty$ we get

$$\lim_{n\to\infty} \varphi_n(u) = \lim_{n\to\infty}\left(1 + \frac{\lambda(e^{iu} - 1)}{n}\right)^n = \exp[\lambda(e^{iu} - 1)] \tag{4.4}$$

which proves the assertion of Proposition 4.1. On the practical side, this proposition is interpreted by saying that the Poisson distribution – besides being often applicable in its own right – can be used as a valid approximation of the binomial distribution when the probability of 'success' p is rather small and n is sufficiently large. In fact it should be noted that all the binomial r.v.s X_n have the same mean $E(X_n) = pn = (\lambda/n)n = \lambda$, thus implying that for large values of n the probability p must be small (incidentally, it is for this reason that the Poisson distribution is often called the distribution of rare events). In this light, as Example 4.1 will show, the parameter λ represents the average number of occurrences of the event under study per measurement unit (of time, length, area, etc., depending on the case). As a general rule of thumb one can use the Poisson distribution to approximate the binomial when either $n \geq 20$ and $p \leq 0.05$ or when $n \geq 100$ and $np \leq 10$; this makes calculations much easier because if we are interested in, say, the probability of 9 successes out of $n = 1000$ trials in a binomial process with $p = 0.006$ (so that $\lambda = np = 6$) it is certainly easier to calculate $(6^9 e^{-6})/9!$ rather than

$$\binom{1000}{9}(0.006)^9(1 - 0.006)^{1000-9}$$

(incidentally, the result of both expressions is 0.0688).

Example 4.1 Two typical cases of Poisson r.v. are as follows. Consider the number of car accidents per month at a given intersection where it is known that, on average, there are 1.7 accidents per month. In this case the month is our measurement unit and the Poisson law can be justified as follows. Divide a month in n intervals, each of which is so small that at most one accident can occur with a probability $p \neq 0$. Then, since it is reasonable to assume that the occurrence of accidents is independent from interval to interval, we are in essence observing a Bernoulli trial where the probability of 'success' p is relatively small if n is large. Also we know that $\lambda = np = 1.7$ and we can, for instance, obtain the probability of zero accidents in a month as $(1.7^0 \mathrm{e}^{-1.7})/0! = 0.183$.

The second example arises from a ballistic problem rather common during II World War. The probability of hitting an airplane in a vulnerable part when shooting with a rifle – that is, a 'success' – is very low, say, $p = 0.001$. However, if an entire military unit shoots, say, $n = 4000$ bullets, one can use the Poisson distribution to determine that the probability of at least two hits is (since $\lambda = np = 4$) $\sum_{x=2}^{4000} 4^x \mathrm{e}^{-4}/x! = 1-(4^0 \mathrm{e}^{-4}/0!)-(4^1 \mathrm{e}^{-4}/1!) = 0.908$ which is rather high and has been confirmed in practice.

Another important limit theorem – which involves D-convergence and points in the direction of the central limit theorem to be considered in a later section – was first partially obtained by deMoivre in the eighteenth century and then completed by Laplace some 60–70 years later. Once again, one considers a sequence of Bernoulli trials and defines the random variables $X_n (n = 1, 2, \ldots)$ which take on the value 0 in case of 'failure' or the value 1 in case of 'success' (recall that the probability of 'success' p does not change from trial to trial). In this light the r.v. $S_n = X_1 + X_2 + \cdots + X_n$ represents the number of successes in n trials and is binomially distributed with mean np and standard deviation $\sqrt{npq}$ (Example 2.8a). With these assumptions we have the deMoivre–Laplace theorem:

Proposition 4.2 *Let S_n be the number of successes in a sequence of Bernoulli trials, then*

$$\lim_{n\to\infty} P\left(a < \frac{S_n - np}{\sqrt{npq}} \le b\right) = \frac{1}{\sqrt{2\pi}} \int_a^b \exp(-z^2/2)\, \mathrm{d}z \tag{4.5}$$

uniformly for all a, b $(-\infty \le a < b \le \infty)$.

The proof is not given here because this proposition is just a particular case of the central limit theorem which will be proven in a later section (Proposition 4.22). Noting that the r.h.s. of eq. (4.5) is $P(a \le Z < b)$ where Z is a standard Gaussian r.v., we can state Proposition 4.2 in words by saying that the sequence of r.v.s $Y_n = (S_n - np)/\sqrt{npq}$ – which, in turn, is obtained by 'standardizing' the sequence of binomial r.v.s S_n – converges in distribution to a standard Gaussian r.v. This result is also frequently expressed by saying that the r.v. Y_n is 'asymptotically standard normal' and sometimes written $Y_n \approx As-N(0,1)$ where $N(0,1)$ denotes the normal probability distribution with zero mean and unit variance (i.e. the standard Gaussian distribution).

In the light of the considerations above, it turns out that – in the limit of large n – the binomial distribution can be approximated either by a Poisson distribution or by a standardized Gaussian. Which one of the two approximations to use depends on the problem at hand; broadly speaking, the Gaussian approximation works well even for moderately large values of n (say $n \geq 20-25$) as long as p is not too close to 0 or 1. If, on the other hand, p is close to 0 or 1, n must be rather large in order to obtain reasonably good results and in these cases the Poisson approximation is preferred. General rules of thumb are often given in textbooks and one finds, for example, that the Gaussian approximation is appropriate whenever (i) $p \pm 2\sqrt{pq/n}$ lies in the interval $(0,1)$ or (ii) $np \geq 5$ if $p \leq 0.5$ or $nq \geq 5$ if $p > 0.5$.

A third important and useful result considers the asymptotic behaviour of Poisson r.v.s. The CF of a Poisson r.v. X is given by eq. (4.2b); as a consequence the CF of the standardized Poisson r.v. $Y = (X - \lambda)/\sqrt{\lambda}$ is given by

$$\varphi_Y(u) = \exp[-\mathrm{i}u\sqrt{\lambda} + \lambda(\mathrm{e}^{\mathrm{i}u/\sqrt{\lambda}} - 1)] \tag{4.6}$$

where eq. (4.6) – since Y and X are linearly related – is obtained by using eq. (3.39b). As $\lambda \to \infty$ we can expand the exponential in parenthesis as $\exp(\mathrm{i}u/\sqrt{\lambda}) = 1 + \mathrm{i}u/\sqrt{\lambda} - u^2/2\lambda + \cdots$ and obtain

$$\lim_{\lambda\to\infty} \varphi_Y(u) = \exp(-u^2/2) \tag{4.7}$$

which, in other words, means that $Y \approx As - N(0,1)$. In the light of Propositions 4.1 and 4.2, this last result is hardly unexpected.

4.2.1 *A few further remarks on weak convergence*

It has been pointed out in the preceding section that weak convergence (or convergence in distribution or in law) concerns the convergence of PDFs and, in general, does not imply the convergence of pmfs or pdfs (when they exist). However, in some cases there is the possibility of establishing 'local' limit theorems for these functions. An example is given by the 'local' version of the DeMoivre–Laplace theorem (see e.g. [9] or [13]) stating that

$$\lim_{n\to\infty} \frac{\sqrt{npq}B_n(m)}{(\sqrt{2\pi})^{-1}\exp(-x^2/2)} = 1 \tag{4.8a}$$

where

$$B_n(m) = \binom{n}{m} p^m (1-p)^{n-m} = \binom{n}{m} p^m q^{n-m}$$
$$x = (m - np)/\sqrt{npq} \tag{4.8b}$$

In words, the result of eq. (4.8a) is expressed by saying that, for any given m, the binomial pmf (multiplied by its standard deviation $\sqrt{npq}$) tends to a standardized Gaussian pdf as n gets larger and larger. As a matter of fact, the approximation is rather good even for relatively small values of n. So, for example, if $n = 25$, $p = 0.2$ and we are interested in $m = 3$, then $\sqrt{npq}B_{25}(3) = 0.2715$ and since $x = -1$ in this case, we get $(\sqrt{2\pi})^{-1}\exp(-x^2/2) = 0.2420$. A graphical representation of this local theorem is given in Figures 4.1 ($n = 25$, $p = 0.2$) and 4.2 ($n = 100$, $p = 0.2$) where one can immediately notice the quality of the approximation: good in the first case and excellent in the second case. The reader should check, however, that larger and larger values of n are needed for a good approximation as p gets close to either 0 or 1.

As stated in the preceding section, when p is close to either 0 or 1 (say $p < 0.1$ or $p > 0.9$) the binomial pdf can be better approximated by a Poisson density. In fact, if we let $n \to \infty$ and $p \to 0$ so that $\lambda = pn$ is finite,

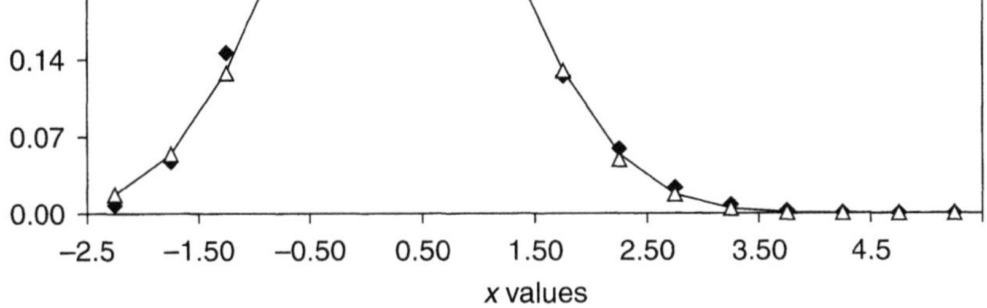

Figure 4.1 Gaussian approx. to binomial ($n = 25, p = 0.2$).

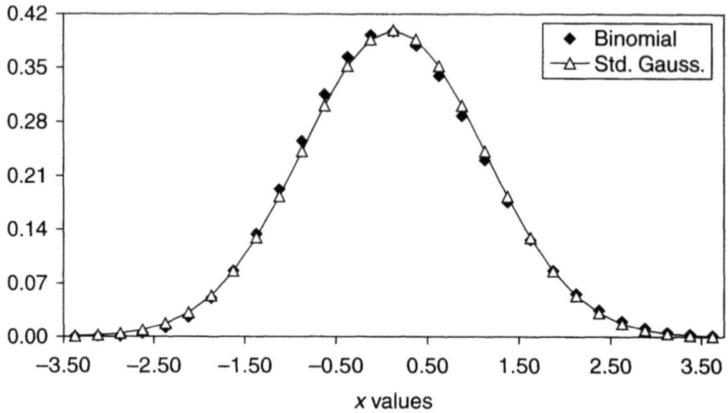

Figure 4.2 Gaussian approx. to binomial ($n = 100, \ p = 0.2$).

then for any fixed value of m

$$\begin{aligned} B_n(m) &= \frac{n!}{m!(n-m)!} p^m (1-p)^{n-m} \\ &= n(n-1)\cdots(n-m+1)\frac{\lambda^m}{m!n^m}\left(1-\frac{\lambda}{n}\right)^{n-m} \\ &= n^m\left(1-\frac{1}{n}\right)\cdots\left(1-\frac{m-1}{n}\right)\frac{\lambda^m}{m!n^m}\frac{(1-\lambda/n)^n}{(1-\lambda/n)^m} \end{aligned}$$

and therefore, since $(1-\lambda/n)^n \to \exp(-\lambda)$ as $n \to \infty$,

$$\lim_{n\to\infty} B_n(m) = \frac{\lambda^m}{m!}\mathrm{e}^{-\lambda} \tag{4.9}$$

The approximations considered here, clearly, are not the only ones. So, for example, it may be reasonable to expect that a distribution arising from an experiment of sampling without replacement can be approximated by a distribution of a similar experiment with replacement if the total number of objects N from which the sample is taken is very large. In fact, as $N \to \infty$ and we extract a finite sample, it no longer matters whether the extraction is done with or without replacement because the probability of 'success' is unaffected by the fact that we replace – or do not replace – the extracted item. In mathematical terms these considerations can be expressed by saying that, under certain circumstances, the so-called hypergeometric distribution – which is relative to sampling without replacement – can be approximated by a binomial distribution (see e.g. [18, Section 3.1.3] or [7, Appendix 1,

Section 17]). However, we do not consider other cases here and, if needed, postpone any further consideration.

4.3 Other types of convergence

Consider a sequence of r.v.s X_n defined on the same probability space (W, S, P). We say that X_n converges in probability to the r.v. X – also defined on (W, S, P) – if for every $\varepsilon > 0$ we have

$$\lim_{n\to\infty} P\{w \in W : |X_n(w) - X(w)| \geq \varepsilon\} = 0 \tag{4.10a}$$

and in this case one often writes $X_n \to X[P]$ and speaks of P-convergence. It is worth noting that convergence in probability is called 'convergence in measure' in mathematical analysis.

In words, eq. (4.10a) states that the probability measure of the set where X_n differs from X by more than any prescribed positive number tends to zero as $n \to \infty$. This, we point out, does not assure that all the values $|X_n(w) - X(w)|$ will be smaller than ε for n larger than a certain N, but only that the probability measure of the event (i.e. set) for which $|X_n(w) - X(w)| \geq \varepsilon$ is very small (zero in the limit). Also, it may be noted that eq. (4.10) can be expressed equivalently by writing

$$\lim_{n\to\infty} P\{w \in W : |X_n(w) - X(w)| \leq \varepsilon\} = 1 \tag{4.10b}$$

and it is immediate to see that $X_n \to X[P]$ if and only if $X_n - X \to 0[P]$ (remember that this is not true in general for convergence in distribution).

In the case of random vectors the condition (4.10a) – or (4.10b) – must hold for all their components and it is understood that the sequence of vectors $\mathbf{X}_n$ and the limit $\mathbf{X}$ must have the same dimension. More specifically, it can be shown that $\mathbf{X}_n \to \mathbf{X}[P]$ if and only if $X_n^{(k)} \to X^{(k)}[P]$ for all k (where k is here the index of component; so, for a m-dimensional vector $k = 1, 2, \ldots, m$).

We turn our attention now on some important results on convergence in probability starting with the following two propositions:

Proposition 4.3 *If $X_n \to X[P]$ and $g : \mathbb{R} \to \mathbb{R}$ is a continuous function, then $g(X_n) \to g(X)[P]$.*

Proposition 4.4(a) *Convergence in probability implies convergence in distribution.*

In fact, we have

$$\begin{aligned} F_n(x) = P(X_n \leq x) &= P(X_n \leq x \cap X > x + \varepsilon) + P(X_n \leq x \cap X \leq x + \varepsilon) \\ &\leq P(|X - X_n| \geq \varepsilon) + P(X \leq x + \varepsilon) = P(|X - X_n| \geq \varepsilon) + F(x + \varepsilon) \end{aligned}$$

where the inequality comes from two facts:

(a) $P(X_n \le x \cap X \le x + \varepsilon) \le P(X \le x + \varepsilon)$ because of the straightforward inclusion $(X_n \le x \cap X \le x + \varepsilon) \subseteq (X \le x + \varepsilon)$, and
(b) $P(X_n \le x \cap X > x + \varepsilon) \le P(|X - X_n| \ge \varepsilon)$ because $(X_n \le x \cap X > x + \varepsilon) \subseteq (|X - X_n| \ge \varepsilon)$. This inclusion is less immediate but the l.h.s. event implies $x < X - \varepsilon$ and, clearly, $X_n \le x$; consequently $X_n < X - \varepsilon$, which, in turn, is included in the event $|X - X_n| \ge \varepsilon$. By a similar line of reasoning we get

$$\begin{aligned} F(x-\varepsilon) = P(X \le x - \varepsilon) &= P(X \le x - \varepsilon \cap X_n > x) \\ &\quad + P(X \le x - \varepsilon \cap X_n \le x) \\ &\le P(|X - X_n| \ge \varepsilon) + P(X_n \le x) = P(|X - X_n| \ge \varepsilon) + F_n(x) \end{aligned}$$

Putting the two pieces together leads to

$$F(x-\varepsilon) - P(|X - X_n| \ge \varepsilon) \le F_n(x) \le P(|X - X_n| \ge \varepsilon) + F(x+\varepsilon)$$

and since $X_n \to X[P]$ then $F_n(x)$ is bracketed between two quantities that – as $\varepsilon \to 0$ – tend to $F(x)$ whenever F is continuous at x. This, in turn, means that $X_n \to X[D]$ and the theorem is proven.

The reverse statement of Proposition 4.4a is not true in general because – we recall – convergence in distribution can occur for r.v.s defined on different probability spaces, a case in which P-convergence is not even defined. However, when the X_n are defined on the same probability space, a partial converse exists:

Proposition 4.4(b) *If X_n converges in distribution to a constant c then X_n converges in probability to c.*

We do not prove the proposition but only point out that:

(i) a r.v. which takes on a constant value c with probability one – that is, such that $P_X(c) = 1$ – is not truly random. Its PDF is $F(x) = 0$ for $x < c$ $F(x) = 1$ for $x \ge c$ and often one speaks of 'degenerate' or 'pseudo' random variable in this case;
(ii) when all the X_n and X are defined on the same probability space and X is not a constant, there are special cases in which the converse of Proposition 4.4 may hold (see [11, Chapter 4]).

The last result on P-convergence we give here is called Slutsky's theorem and its proof can be found, for example, in Ref. [1]

Proposition 4.5 *If $X_n \to X$ [D] and $Y_n \to c$ [D] (and therefore $Y_n \to c$ [P]), then*

(a) $X_n + Y_n \to X + c$ [D]
(b) $X_n Y_n \to cX$ [D]
(c) $X_n / Y_n \to X/c$ [D] *if* $c \neq 0$.

Turning now to another important notion of convergence, we say that the sequence of r.v.s X_n converges almost-surely (some authors say 'with probability 1') to X if

$$P\left\{w \in W : \lim_{n\to\infty} X_n(w) = X(w)\right\} = 1 \tag{4.11}$$

and we will write $X_n \to X$ [a.s.] or $X_n \to X[P - \text{a.s.}]$ if the measure needs to be specified. Clearly, $X_n \to X$ [a.s.] if and only if $X_n - X \to 0$ [a.s.]. Definition 4.11 implies that the set N of all w where $X_n(w)$ fails to converge to $X(w)$ is such that $P(N) = 0$ and that, on the other hand, $X_n(w) \to X(w)$ for all $w \in N^c$ where, clearly, $P(N^c) = 1$. Given a measure P – and a probability is a finite, non-negative measure – in mathematical analysis one speaks of 'convergence almost-everywhere' (a.e.) when condition (4.11) holds; therefore a.s.-convergence is just the probabilistic name given to the notion of a.e.-convergence of advanced calculus. In general, there is no relation between a.e.-convergence and convergence in measure (eq. (4.10)); however, the fact that P is a finite measure has an important consequence for our purposes:

Proposition 4.6 *Almost-sure convergence implies convergence in probability (and therefore, by Proposition 4.4, convergence in distribution).*

This result is a consequence of the following criterion for a.s.-convergence: the sequence X_n converges almost surely to X if and only if for every $\varepsilon > 0$

$$\lim_{n\to\infty} P\left[\bigcup_{k=n}^{\infty}\{|X_k - X| \geq \varepsilon\}\right] = 0 \tag{4.12a}$$

or, equivalently,

$$\lim_{n\to\infty} P\left[\bigcap_{k=n}^{\infty}\{|X_k - X| < \varepsilon\}\right] = 1 \tag{4.12b}$$

In fact, if (4.12a) holds then eq. (4.10) follows by virtue of the fact that the probability of a union of events is certainly not less than the probability of each one of the individual events in the union. The proof of the criterion is more involved and is not given here; the interested reader may refer, for example, to [16] or [17]. Regarding the converse of Proposition 4.6 – which is not, in general, true – a remark is worthy of notice: it can be shown that if $X_n \to X[P]$ then there exists a subsequence X_{n_k} of X_n such that $X_{n_k} \to X$ [a.s.] as $k \to \infty$.

Proposition 4.7 *If* $X_n \to X$ [a.s.] *and* g *is a continuous function, then* $g(X_n) \to g(X)$ [a.s.].

In fact, for every fixed w such that $X_n(w) \to X(w)$ then $Y_n(w) \equiv g(X_n(w)) \to g(X(w)) \equiv Y(w)$ because of the continuity of g. Therefore $\{w: X_n(w) \to X(w)\} \subseteq \{w: Y_n(w) \to Y(w)\}$ so that $P\{w: Y_n(w) \to Y(w)\} \geq P\{w: X_n(w) \to X(w)\}$ and the theorem follows.

The last comment we make here on a.s.-convergence regards random vectors. As for P-convergence, a sequence of m-dimensional random vectors $\mathbf{X}_n = (X_n^{(1)}, \ldots, X_n^{(m)})$ converges a.s. to the m-dimensional vector $\mathbf{X} = (X^{(1)}, \ldots, X^{(m)})$ if and only if $X_n^{(k)} \to X^{(k)}$ [a.s.] for all $k = 1, 2, \ldots, m$.

Before turning to the collection of results known as 'law of large numbers', we close this section by introducing another type of convergence. A sequence of r.v.s X_n is said to converge to X 'in the kth mean' ($k = 1, 2, \ldots$) if

$$\lim_{n\to\infty} E(|X_n - X|^k) = \lim_{n\to\infty} \int_W |X_n - X|^k \, dP = 0 \tag{4.13}$$

and we will write $X_n \to X$ $[M_k]$. In the above definition it is assumed that all the X_n and X are such that $E(X_n^k) < \infty$ and $E(X^k) < \infty$ because these conditions imply the existence of the expectation in eq. (4.13). In fact, from the inequality $|X_n - X|^k \leq 2^k(|X_n|^k + |X|^k)$ we can pass to expectations to get $E(|X_n - X|^k) \leq 2^k E(|X_n|^k) + 2^k E(|X|^k)$ so that the l.h.s. is finite whenever the r.h.s. is. Also, it is easy to see that $X_n \to X$ $[M_k]$ if and only if $X_n - X \to 0$ $[M_k]$.

The most important special cases of (4.13) in applications are $k = 1$ – the so-called 'convergence in the mean' – and $k = 2$, called 'convergence in the quadratic mean'. This latter type plays a role in probability when only 'second-order data' are available, that is, when the only information is given by the means $m_n = E(X_n)$ and covariances $K_{ij} (i, j = 1, \ldots, n)$ and one cannot determine whether the sequence converges in any one of the modes considered before. However, the following result holds:

Proposition 4.8 *If* $X_n \to X[M_k]$ *– with* k *being any one integer – then* $X_n \to X$ $[P]$ *and therefore (Proposition 4.4)* $X_n \to X$ $[D]$.

In fact, consider Chebyshev's inequality (eq. (2.36a)) applied to the r.v. $X_n - X$; for every $\varepsilon > 0$ we have $P(|X_n - X| \geq \varepsilon) \leq E(|X_n - X|^k)/\varepsilon^k$ and therefore the l.h.s. tends to zero whenever the r.h.s. does. So, in particular, if a sequence converges in the mean or in the quadratic mean then convergence in probability and convergence in distribution follow. Furthermore, by virtue of Proposition 2.12, it is immediate to show that convergence in the quadratic mean implies convergence in the mean or, more generally:

Proposition 4.9 *Convergence in the* kth *mean implies convergence in the* jth *mean for all integers* $j \leq k$.

4.3.1 *Additional notes on convergences*

In the preceding section we have determined the following relations:

(a) a.s.-convergence is stronger than P-convergence which, in turn, is stronger than D-convergence unless the limit is a constant random variable.
(b) M_k-convergence (for any one integer k) implies P-convergence and therefore D-convergence.

At this point one may ask, for instance, about the relation between M_k and a.s.-convergence. The answer is that, in general, without additional assumptions, there are no relations other than the ones given above. An example is given by the celebrated Lebesgue dominated convergence theorem which, for our purposes, can be stated as follows

Proposition 4.10 *Let* $X_n \to X$ [a.s.] *or* $X_n \to X$ $[P]$ *and let* Y *be a r.v. such that* $E(Y) < \infty$ *(i.e. with finite mean) and* $|X_n(w)| \leq Y(w)$ *for each* n *and for almost all* $w \in W$. *Then* $X_n \to X$ $[M_1]$. *(see Ref. [8] or [15]).*

Note that the expression $|X_n(w)| \leq Y(w)$ for almost all $w \in W$ brings into play the measure P and means that the set N where the inequality does not hold is such that $P(N) = 0$ (again, this is the 'almost everywhere' notion of mathematical analysis).

Another important result establishes a relation between D- and *a.s.*-convergence. This is due to Skorohod and, broadly speaking, states that convergence in distribution can be turned into almost sure convergence by appropriately changing probability space.

Proposition 4.11 (Skorohod's theorem) *Let* X_n *and* X *be r.v.s defined on a probability space* (W, S, P) *and such that* $X_n \to X$ $[D]$. *Then, it is possible to construct a probability space* $(\widehat{W}, \widehat{S}, \widehat{P})$ *and random variables* $\widehat{X}_n$ *and* $\widehat{X}$ *such that* $\widehat{P}(\widehat{X} \leq x) = P(X \leq x)$, $\widehat{P}(\widehat{X}_n \leq x) = P(X_n \leq x)$ *for* $n = 1, 2, \ldots$ *(i.e.* $\hat{F}(x) = F(x)$ *and* $\hat{F}_n(x) = F_n(x)$ *for all* n*) and* $\widehat{X}_n \to \widehat{X}$ $[\widehat{P} - \text{a.s.}]$.

We do not prove the theorem here but it is worth noting that, in essence, Proposition 4.11 is due to the fact that any PDF $F : \mathbb{R} \to [0, 1]$ can be 'inverted' to obtain a r.v. defined on the interval $U = [0, 1]$ whose PDF is F. In this light, it turns out that $(\widehat{W}, \widehat{S}, \widehat{P}) = (U, \mathbb{B}(U), \mu)$ – where μ is the Lebesgue measure. For more details the interested reader can refer, for example, to [1, 2] or [19].

A third remark of interest is that P-, a.s.- and M_k-convergence can all be established by the well-known Cauchy criterion of mathematical analysis. So, for example, if a sequence X_n satisfies the Cauchy criterion in probability, that is,

$$\lim_{m,n\to\infty} P(|X_m - X_n| \geq \varepsilon) = 0 \tag{4.14}$$

(which can also be written $X_m - X_n \to 0$ $[P]$ as $m,n \to \infty$), then there exists a r.v. X such that $X_n \to X$ $[P]$. The fact that $X_n \to X$ $[P]$ implies eq. (4.14) is clear; therefore it can be said that the Cauchy criterion (4.14) is a necessary and sufficient condition for the sequence X_n to converge (in probability) to a r.v. X defined on the same probability space. Similarly, it can be shown that $|X_m - X_n| \to 0$ [a.s.] implies that there exists X such that $X_n \to X$ [a.s.]; consequently, by the same reasoning as above $X_n \to X$ [a.s.] if and only if $|X_m - X_n| \to 0$ [a.s.]. By the same token, $X_n \to X$ $[M_k]$ if and only if the Cauchy criterion $E(|X_m - X_n|^k) \to 0(m,n \to \infty)$ in the kth mean holds. In mathematical terminology, these results can be expressed by saying that the 'space' of random variables defined on a probability space (W, S, P) is complete with respect to P, a.s. and M_k convergence. Moreover, if we consider as equal any two r.v.s which are almost everywhere equal (with respect to the measure P) the spaces of r.v.s. with finite kth order moment ($k = 1, 2, \ldots$) are the so-called L^k spaces of functional analysis. It is well known, in fact, that defining the norm $\|X\|_k = \{E(|X|^k)\}^{1/k}$ these are Banach spaces (i.e. complete normed spaces) and, in particular, the space L^2 is a Hilbert space. Although it is beyond our scopes, this aspect of probability theory has far-reaching consequences in the light of the fact that the study of Banach and Hilbert spaces is a vast and rich field of mathematical analysis in its own right.

4.4 The weak law of large numbers (WLLN)

Broadly speaking, the so-called 'law of large numbers' (LLN) deals with the asymptotic behaviour of the arithmetic mean of a sequence of random variables. Since the term 'asymptotic behaviour' implies some kind of limit and therefore a notion of convergence, it is customary to distinguish between the 'weak' law of large numbers (WLLN) and 'strong' law of large numbers (SLLN), where in the former case the convergence is in the probability sense while in the latter almost sure convergence is involved. Clearly, the attributes of 'weak' and 'strong' are due to the fact that a.s.-convergence is stronger than P-convergence and therefore the SLLN implies the WLLN.

In order to cast these ideas in mathematical form, let us consider the WLLN first and start with a general result which is a consequence of Chebychev's inequality. For $n = 1, 2, \ldots$ consider a sequence $\{Y_n\}$ of r.v.s with finite means $E(Y_n)$ and standard deviations $\sigma_n = \sqrt{\mathrm{Var}(Y_n)}$. Then our first statement is:

Proposition 4.12 *If the numerical sequence of standard deviations is such that $\sigma_n \to 0$ as $n \to \infty$, then for every $\varepsilon > 0$*

$$\lim_{n\to\infty} P(|Y_n - E(Y_n)| \geq \varepsilon) = 0 \tag{4.15}$$

By setting $b = \varepsilon$, the proof follows immediately from the first of eq. (2.36b). Now, given a sequence of r.v.s X_k defined on a probability space (W, S, P) we can define, for every $n = 1, 2, \ldots$, the new r.v.

$$S_n = X_1 + X_2 + \ldots + X_n \tag{4.16}$$

with mean $E(S_n)$ and variance $\mathrm{Var}(S_n)$. (Note that $E(S_n) = \sum_{k=1}^{n} E(X_k)$ while, in the general case, eq. (2.35b) gives $\mathrm{Var}(S_n)$ in terms of the variances and covariances of the original variables X_k. If these variables are independent or uncorrelated then $\mathrm{Var}(S_n) = \sum_{k=1}^{n} \mathrm{Var}(X_k)$.) With these definitions in mind, the following propositions hold:

Proposition 4.13 (Markov's WLLN) *If $\mathrm{Var}(S_n)/n^2 \to 0$ as $n \to \infty$ then*

$$\lim_{n\to\infty} P\left(\left|\frac{S_n - E(S_n)}{n}\right| \geq \varepsilon\right) = 0 \tag{4.17}$$

Proposition 4.14(a) (Chebychev's WLLN) *If the variables X_k are independent or uncorrelated and there exists a finite, positive constant C such that $\mathrm{Var}(X_k) < C$ for all k (in other words, this latter condition can be expressed by saying that the variances $\mathrm{Var}(X_k)$ are 'uniformly bounded'), then eq. (4.17) holds.*

The proof of Proposition 4.13 is almost immediate. If we set $Y_n = S_n/n$ then, by hypothesis, $\mathrm{Var}(Y_n) = \mathrm{Var}(S_n)/n^2 \to 0$ as $n \to \infty$ and $E(Y_n) = E(S_n)/n$. In this light, Proposition 4.13 is a consequence of Proposition 4.12. For Proposition 4.14 we note first that $\mathrm{Var}(S_n) = \sum_{k=1}^{n} \mathrm{Var}(X_k) < nC$, where the equality holds because of independence (or uncorrelation). Consequently, $\mathrm{Var}(S_n)/n^2 < C/n$, and since $C/n \to 0$ as $n \to \infty$ the result follows by virtue of Proposition 4.13.

At this point, some remarks are in order. First of all, we note that eq. (4.17) can be rewritten equivalently as $(S_n - E(S_n))/n \to 0[P]$ or $S_n/n \to E(S_n)/n[P]$, where S_n/n is the arithmetic mean of the r.v.s $X_1, X_2, \ldots, X_n$. So, if the X_k are such that $E(X_k) = \mu$ for all k, then $E(S_n) = n\mu$ and

$$S_n/n \to \mu\ [P] \tag{4.18}$$

meaning that for large n the arithmetic mean of n independent r.v.s (each with finite expectation μ and with uniformly bounded variances) is very likely to be close to μ. This is what happens, for instance, when we repeat a given experiment a large number of times. In this case we 'sample' n times a given r.v. X – which is assumed to have finite mean $E(X) = \mu$ and variance $Var(X) = \sigma^2$ – so that $X_1, X_2, \ldots, X_n$ are independent r.v.s distributed as X. Then, by calculating the arithmetic mean $(X_1 + X_2 + \cdots + X_n)/n$ of our n observations we expect that

$$\frac{S_n}{n} = \frac{X_1 + X_2 + \cdots + X_n}{n} \cong \mu \tag{4.19}$$

We will have more to say about this in future chapters but, for the moment, we note that a typical example in this regard is the measuring process of an unknown physical quantity Q: we make n independent measurements of the quantity, calculate the mean of these observed values and take the result as a good (if n is sufficiently large) estimate of the 'true value' Q. Note that the assumptions of Proposition 4.14 are satisfied because all the X_k have the same distribution as X so that, in particular, $E(X_k) = \mu$ (if the measurements have no systematic error) and $\text{Var}(X_k) = \sigma^2$ (and since σ is a finite number, the variances are uniformly bounded).

The relative frequency interpretation of the probability of an event A (recall Section 1.3) is also dependent on the LLN. In fact, by performing n times an experiment in which A can occur, the relative frequency $f(A)$ of A is

$$f(A) = \frac{1}{n}\sum_{k=1}^{n} I_k \tag{4.20}$$

where I_k is the indicator function of event A in the kth repetition of the experiment. As n gets larger and larger, it is observed that $f(A)$ tends to stabilize in the vicinity of a value – for example, 0.50 in the tossing of a fair coin or, say, 0.03 for the fraction of defective items in the daily production of a given industrial process – which, in turn, is postulated to be the probability of A. In this light, it is clear that we cannot rigorously prove or disprove the existence, in the real world, of such a limiting value because an infinite number of trials is impossible. The best we can do is to build up confidence in our assumptions and check them against real observations; continued success tends to increase our confidence, thus leading us to believe in the adequacy of the postulate.

Returning to our main discussion we note that a special case of Proposition 4.14 is given by the celebrated Bernoulli theorem whose basic assumption is that we perform a sequence of Bernoulli trials and p is the probability of success in each trial. If $X_k = I_k$ – the indicator function of a success in the kth trial – the sum S_n is the total number of successes in

n trials and is binomially distributed with (Example 2.8) $E(S_n) = np$ and $\text{Var}(S_n) = np(1-p) = npq$. Then, Bernoulli's theorem asserts:

Proposition 4.14(b) (Bernoulli's WLLN) *With the above assumptions*

$$\lim_{n\to\infty} P\left(\left|\frac{S_n}{n} - p\right| < \varepsilon\right) = 1 \tag{4.21}$$

or, equivalently, $S_n/n \to p[P]$.

The proof follows from Markov's theorem (Proposition 4.13) once we note that $\text{Var}(S_n)/n^2 = pq/n \to 0$ as $n \to \infty$. Now, although the proof of the theorem may seem almost trivial, we must keep in mind that it was the first limit theorem to be proved (in the book Ars Conjectandi published in 1713), and therefore Bernoulli did not have the mathematical resources at our disposal. Moreover, since the theorem states that the average number of successes in a long sequence of trials is close to the probability of success on any given trial, its historical importance lies in the fact that this is the first step in the direction of removing the restriction of 'equally likely outcomes' – necessary in the 'classical' notion of probability – in defining the probability of an event. As a consequence, it provides mathematical support to the idea that probabilities can be determined as relative frequencies in a sufficiently long sequence of repeated trials.

The different forms of the WLLN given so far assume that all the variables X_i have finite variance. Khintchine's theorem shows that this is not necessary if the variables are independent and have the same distribution.

Proposition 4.15 (Khintchine's WLLN) *If the r.v.s X_k are independent and identically distributed (iid) with finite first moment $E(X_k) = \mu$ then $S_n/n \to \mu\ [P]$.*

In order to prove the theorem we can use characteristic functions to show that $S_n/n \to \mu[D]$. This, by virtue of Proposition 4.4(b), implies convergence in probability. Let $\varphi(u)$ be the common CF of the variables X_k, then we can write the MacLaurin expansion $\varphi(u) = \varphi(0) + iuE(X_k) + \cdots = 1 + iu\mu + \cdots$ (see Proposition 2.18(a) and the first of eq. (2.47b)) where the excluded terms tend to zero as $u \to 0$. Then, if we call $\psi(u) = E[\exp(iuS_n)]$ the CF of S_n we have

$$E[\exp(iuS_n/n)] = \psi(u/n) = \prod_{k=1}^{n} \varphi(u/n) = \{\varphi(u/n)\}^n = (1 + iu\mu/n + \cdots)^n$$

where we used independence in the second equality. Now, as $n \to \infty$, the last expression on the r.h.s. tends to $\exp(iu\mu)$ which, in turn, is the CF of a pseudo-r.v. μ. Consequently, $S_n/n \to \mu\ [D]$ and therefore

$S_n/n \to \mu[P]$. A different proof of this theorem is based on the so-called 'method of truncation' and can be found, for example, in [2] or [9].

At this point it could be asked if there is a necessary and sufficient condition for the WLLN to hold. In fact, all the results above provide sufficient conditions and examples can be given of sequences which obey the WLLN but do not verify the assumptions of any one of the theorems above. Such a condition exists and is given in the next theorem due to Kolmogorov.

Proposition 4.16 (Kolmogorov's WLLN) *A sequence X_k of r.v.s with finite expectations $E(X_k)$ satisfies eq. (4.17) – that is, the WLLN – if and only if*

$$\lim_{n\to\infty} E\left\{\frac{\Lambda_n^2}{1+\Lambda_n^2}\right\} = 0 \tag{4.22}$$

where $\Lambda = [S_n - E(S_n)]/n$.

We do not prove this proposition here and the interested reader may refer, for example, to [9]. However, it is worth noting that the theorem requires neither independence nor the existence of finite second-order moments. Also, since Kolmogorov's theorem expresses an 'if and only if' statement, it can be said that the various conditions of the propositions above are all sufficient conditions for (4.22) to hold, meaning that they imply (but are not implied by) eq. (4.22). In fact, for example, in case of finite variances we have

$$\frac{\Lambda_n^2}{1+\Lambda_n^2} \le \Lambda_n^2 = \frac{1}{n^2}(S_n - E(S_n))^2$$

so that taking expectations on both sides it follows that (4.22) holds whenever Markov's condition on variances (Proposition 4.13) holds.

4.5 The strong law of large numbers (SLLN)

As stated in the preceding section, the type of convergence involved in the different forms of the SLLN is a.s.-convergence, which, in turn, implies P- and D-convergence. Being a stronger statement than the WLLN, the mathematical proofs of the SLLN are generally longer and more intricate than in the weak case; for this reason we will mainly limit ourselves to the results. The reader interested in the proofs of the theorems can find them in the references at the end of the chapter.

Historically, the first statement of SLLN is due to Borel and is somehow a stronger version of Bernoulli's theorem (Proposition 4.15):

Proposition 4.17 (Borel's SLLN) *Let $X_k = I_k$ $(k = 1, 2, \ldots)$ be the indicator function of a success in the* k*th trial in a sequence of independent trials* *and let p be the probability of success in each trial. Then* $S_n/n \to p$ [a.s.], *where, as before,* $S_n = X_1 + X_2 + \cdots + X_n$.

Borel's theorem, in turn, is a special case of the more general result due to Kolmogorov:

Proposition 4.18 (Kolmogorov's SLLN) *Let* X_k *be a sequence of independent r.v.s with finite variances* $\mathrm{Var}(X_k)$ *such that*

$$\sum_{k=1}^{\infty} \frac{\mathrm{Var}(X_k)}{k^2} < \infty \tag{4.23}$$

(i.e. the series on the l.h.s. converges). Then, as $n \to \infty$, *the SLLN holds, that is,*

$$\frac{S_n}{n} \to \frac{E(S_n)}{n} \text{ [a.s.]} \tag{4.24a}$$

or, equivalently, $[S_n - E(S_n)]/n \to 0$ [a.s.].

Three corollaries to this proposition are:

(a) If the variables of the sequence are independent and have uniformly bounded variances – that is, $\mathrm{Var}(X_k) < C$ for all k – then eq. (4.24) holds.
In particular,
(b) If the variables of the sequence are independent and have the same mean μ and variance σ^2 then

$$S_n/n \to \mu \text{ [a.s.]} \tag{4.24b}$$

In Ref. [3] one can find a slightly different version of this result stating that if the variables are identically distributed with common finite mean μ and variance σ^2 and are uncorrelated – that is, $\mathrm{Cov}(X_i X_j) = 0$ for $i \neq j$ – then eq. (4.24b) holds. The usefulness of this theorem lies in the fact that non-correlation is generally easier to test than independence.
(c) If the X_k are independent and have uniformly bounded fourth central moments – that is, for all k we have $E[(X_k - \mu_k)^4] \leq C$ (where $\mu_k = E(X_k)$) for some positive constant C – then eq. (4.24) holds.

This last result is due to Cantelli and is a consequence of Liapunov's inequality (Proposition 2.12); in fact, setting $Z_k = X_k - \mu_k$ for simplicity of notation, we have $\mathrm{Var}(X_k) = E\left(Z_k^2\right)$, and the variances are uniformly bounded because

$$\{E(Z_k^2)\}^{1/2} \le \{E(Z_k^4)\}^{1/4} \le C^{1/4}$$

As in the case of the WLLN, the condition of finite variances can be relaxed if one considers a sequence of iid variables with finite mean μ. Moreover, it is worth noting that we have an 'if and only if' statement

Proposition 4.19 *Let X_k be a sequence of iid random variables. Then the existence of finite first-order moment μ is a necessary and sufficient condition for the SLLN (eq. (4.24b)) to hold.*

Although not given here, the proofs of some of the theorems above use two results which are worthy of mention in their own right because of their importance in many aspects of probability theory. These results are known as Kolmogorov's inequality and the Borel–Cantelli lemma.

Proposition 4.20 (Kolmogorov's inequality) *Let $X_i (i = 1, 2, \ldots, n)$ be a finite collection of independent (not necessarily identically distributed) random variables with finite variances and let $S_k = X_1 + X_2 + \cdots + X_k$ for $1 \le k \le n$. Then, for each $b > 0$*

$$P\left(\max_{1\le k\le n} |S_k - E(S_k)| \ge b\right) \le \frac{\mathrm{Var}(S_n)}{b^2} \tag{4.25}$$

Note that if $n = 1$ eq. (4.25) reduces to Chebishev's inequality (2.36b).

Proposition 4.21 *Borel–Cantelli lemma consists of two parts:*

(a) Let (W, S, P) be a probability space and $A_1, A_2, \ldots$ be a sequence of events (i.e. $A_n \in S$ for all $n = 1, 2, \ldots$). If $\sum_{n=1}^{\infty} P(A_n) < \infty$, then

$$P\left(\limsup_{n\to\infty} A_n\right) = 0 \tag{4.26a}$$

where, we recall from Appendix A, $\limsup_{n\to\infty} A_n = \cap_{n=1}^{\infty}\left(\cup_{k=n}^{\infty} A_k\right)$ is itself an event which, by definition, occurs if and only if infinitely many of the A_ns occur.

(b) If $A_1, A_2, \ldots$ are mutually independent and $\sum_{n=1}^{\infty} P(A_n) = \infty$, then

$$P\left(\limsup_{n\to\infty} A_n\right) = 1 \tag{4.26b}$$

To prove the first part of the lemma, set $E_n = \cup_{k=n}^{\infty} A_k$. Then $E_1 \supset E_2 \supset E_3 \supset \cdots$ is a decreasing sequence of events and the theorem follows from the chain of relations

$$P\left(\limsup_{n\to\infty} A_n\right) = P\left(\lim_{n\to\infty} E_n\right) = \lim_{n\to\infty} P(E_n) \le \lim_{n\to\infty} \sum_{k=n}^{\infty} P(A_k) = 0$$

where we used the definition of limit of a decreasing sequence of sets (see Appendix A, Section A1) first and then the continuity and subadditivity properties of probability. For the second part of the lemma we can write

$$\begin{aligned} P(E_n^C) &= P\left(\bigcap_{k=n}^{\infty} A_k^C\right) = \lim_{m\to\infty} P\left(\bigcap_{k=n}^{m} A_k^C\right) = \lim_{m\to\infty} \prod_{k=n}^{m} P(A_k^C) \\ &= \lim_{m\to\infty} \prod_{k=n}^{m} [1 - P(A_k)] \le \lim_{m\to\infty} \prod_{k=n}^{m} \exp[-P(A_k)] \\ &= \lim_{m\to\infty} \exp\left[-\sum_{k=n}^{m} P(A_k)\right] = 0 \end{aligned}$$

where we used, in the order, De Morgan's law (i.e. eq. A.6), the independence of the A_k and the inequality $1 + x \le e^x$ (which holds for any real number x). Moreover, the last equality holds because $\sum_{k=n}^{\infty} P(A_k) = \infty$. From the relations above, part (b) of the lemma follows from the fact that $P(E_n) = 1 - P(E_n^C) = 1$ for each n.

4.6 The central limit theorem

In problems where probabilistic concepts play a part it is often reasonable to assume that the unpredictability may be due to the overall effect of many random factors and that each one of them has only a small influence on the final result. Moreover, these factors – being ascribable to distinct and logically unrelated causes – can be frequently considered as mutually independent. Since our interest lies in the final result and not in the individual factors themselves – which, often, are difficult or even impossible to identify – it becomes important to study the existence of limiting probability distributions when an indefinitely large number of independent random effects combine to yield an observable outcome. Needless to say, the ubiquitous Gaussian distribution is one of these limits and the mathematical results formalizing this fact – that is, convergence to a Gaussian distribution – go under the general name of 'central limit theorem' (CLT). The various forms of the theorem differ in the assumptions made on the probabilistic nature of the causes affecting the final result.

In order to cast the above ideas in mathematical form, we consider a sequence $X_1, X_2, \ldots$ of independent random variables with finite means $E(X_1) = m_1, E(X_2) = m_2, \ldots$ and variances $\mathrm{Var}(X_1) = \sigma_1^2, \mathrm{Var}(X_2) = \sigma_2^2, \ldots$; also, as in the previous sections, we denote by S_n the sum $S_n = X_1 + X_2 + \cdots + X_n$ whose mean and variance are $E(S_n) = \sum_{k=1}^n m_k$ and $\mathrm{Var}(S_n) = \sum_{k=1}^n \sigma_k^2$, respectively. The simplest case is when the variables X_k are iid; then, by calling $m \equiv m_1 = m_2 = \cdots$ the common mean and $\sigma^2 \equiv \sigma_1^2 = \sigma_2^2 = \cdots$ the common variance, we have $E(S_n) = nm$ and $\mathrm{Var}(S_n) = n\sigma^2$. Since these quantities both diverge as $n \to \infty$, it cannot be expected that the sequence S_n can converge in distribution to a random variable with finite mean and variance (unless, of course, the case $m = 0$ and $\sigma = 0$). However, we can turn our attention to the 'standardized' sequence Y_n defined by

$$Y_n = \frac{S_n - E(S_n)}{\sqrt{\mathrm{Var}(S_n)}} = \frac{S_n - nm}{\sigma\sqrt{n}} \tag{4.27}$$

whose mean and variance are, respectively, 0 and 1. In this light, the first form of the CLT – also known as Lindeberg–Levy theorem – is as follows:

Proposition 4.22 (Lindeberg–Levy: CLT for iid variables) *Let $X_k (k = 1, 2, \ldots)$ be iid random variables with finite mean m and variance σ^2; then $Y_n \to Z[D]$ as $n \to \infty$, where the symbol Z denotes the standardized Gaussian r.v. whose PDF is*

$$F_Z(x) = \frac{1}{\sqrt{2\pi}} \int_{-\infty}^{x} \exp(-t^2/2)\, dt$$

The proposition can be proven by recalling Levy's theorem (Proposition 2.24) and using characteristic functions. In fact, by introducing the iid r.v.s (with zero mean and unit variance) $U_j = (X_j - m)/\sigma$ for $j = 1, 2, \ldots$, we have

$$Y_n = \frac{S_n - nm}{\sigma\sqrt{n}} = \frac{1}{\sqrt{n}} \sum_{j=1}^{n} U_j$$

Now, denoting by $\varphi(u)$ the common CF of the variables U_j, the existence of finite mean and variance allows one to write the MacLaurin expansion

$$\varphi(u) = \varphi(0) + u\varphi'(0) + \frac{u^2}{2}\varphi''(0) + \cdots = 1 - \frac{u^2}{2} + \cdots$$

where the dots indicate higher order terms that tend to zero more rapidly than u^2 as $u \to 0$. Then, since by virtue of independence the CF ψ_n of Y_n is given by $\psi_n(u) = \{\varphi(u/\sqrt{n})\}^n$, we have

$$\psi_n(u) = \{\varphi(u/\sqrt{n})\}^n = \left(1 - \frac{u^2}{2n} + \cdots\right)^n$$

so that letting $n \to \infty$ we get $\lim_{n\to\infty} \psi_n(u) = \exp(-u^2/2)$ (the technicality of justifying the fact that we neglect higher order terms can be tackled by passing to natural logarithms; for more details the reader can refer to [19, Chapter VI, Section 7]). This limiting function is precisely the CF of a standardized Gaussian r.v., therefore proving the assertion $Y_n \to Z \approx N(0,1)[D]$ which, more explicitly, can also be expressed as

$$\lim_{n\to\infty} P(a < Y_n \le b) = \frac{1}{\sqrt{2\pi}} \int_a^b \exp(-x^2/2)\,dx \tag{4.28a}$$

for all a, b such that $-\infty \le a < b \le \infty$. Equivalently, by taking $a = -1$ and $b = 1$ we can also write

$$\lim_{n\to\infty} P\left(\left|\frac{S_n}{n} - m\right| < \frac{\sigma}{\sqrt{n}}\right) = \sqrt{\frac{2}{\pi}} \int_0^1 \exp(-x^2/2)\,dx \tag{4.28b}$$

which, for large n, can also be interpreted as an estimate on the probability that the arithmetic mean S_n/n (see also the following remark (c)) takes values within an interval of length $2\sigma/\sqrt{n}$ centered about the mean m.

At this point, a few remarks are in order:

(a) The DeMoivre–Laplace theorem (Proposition 4.2) is a special case of CLT of Proposition 4.22. In DeMoivre–Laplace case, in fact, the variables X_j are all binomially distributed with mean p and variance $pq = p(1-q)$. Consequently, the mean and variance of S_n – the number of successes in n independent trials – are np and npq, respectively, so that eq. (4.28) reduces to eq. (4.5).

(b) If the X_j are (independent) Poisson r.v.s with parameter λ, then – by virtue of the 'self-reproducing property' of Poisson variables pointed out in Section 4.2 – the variable S_n is also Poisson distributed with parameter $\Lambda = n\lambda$ and the CF ψ_n of the variable Y_n is obtained by simply substituting Λ in place of λ in eq. (4.6). Then $\psi_n(u) \to \exp(-u^2/2)$ as $n \to \infty$, therefore leading to another important special case of Proposition 4.22.

(c) In different words, the statement of Proposition 4.22 can be expressed by saying that the variable S_n is asymptotically Gaussian with mean nm

and variance $n\sigma^2$. This, in turn, implies that the arithmetic mean

$$\overline{X}_n = \frac{1}{n}\sum_{k=1}^{n} X_k = \frac{S_n}{n} \tag{4.29}$$

is itself a r.v. which is asymptotically normal with mean $E(\overline{X}_n) = m$, variance $\mathrm{Var}(\overline{X}_n) = \sigma^2/n$ and standard deviation $\sqrt{\mathrm{Var}(\overline{X}_n)} = \sigma/\sqrt{n}$. This fact, we will see, often plays an important role in cases where a large number of elements is involved. In particular, it is at the basis of Gauss' theory of errors where the experimental value of the (unknown) quantity, say Q, is 'estimated' by calculating the arithmetic mean of many repeated measurements under the assumption that the errors are iid random variables with zero mean and finite variance σ^2. Then – besides relying on the SLLN stating that $\overline{X}_n \to Q$ [a.s.] – if n is sufficiently large we can also use the Gaussian distribution to make probability statements regarding the accuracy of our result. More about these and other statistical applications is delayed to later chapters.

(d) Berry–Esseen inequality: Since for large values of n the standardized Gaussian can be considered as an approximation of the PDF of the variable Y_n, the question may arise on how good is this estimate as a function of n. Now, besides the practical fact – also supported by the results of many computer simulations – that the approximation is generally rather good for $n \geq 10$, a more definite answer can be obtained if one has some additional information on the X variables. If, for example, it is known that these variables have a finite third-order absolute central moment – that is, $E(|X - m|^3) < \infty$ – a rather general result is given by Berry–Esseen inequality which states that for all x

$$|F_n(x) - F_Z(x)| \leq C\frac{E(|X - m|^3)}{\sigma^3\sqrt{n}} \tag{4.30}$$

where we called $F_n(x) = P(Y_n \leq x)$ the PDF of Y_n, $F_Z(x)$, as above, is the standardized Gaussian PDF and C is a constant whose current best estimate is $C = 0.798$ (see Refs [11–14]).

Although Lindeberg–Levy theorem is important and often useful, the requirement of iid random variables is too strict to justify all the cases in which the Gaussian approximation seems to apply. In fact, other forms of the CLT show that the assumption of identically distributed variables can be relaxed without precluding the convergence to the Gaussian distribution. Retaining the assumption of independence, a classical result in this direction is Lindeberg's theorem. We state it without proof and the interested reader can refer, for instance, to [1, 9] or [19].

Proposition 4.23 (Lindeberg's CLT) *Let $X_1, X_2, \ldots$ be a sequence of independent random variables with finite means $E(X_k) = m_k$ and variances $\mathrm{Var}(X_k) = \sigma_k^2$ $(k = 1, 2, \ldots)$ and let $F_k(x)$ be the PDF of X_k. If, for every $\varepsilon > 0$ (ε enters in the domain of integration, see eq. (4.31b)) the Lindeberg condition*

$$\lim_{n\to\infty} \frac{1}{\mathrm{Var}(S_n)} \sum_{k=1}^{n} \int_{C_k} (x - m_k)^2 \, dF_k(x) = 0 \tag{4.31a}$$

holds, then $Y_n \to Z \approx N(0,1)[D]$, that is, the variable $Y_n = [S_n - E(S_n)]/\sqrt{\mathrm{Var}(S_n)}$ converges in distribution to a standardized Gaussian r.v. Z.

Two remarks on notation:

(i) Clearly $E(S_n) = \sum_{k=1}^{n} m_k$, $\mathrm{Var}(S_n) = \sum_{k=1}^{n} \sigma_k^2$ and $\sqrt{\mathrm{Var}(S_n)}$ is the standard deviation of S_n. In the following, for brevity these last two parameters will often be denoted by V_n^2 and V_n, respectively.
(ii) The domain of integration C_k in condition (4.31) is the set defined by

$$C_k = \{x : |x - m_k| \geq \varepsilon V_n\} \tag{4.31b}$$

Basically, the Lindeberg condition is an elaborate – and perhaps rather intimidating-looking – way of requiring that the contribution of each individual X_k to the total be small (recall the discussion at the beginning of this section). In fact, since the variable Y_n is the sum of n ratios, that is,

$$Y_n = \frac{S_n - E(S_n)}{V_n} = \sum_{k=1}^{n} \frac{X_k - m_k}{V_n}$$

the condition expresses the fact that each individual summand must be uniformly small or, more precisely, that for every $\varepsilon > 0$

$$\lim_{n\to\infty} P\left(\frac{|X_k - m_k|}{V_n} \geq \varepsilon \right) = 0 \tag{4.32}$$

that is, $V_n^{-1}|X_k - m_k| \to 0[P]$, which holds whenever eq. (4.31a) holds since

$$\varepsilon^2 P(|X_k - m_k| \geq \varepsilon V_n) = \varepsilon^2 \int_{C_k} dF_k \leq \frac{1}{V_n^2} \int_{C_k} (x - m_k)^2 \, dF_k$$
$$\leq \frac{1}{V_n^2} \sum_{k=1}^{n} \int_{C_k} (x - m_k)^2 \, dF_k \to 0$$

where the first inequality is due to the fact that the domain of integration C_k includes only those x such that $|x - m_k| \geq \varepsilon V_n$, that is, $(x - m_k)^2 \geq \varepsilon^2 V_n^2$. Property (4.32) is sometimes called 'uniform asymptotic negligibility' (uan).

As it should be expected, the iid case of Lindeberg–Levy theorem is just a special case of Proposition 4.23. In fact, if the X_k are iid variables with finite means m and variances σ^2, then the sum in (4.31) is simply a sum of n identical terms resulting in

$$\frac{1}{\sigma^2} \int\limits_{\{|x-m| \geq \varepsilon\sigma\sqrt{n}\}} (x - m)^2 \, \mathrm{d}F$$

which, in turn, must converge to zero because $\{x : |x - m_k| \geq \varepsilon\sqrt{n}\} \to \emptyset$ as $n \to \infty$. A second special case of Proposition 4.23 occurs when the X_k are uniformly bounded – that is, $|X_k| \leq M$ for all k – and $V_n^2 \to \infty$ as $n \to \infty$. Then

$$\int\limits_{C_k} (x - m_k)^2 \, \mathrm{d}F_k = \int\limits_{\mathbb{R}} I_{C_k}(x - m_k)^2 \, \mathrm{d}F_k \leq (2M)^2 P\{|x - m_k| \geq \varepsilon V_n\}$$
$$\leq \frac{(2M)^2 \sigma_k^2}{\varepsilon^2 V_n^2}$$

where Chebyshev's inequality (eq. (2.36b)) has been taken into account in the second inequality. From the relations above the Lindeberg condition follows because

$$\frac{1}{V_n^2} \sum_{k=1}^{n} \int\limits_{C_k} (x - m_k)^2 \, \mathrm{d}F_k \leq \frac{(2M)^2}{\varepsilon^2 V_n^2} \to 0$$

as $n \to \infty$. A third special case of Lindeberg theorem goes under the name of Liapunov's theorem which can be stated as follows

Proposition 4.24(a) (Liapunov's CLT) *Let X_k be a sequence of independent r.v.s with finite means m_k and variances $\sigma_k^2 (k = 1, 2, \ldots)$. If, for some $\alpha > 0$,*

$$\frac{1}{V_n^{2+\alpha}} \sum_{k=1}^{n} \mathrm{E}(|X_k - m_k|^{2+\alpha}) \to 0 \tag{4.33a}$$

as $n \to \infty$, then $Y_n \to Z[D]$.

In fact, Lindeberg's condition follows owing to the relations

$$\frac{1}{V_n^2}\sum_k \int_{C_k} (x-m_k)^2\,\mathrm{d}F_k \le \frac{1}{\varepsilon^\alpha V_n^{2+\alpha}}\sum_k \int_{C_k} |x-m_k|^{2+\alpha}\,\mathrm{d}F_k$$

$$\le \frac{\sum_k E(|X_k-m_k|^{2+\alpha})}{\varepsilon^\alpha V_n^{2+\alpha}}$$

where the first inequality holds because $|x-m_k| \ge \varepsilon V_n$.

Liapunov's theorem is sometimes given in a slightly less general form by requiring that $\rho_k = E(|X_k - m_k|^3) < \infty$, that is, that all the X_k have finite third-order central absolute moment. Then $Y_n \to Z[D]$ if

$$\lim_{n\to\infty}\frac{1}{V_n}\left(\sum_{k=1}^{n}\rho_k\right)^{1/3} = 0 \tag{4.33b}$$

At this point it is worth noting that eq. (4.31) is a sufficient but not necessary condition for convergence in distribution to Z. This means that there exist sequences of independent r.v.s which converge (weakly) to Z without satisfying Lindeberg's condition. However, it turns out that for those sequences X_k (of independent r.v.s) such that

$$\lim_{n\to\infty}\max_{k\le n}\frac{\sigma_k^2}{V_n^2} = 0 \tag{4.34}$$

eq. (4.31) is a necessary and sufficient condition for weak convergence to Z. This is expressed in the following proposition

Proposition 4.24(b) (Lindeberg–Feller CLT) *Let $X_1, X_2, \ldots$ be as in Proposition 4.23. Then the Lindeberg condition (4.31) holds if and only if $Y_n \to Z[D]$ and eq. (4.34) holds.*

A slightly different version of this theorem replaces eq. (4.34) by the uan condition of eq. (4.32). The interested reader can find both the statement and the proof of this theorem in Ref. [1].

4.6.1 Final remarks

We close this chapter with a few complementary remarks which, although outside our scopes, can be useful to the reader interested in further analysis. The different forms of CLT given above consider D-convergence which, we recall, is a statement on PDFs and, in general, implies nothing on the convergence properties of pmfs or – when they exist – pdfs. However, in

Section 4.2.1 we mentioned the local deMoivre–Laplace theorem where a sequence of (discrete) Bernoulli pmfs converges to a standardized Gaussian pdf. This is a special case of 'lattice distributions' converging to the standardized Gaussian pdf. Without entering into details we only say here that a random variable is said to have a 'lattice distribution' if all its values can be expressed in the form $a + hk$ where a, h are two real numbers, $h > 0$ and $k = 0, \pm 1, \pm 2, \ldots$. Bernoulli's and Poisson's distributions are just two examples among others. Under the assumption of iid variables with finite means and variances, a further restriction on h (the requirement of being 'maximal') provides a necessary and sufficient condition for the validity of a 'local' version of the CLT. The definition of maximality for h, the theorem itself and its proof can be found in Chapter 8 of [9]. Also, in the same chapter, the following result for continuous variables is given:

Proposition 4.25 *Let $X_1, X_2, \ldots$ be iid variables with finite means m and variances σ^2. If, starting from a certain integer $n = n_0$ the variable $Y_n = (\sigma\sqrt{n})^{-1}[S_n - nm]$ has a density $f_n(x)$, then*

$$f_n(x) - \frac{1}{\sqrt{2\pi}} \exp(-x^2/2) \to 0$$

uniformly for $-\infty < x < \infty$ if and only if there exists n_1 such that $f_{n_1}(x)$ is bounded.

A second aspect to consider is whether the Gaussian is the only limiting distribution for sums of independent random variables. The answer to this question is no. In fact, a counterexample has been given in Proposition 4.1 stating that the Poisson distribution is a limiting distribution for binomial r.v.s. Moreover, even in the case of iid variables the requirement of finite means and variances may not be met. So, in the light of the fact that a so-called Cauchy r.v., whose pdf is

$$f(x) = \frac{1}{\pi(1 + x^2)} \tag{4.35a}$$

has not a finite variance, one might ask, for example, if (4.35) could be a limiting distribution or, conversely, what kind of distribution – if any – is the limit of a sequence of independent r.v.s X_k distributed according to (4.35a). Incidentally, we note that (i) the PDF and CF of a Cauchy r.v. are, respectively

$$\begin{aligned} F(x) &= \frac{1}{\pi} \arctan x + \frac{1}{2} \\ \phi(u) &= \exp(-|u|) \end{aligned} \tag{4.35b}$$

(ii) in (4.35a), failure to converge to the Gaussian distribution is due to the presence of long 'inverse-power-law' tails as $|x| \to \infty$. These broad tails, however, do not preclude the existence of a limiting distribution.

Limiting problems of the types just mentioned led to the identification of the classes of 'stable' (or Levy) distributions and of 'infinitely divisible' distributions, where the latter class is larger and includes the former. As it should be expected, the Gaussian, the Poisson and the Cauchy distributions are infinitely divisible (the Gaussian and Cauchy distributions, moreover, belong to the class of stable distributions). For the interested reader, more on this topic can be found, for example, in [1, 9, 10, 21].

The third and last remark is on the multi-dimensional CLT for iid random vectors which, in essence, is a straightforward extension of the one-dimensional case. In fact, just as the sum of a large number of iid variables is approximately Gaussian under rather wide conditions, similarly the sum of a large number of iid vectors is approximately Gaussian (with the appropriate dimension). In more mathematical terms, we have the following proposition:

Proposition 4.26 *Let* $\mathbf{X}_1 = (X_1^{(1)}, \ldots, X_1^{(k)}), \mathbf{X}_2 = (X_2^{(1)}, \ldots, X_2^{(k)}), \ldots$ *be k-dimensional iid random vectors with finite mean* $\mathbf{m}$ *and covariance matrix* $\mathbf{K}$. *Denoting by* $\mathbf{S}_n$ *the vector sum*

$$\mathbf{S}_n = \sum_{j=1}^{n} \mathbf{X}_j = \left(\sum_{j=1}^{n} X_j^{(1)}, \sum_{j=1}^{n} X_j^{(2)}, \ldots, \sum_{j=1}^{n} X_j^{(k)} \right) \tag{4.36}$$

then the sequence $(\mathbf{S}_n - n\mathbf{m})/\sqrt{n}$ *converges weakly to* $\mathbf{Z}$, *where* $\mathbf{Z}$ *is a k-dimensional Gaussian vector with mean* $\mathbf{0}$ *and covariance matrix* $\mathbf{K}$.

4.7 Summary and comments

In experiments involving elements of randomness, long-term regularities tend to become clearer and clearer as the number of trials increases and one of the great achievements of probability theory consists in having established the general conditions under which these regularities occur.

On mathematical grounds, a tendency towards something implies some kind of limit, although – as is the case in probability – this is not necessarily the familiar limit of elementary calculus. In this light, Sections 4.2 and 4.3 define a number of different types of convergence by also giving their main individual properties and, when they exist, their mutual relations. Both sections have a subsection – 4.2.1 and 4.3.1, respectively – where additional remarks are made and further details are considered.

In essence, the main types of convergences used in probability theory are: convergence in distribution (or weak convergence), convergence in probability, almost-sure convergence and convergence in the kth median

($k = 1, 2, \ldots$). Respectively, they are denoted in this text by the symbols D, P, a.s. and M_k convergence and the main mutual relations are as follows:

$$\text{(i) } M_k \Rightarrow M_j (j \leq k), \quad \text{(ii) } M_1 \Rightarrow P \Rightarrow D, \quad \text{(iii) a.s.} \Rightarrow P \Rightarrow D.$$

The relation between M_k- and a.s.-convergence is considered in Proposition 4.10 and partial converses of the implications above, when they exist, are also given.

With these notions of convergence at our disposal, we then investigate the results classified under the name LLN, which concern the asymptotic behaviour of the arithmetic mean of a sequence of random variables. In this regard, it is customary to distinguish between weak (WLLN) and strong (SLLN) law of large numbers depending on the type of convergence involved in the mathematical formulation – that is, P- or a.s.–convergence, respectively. The names 'weak' and 'strong' follow from implication (iii) above and the two laws are considered in Sections 4.4 (WLLN) and 4.5 (SLLN).

So, in Section 4.4 we find, for instance, Markov's, Chebychev's, Khintchine's and Kolmogorov's WLLN, the various form differing on the conditions satisfied by the sequence of r.v.s involved (e.g. independence, independence and equal probability distributions, finite variances, etc.). Also, it is shown that Bernoulli's WLLN – one of the oldest results of probability theory – is a consequence of the more general (and more recent) result due to Markov. It should be noted that most of the above results provide sufficient conditions for the WLLN to hold and only Kolmogorov's theorem is an 'if and only if' statement.

Section 4.5 on the SLLN is basically similar to Section 4.4; various forms of SLLN are given and a noteworthy result is expressed by Proposition 4.19 which shows that for iid r.v.s (a frequently encountered case in applications) the existence of a finite first-order moment is a necessary and sufficient condition for the WLLN to hold.

In Section 4.5, moreover, we also give two additional results: (a) Borel–Cantelli lemma and (b) Kolmogorov's inequality. These are two fundamental results of probability theory in general. The main reason why they are included in this section is because they play a key part in the proofs of the theorems on the SLLN, but it must be pointed out that their importance lies well beyond this context.

Having established the conditions under which the LLN holds, Section 4.6 turns to one of the most famous results of probability, the so-called CLT which concerns the D-convergence of sequences of (independent) random variables to the normal distribution. We give two forms of this result: Lindeberg–Levy CLT and Lindeberg's CLT, where this latter result shows that – provided that the contribution of each individual r.v. is 'small' – the assumption of identically distributed variables is not necessary. A number of special cases of the theorem are also considered.

The chapter ends with Section 4.6.1 including some additional remarks worthy of mention in their own right. First, some comments are made on 'local' convergence theorems, where the term means the convergence of pdfs (or pmfs for discrete r.v.s) to the normal pdf. We recall, in fact, that *D*-convergence is a statement on PDFs and does not necessarily imply the convergence of densities.

Second, owing to the popularity of the CLT, one might be tempted to think that the normal distribution is the only limiting distribution of sequences of r.v.s. As a matter of fact, it is important to point out that it is not so because there exists a whole class of limiting distributions and the normal is just a member of this class. The subject, however, is outside the scope of the book and references are given for the interested reader. The third and final remark is an explicit statement of the multi-dimensional CLT for iid random vectors.

References

[1] Ash, R.B., Doléans-Dade, C., *'Probability and Measure Theory'*, Harcourt Academic Press, San Diego (2000).

[2] Boccara, N., *'Probabilités'*, Ellipses – Collection Mathématiques Pour l'Ingénieur (1995).

[3] Brémaud, P., *'An Introduction to Probabilistic Modeling'*, Springer-Verlag, New York (1988).

[4] Breiman, L., *'Probability'*, SIAM – Society for Industrial and Applied Mathematics, Philadelphia (1992).

[5] Cramer, H., *'Mathematical Methods of Statistics'*, Princeton Landmarks in Mathematics, Princeton University Press,19th printing (1999).

[6] Dall'Aglio, G., *'Calcolo delle Probabilità'*, Zanichelli, Bologna (2000).

[7] Duncan, A.J., *'Quality Control and Industrial Statistics'*, 5th edn., Irwin, Homewood, Illinois.

[8] Friedman, A., *'Foundations of Modern Analysis'*, Dover Publications, New York (1982).

[9] Gnedenko, B.V., *'Teoria della Probabilità'*, Editori Riuniti, Roma (1987).

[10] Gnedenko, B.V., Kolmogorov, A.N., *'Limit Distributions for Sums of Independent Random Variables'*, Addison-Wesley, Reading MA (1954).

[11] Heathcote, C.R., *'Probability: Elements of the Mathematical Theory'*, Dover Publications, NewYork (2000).

[12] Jacod J., Protter, P., *'Probability Essentials'*, 2nd edn., Springer-Verlag, Berlin (2003).

[13] Karr, R.A., *'Probability'*, Springer Texts in Statistics, Springer-Verlag, New York (1993).

[14] Klimov, G., *'Probability Theory and Mathematical Statistics'*, Mir Publishers, Moscow (1986).

[15] McDonald, J.N., Weiss, N.A., *'A Course in Real Analysis'*, Academic Press, San Diego (1999).

[16] Monti, C.M., Pierobon, G., *'Teoria della Probabilità'*, Decibel editrice, Padova (2000).

[17] Pfeiffer, P.E., '*Concepts of Probability Theory*', 2nd edn., Dover Publications, New York (1978).
[18] Biswas, S., '*Topics in Statistical Methodology*', Wiley Eastern Limited, New Delhi (1991).
[19] Taylor, J.C., '*An Introduction to Measure and Probability*', Springer-Verlag, New York (1997).
[20] Ventsel, E.S., '*Teoria delle Probabilità*', Mir Publisher, Moscow (1983).
[21] Wolfgang, P., Baschnagel, J., '*Stochastic Processes, from Physics to Finance*', Springer-Verlag, Berlin (1999).

Part II

Mathematical statistics

5 Statistics: preliminary ideas and basic notions

5.1 Introduction

With little doubt, the theory of probability considered in the previous chapters is an elegant and consistent mathematical construction worthy of study in its own right. However, since it all started out from the need to obtain answers and/or make predictions on a number of practical problems, it is reasonable to expect that the abstract objects and propositions of the theory must either have their counterparts in the physical world or express relations between real-world entities.

As far as our present knowledge goes, real-world phenomena are tested by observation and experiment and these activities, in turn, produce a set – or sets – of data. With the hope to understand the phenomena under investigation – or at least of some of their main features – we 'manipulate' these data in order to extract the useful information. In experiments where elements of randomness play a part, the manipulation process is the realm of 'Statistics' which, therefore, is a discipline closely related to probability theory although, in solving specific problems, it uses techniques and methods of its own.

Broadly speaking, the main purposes of statistics are classified under three headings: description, analysis and prediction. In most cases, clearly, the distinction is not sharp and these classes are introduced mainly as a matter of convenience. The point is that, in general, the individual data are not important in themselves but they are considered as a means to an end: the measure of a certain physical property of interest, the test of a hypothesis or the prediction of future occurrences under given conditions.

Whatever the final objectives of the experiment, statistical methods are techniques of 'inductive inference' in which a particular set (or sets) of data – the so-called 'realization of the sample' – is used to draw inferences of general nature on a 'population' under study. This process is intrinsically different and must be distinguished from 'deductive inference' where conclusions based on partial information are always correct, provided that the original information is correct. For example, in basic geometry the examination of particular cases leads to the deduction that the sum of the angles of a triangle equals 180 degrees, a conclusion which is always correct within the framework of plane geometry. By contrast, inductive inferences drawn from incomplete information may be wrong even if the original information is not. In the field of statistics, this possibility is often related to the (frequently overlooked) process of data collection on one hand – it is evident that insufficient or biased data and/or the failure to consider an important influencing factor in the experiment may lead to incorrect conclusions – and, on the other hand, to the fact that in general we can only make probabilistic statements and/or predictions. By their own nature, in fact, statements or predictions of this kind always leave a 'margin of error' even if the data have been properly collected. To face this problem, it is necessary to design the experiment in such a way as to reduce this uncertainty to values which may be considered acceptable for the situation at hand. Statistics itself, of course, provides methods and guidelines to accomplish this task but the analyst's insight of the problem is, in this regard, often invaluable. Last but not least, it should always be kept in mind that an essential part of any statistical analysis consists in a quantitative statement on the 'goodness' of our inferences, conclusions and/or results.

A final remark is not out of place in these introductory notes. It is a word of caution taken from Mandel's excellent book [20]. In the light of the fact that statistical results are often stated in mathematical language, Mandel observes that 'the mathematical mode of expression has both advantages and disadvantages. Among its virtues are a large degree of objectivity, precision and clarity. Its greatest disadvantage lies in its ability to hide some very inadequate experimentation behind a brilliant facade'. In this regard, it is surely worth having a look at Huff's fully enjoyable booklet [15].

5.2 The statistical model and some notes on sampling

As explained in Chapter 4, the mathematical models of probability are based on the notion of probability space (W, S, P), where W is a non-empty set, S a σ-algebra of subsets of W (the 'events') and P is a probability function defined on S. Moreover, an important point is that one generally considers – more or less implicitly – P to be fully defined.

In practice, however, P is seldom known fully and there exists some degree of uncertainty attached to it. Depending on the problem, the degree of uncertainty may vary from a situation of complete indeterminacy – where P could be any probability function that can be defined on S – to cases of partial indeterminacy in which P is known to belong to a given class but we lack some information which, were it available, would specify P completely. In general terms, the goal of statistics is to reduce the uncertainty in order to gain information and/or make predictions on the phenomena under investigation. This task, as observed in the introduction, is accomplished by using and 'manipulating' the data collected in experiment(s).

In more mathematical terms, the general idea is as follows. We perform an experiment consisting of n trials by assuming that the result $x_i (i = 1, 2, \ldots, n)$ of the ith trial is associated to a random variable X_i. By so doing, we obtain a set of n observations $(x_1, x_2, \ldots, x_n)$ – the so-called *realization of the sample* – associated to the set of r.v.s $(X_1, X_2, \ldots, X_n)$ which, in turn, is called a *random sample* (of size n). Both quantities can be considered as n-dimensional vectors and denoted by $\mathbf{x}$ and $\mathbf{X}$, respectively. Also, we call sample space the set Ξ of all possible values of $\mathbf{X}$ and this, depending on $\mathbf{X}$, may be the whole $\mathbb{R}^n$ or part of it (if $\mathbf{X}$ is continuous) or may consist in a finite or countable number of points of $\mathbb{R}^n$ (if $\mathbf{X}$ is discrete). With the generally implicit assumption that there exists a collection of subsets of Ξ forming a σ-algebra – which is always the case in practice – one defines (Ξ, Π) as the *statistical model* of the experiment, where here Π denotes the class of possible candidates of probability functions pertaining to the sample $\mathbf{X}$. Clearly, one of the elements of Π will be the (totally or partially unknown) 'true' probability function $P_{\mathbf{X}}$.

Now, referring back for a moment to Chapters 2 and 3, we recall that a random vector $\mathbf{X}$ on (W, S, P) defines a PDF $F_{\mathbf{X}}$ which, in turn, completely determines both the 'induced' probability $P_{\mathbf{X}}$ and the original probability P. Any degree of uncertainty on P, therefore, will be reflected on $F_{\mathbf{X}}$ (or, as appropriate, on the pmf $p_{\mathbf{X}}$ or pdf $f_{\mathbf{X}}$, when they exist) so that, equivalently, we can say that our statistical model is defined by (Ξ, Φ), where Φ is a class of PDFs such that $F_{\mathbf{X}}(x) = P_{\mathbf{X}}(X_1 \le x_1, \ldots, X_n \le x_n) \in \Phi$.

A particular but rather common situation occurs when the experiment consists in n independent repetitions of the same trial (e.g. tossing a coin, rolling a die, measuring n times a physical quantity under similar conditions, etc.). In this case the components of the sample $X_1, X_2, \ldots, X_n$ are iid random variables, that is, they are mutually independent and are all distributed like some r.v. X so that $F_{X_i}(x_i) = F_X(x_i)$ for all $i = 1, 2, \ldots, n$. The variable X is often called the *parent* random variable and the set R_X of all possible values of X is called the *population*; also, with this terminology, one can call $\mathbf{X}$ a 'sample (of size n) from the population R_X'. So, depending on the problem at hand, there are two possibilities

(1) the PDF F_X is totally unknown (and therefore, *a priori*, Φ may include any PDF), or
(2) the general type of F_X is known – or assumed to be known – but we lack information on a certain parameter θ whose 'true' value may vary within a certain set Θ (note that, in general, θ may be a scalar – then $\Theta \subseteq \mathbb{R}$ – or a k-dimensional vector ($k = 2, 3, \ldots$), and then $\Theta \subseteq \mathbb{R}^k$).

In case (1) our interest may be (1a) to draw inferences on the type of PDF underlying the phenomena under study or (1b) to draw inferences which do not depend on the specific distribution of the population from which the sample is taken. Statistical techniques that are – totally or partially – insensitive to the type of distribution and can be applied ignoring this aspect are called *non-parametric* or *distribution-free* methods.

In case (2) we speak of *parametric model* and the class Φ is of the form

$$\Phi = \{F(x;\theta) : \theta \in \Theta\} \tag{5.1}$$

where, for every fixed $\theta \in \Theta$, $F(x;\theta)$ is a well-defined PDF (the semicolon between x and θ denotes that F is a function of x with θ as a parameter, and not a function of two variables). Clearly, one denotes by P_θ the probability function associated to $F(x;\theta)$. In most applications, parametric models are either discrete of absolutely continuous, depending on the type of PDFs in Φ. It is evident that in these cases the model can be specified by means of a class of pmfs or pdfs, respectively.

Example 5.1 (Parametric models) Two examples may be of help to clarify the theoretical discussion above.

(i) Consider the experiment of tossing n times a coin whose bias, if any, is unknown. Assuming that we call a head a 'success', the natural parent r.v. X associated with the experiment assigns the value 1 to a success and 0 to a failure (tail). Since the experiment is a Bernoulli scheme, the distribution of X will clearly be a binomial pmf (eq. (2.41a)) whose probability of success, however, is not known. Then, our statistical model consists of two sets: Ξ, which includes any possible sequence (of n elements) of 1s and 0s, and Φ which includes all the pmfs of the type

$$p(x;\theta) = \binom{n}{x}\theta^x(1-\theta)^{n-x} \tag{5.2}$$

where x denotes the number of successes ($x = 0, 1, \ldots, n$), and $\theta \in \Theta = [0, 1]$ because the probability of success can be any number between 0 and 1 (0 and 1 representing the case of totally biased coin). Performing the experiment once leads to a realization of the sample – that is, one element of Ξ – which form our experimental data. Statistics, using these data, provides methods of evaluating – estimating is the correct term – the unknown parameter θ, that is, to make inferences on how much biased is the coin.

(ii) From previous information it is known that the length of the daily output – say, 5000 pieces – of a machine designed to cut metal rods in pieces of nominal length = 1.00 m, follows a Gaussian distribution. The mean and variance of the distribution, however, are unknown. In this case X is the length of a rod and Φ consists of the density functions

$$f(x;\theta_1,\theta_2) = \left(\sqrt{2\pi}\theta_2\right)^{-1} \exp\left[-(x-\theta_1)^2/2\theta_2^2\right] \tag{5.3}$$

where, in principle, $\theta_1 \in \Theta_1 = (-\infty, \infty)$ and $\theta_2 \in \Theta_2 = (-\infty, \infty)$ but of course – being θ_1, θ_2 the mean and standard deviation of the process – the choice can in reality be restricted to much smaller intervals of possible values. By selecting n, say $n = 50$, pieces of a daily production and by accurately measuring their lengths we can estimate the two parameters, thus drawing inferences on the population (the lengths of the 5000 daily pieces). The realization of the sample are our experimental data, that is, the 50 numbers $(x_1, x_2, \ldots, x_{50})$ resulting from the measurement.

Although the type of parametrization is often suggested by the problem, it should be noted that it is not unique. In fact, if $h : \Theta \to \Psi$ is a one-to-one function, the model (5.1) can be equivalently written as

$$\Phi = \{F(x; \psi) : \psi \in \Psi\} \tag{5.4}$$

where $\Psi = \{\psi : \psi = h(\theta), \theta \in \Theta\}$ and the choice between (5.1) and (5.2) is generally a matter of convenience. One word of caution, however, is in order: some inferences are not invariant under a change of parametrization, meaning in other words that there are statistical techniques which are affected by the choice of parametrization. This point will be considered in due time if and whenever needed in the course of future discussions.

It is worth at this point to pay some attention to the process by which we collect our data, that is, the so-called procedure of sampling. Its importance lies in the fact that inappropriate sampling may lead to wrong conclusions because our inferences cannot be any better than the data from which they originate. If the desired information is not implicitly contained in the data, it will never come out – no matter how sophisticated the statistical technique we adopt. Moreover, if needed, a good set of data can be analysed more that once by using different techniques, while a poor set of data is either hopeless or leads to conclusions which are too vague to be of any practical use.

At the planning stage, therefore – after the goal of the experiment has been clearly stated – there are a certain number of questions that need an answer. Two of these, the most intuitive, are: 'how do we select the sample?' and 'of what size?'. In regard to the first question the basic prerequisite is that the sample must be drawn at random. A strict definition of what exactly constitutes a 'random sample' is rather difficult to give but, luckily, it is often easier to spot signs of the contrary and decide that a given procedure should be discarded because of non-randomness. The main idea, clearly, is to avoid any source of bias and make our sample, as it is often heard, 'representative' of the population under study. In other words, we must adopt a sampling method that will give every element of the population an equal chance of being drawn.

In this light, the two simplest sampling schemes are called 'simple sampling (with replacement)' and 'sampling without replacement'. In both cases the sampling procedure is very much like drawing random tickets from an urn: in the first case we draw a ticket, note the value inscribed, and replace the ticket in the urn while in the second case we do not replace the ticket before the next drawing. It is worth noting that the main difference is that sampling without replacement, in general, cannot be considered as a repetition of a random experiment under uniform conditions because the composition of the population changes from one drawing to another. However, if (i) the population is infinite or (ii) very large and/or our sample consists in a small fraction of it (as a rule of thumb, at most 5%), the two schemes are essentially the same because removal of a few items does not significantly alter – case (ii) – the composition of the population or – case (i) – does not alter it at all.

The two sampling schemes mentioned above are widely used although, clearly, they are not the only ones and sometimes more elaborate techniques are used in specific applications. For our purposes, we will generally assume the case of simple sampling, unless otherwise explicitly stated in the course of future discussions. In regard to random samples, a final point worthy of mention is that, for finite populations, the use of (widely available) tables of random numbers is very common among statisticians. The members of the population are associated with the set of random numbers or some subset thereof; then a sample is taken from this set – for example, by blindly putting a pencil down on the table and picking n numbers in that section of the table – and the corresponding items of the population are selected. In case of sampling without replacement we must disregard any number that has already appeared.

The number n is the size of the sample which, as noted above, is one of the main points to consider at the planning stage because its value – directly or indirectly – affects the quality and accuracy of our conclusions. However, since the role of the sample size will become clearer as we proceed, further considerations will be made in due time.

5.3 Sample characteristics

It often happens that an experiment consists in performing n independent repetitions of a trial to which a one-dimensional parent r.v. X, with PDF F_X, is attached. Then, the sample is the sequence of iid r.v.s $X_1, \ldots, X_n$ and the realization of the sample will be is a sequence $x_1, x_2, \ldots, x_n$ of n observed values of X. Recalling from Section 2.3.2 the numerical descriptors of a r.v. – that is, mean, variance, moments, central moments, etc. – we can define the sample, or statistical, counterparts of these quantities. So, the ordinary (i.e. non-central) sample moment and sample central moment of order k ($k = 1, 2, \ldots$) are

$$A_k = \frac{1}{n}\sum_{i=1}^{n} X_i^k \tag{5.5}$$

$$C_k = \frac{1}{n}\sum_{i=1}^{n} (X_i - A_1)^k \tag{5.6}$$

respectively, where A_1 is the sample mean and C_2 is the sample variance. The specific values of these quantities obtained from a realization of the sample $x_1, x_2, \ldots, x_n$ will be denoted by the corresponding lowercase letters, that is, a_k and c_k, respectively. So, for instance, if we repeat the experiment (i.e. other n trials) a second time – thus obtaining a new set $x'_1, x'_2, \ldots, x'_n$ of observed values – we will have, in general $a'_k \neq a_k$ and $c'_k \neq c_k (k = 1, 2, \ldots)$.

At this point, in order not to get lost in symbols, a few comments on notation are necessary:

(a) the sample characteristics are denoted by A_k and C_k to distinguish them from their population (or theoretical) counterparts $E(X^k)$ – with the mean $E(X) = \mu$ as a special case – and $E[(X - \mu)^k]$ – with the variance $E[(X - \mu)^2] = \sigma^2$ (or $\mathrm{Var}(X)$) as a special case. The parent r.v. X and the sample size n to which A_k and C_k refer are often clear from the context and therefore will be generally omitted unless necessary either to avoid confusion or to make a point.

(b) Since some population characteristics are given special Greek symbols – for example, μ, σ^2 and the standard deviation σ – it is customary to indicate their sample counterparts by the corresponding uppercase italic letters, that is, M, S^2 and S, respectively. So, in the light of eqs (5.5) and (5.6) we have

$$M = A_1; \quad S^2 = C_2 = n^{-1} \sum_i (X_i - M)^2 \text{ and, clearly, } S = \sqrt{S^2}.$$

(c) Italic lowercase letters, m, s^2 and s, denote the specific realization of the sample characteristic obtained as a result of the experiment, that is:

$$m = n^{-1} \sum_i x_i; \quad s^2 = c_2 = n^{-1} \sum_i (x_i - m)^2 \text{ and } s = \sqrt{s^2}$$

(d) Greek letters will be often used for higher-order population characteristics. So, α_k and μ_k will denote respectively the (population) ordinary and central moments of order k. In this light, clearly, $\alpha_1 = \mu$ and $\mu_2 = \sigma^2$ but for these lower-order moments the notation μ and σ^2 (or $\mathrm{Var}(X)$) will generally be preferred.

The main difference to be borne in mind is that the population characteristics are fixed (though sometimes unknown) constants while the sample characteristics are conceived as random variables whose realizations are obtained by actually performing the experiment. More generally, since A_k and C_k are just special cases of (measurable) functions of $X_1, \ldots, X_n$, the above considerations apply to any (measurable) function $G(X_1, \ldots, X_n)$ of the sample. Any function of this type which contains no unknown parameters is often called a *statistic*. So, for instance, the C_k defined in (5.6) are statistics while the quantities $n^{-1} \sum_i (X_i - \mu)^k$ are not if μ is unknown.

Returning to the main discussion, a first observation to be made is that the relations between theoretical moments given in previous chapters still hold true for their sample counterparts and for their realizations as well. Then, for example, by appropriately changing the symbols, eq. (2.34) becomes

$$S^2 = A_2 - A_1^2 = A_2 - M^2 \tag{5.7}$$

or, more generally, the relation (2.33) between central and ordinary moments is

$$C_k = \sum_{j=0}^{k} \frac{(-1)^j k!}{j!(k-j)!} M^j A_{k-j} \tag{5.8}$$

and similar equations hold between a_k and c_k. Moreover, conceiving the sample characteristics as random variables implies that they will have a probability distribution in their own right which, as should be expected, will be determined by F_X. Whatever these distributions may be, the consequence is that it makes sense to speak, for instance, of the mean, variance and, in general, of moments of the sampling moments. Let us start by considering the mean $E(M)$ of the sample mean M. Since $E(X_i) = \mu$ for all $i = 1, \ldots, n$, the properties of expectation give

$$E(M) = \frac{1}{n} E\left(\sum_i X_i\right) = \frac{1}{n} \sum_i E(X_i) = \mu \tag{5.9}$$

The variance of M, in turn, can be obtained by using eq. (3.116) and the independence of the X_i. Therefore

$$\mu_2(M) = \text{Var}(M) = \frac{1}{n^2} \sum_i \text{Var}(X_i) = \frac{1}{n^2} n \,\text{Var}(X) = \frac{\sigma^2}{n} \tag{5.10}$$

and the standard deviation is $\sigma_M = \sigma/\sqrt{n}$. Similarly, it is left to the reader to show that the third- and fourth-order central moments of M are given by

$$\begin{aligned} \mu_3(M) &\equiv E\left[(M-\mu)^3\right] = \frac{\mu_3}{n^2} \\ \mu_4(M) &\equiv E\left[(M-\mu)^4\right] = \frac{\mu_4}{n^3} + \frac{3(n-1)}{n^3}\sigma^4 \end{aligned} \tag{5.11a}$$

and so on, with more tedious calculations, for the fifth, sixth order, etc. It is useful, however, to know the order of magnitude of the leading term in these central moments of M; for even and odd moments we have

$$\begin{aligned} \mu_{2m}(M) &= O(n^{-m}), \\ \mu_{2m-1}(M) &= O(n^{-m}), \end{aligned} \qquad m = 1, 2, \dots \tag{5.11b}$$

respectively, where the symbol $O(q)$ is well known from analysis and means 'of the same order of magnitude' of the quantity q in parenthesis. Equation (5.11b) can be checked by looking at (5.10) and (5.11a).

Next, turning our attention to the sample variance S^2, eq. (5.7) gives $E(S^2) = n^{-1}E\left(\sum_i X_i^2\right) - E(M^2) = E(X^2) - E(M^2)$. Then, using eq. (2.34) for both r.v.s X and M we get its mean as

$$E(S^2) = \sigma^2 + \mu^2 - \sigma_M^2 - \mu^2 = \sigma^2 - \frac{\sigma^2}{n} = \frac{n-1}{n}\sigma^2 \tag{5.12}$$

The calculation of the variance of S^2 is a bit more involved. Defining $Y_i = X_i - \mu (i = 1, 2, \dots, n)$ we get $S^2 = n^{-1}\sum_i (Y_i - \bar{Y})^2$, where $\bar{Y} = n^{-1}\sum_j Y_j$. Then we can write

$$\begin{aligned} S^2 &= \frac{1}{n}\sum_i (Y_i - \bar{Y})^2 = \frac{1}{n}\left[\sum_i Y_i^2 - \frac{1}{n}\left(\sum_i Y_i\right)^2\right] \\ &= \frac{1}{n}\left[\sum_i Y_i^2 - \frac{1}{n}\sum_i Y_i^2 - \frac{2}{n}\sum_{i<j} Y_i Y_j\right] \\ &= \frac{n-1}{n^2}\sum_i Y_i^2 - \frac{2}{n^2}\sum_{i<j} Y_i Y_j \end{aligned}$$

Squaring this quantity and taking its expectation gives

$$\begin{aligned} E[(S^2)^2] = {} & \left(\frac{n-1}{n^2}\right)^2 E\left[\left(\sum_i Y_i^2\right)^2\right] \\ & + \frac{4}{n^4} E\left[\left(\sum_{i<j} Y_i Y_j\right)^2\right] - \frac{4(n-1)}{n^4} E\left[\sum_r Y_r^2 \sum_{i<j} Y_i Y_j\right] \end{aligned}$$

Now, taking independence into account plus the fact that $E(Y_i) = 0$ for all i, the last term on the r.h.s. is zero while the first and second term lead to

$$\left(\frac{n-1}{n^4}\right)^2 E\left(\sum_i Y_i^4 + 2\sum_{i<j} Y_i^2 Y_j^2\right)$$

$$\frac{4}{n^4}E\left(\sum_{i<j} Y_i^2 Y_j^2\right)$$

respectively. Therefore

$$\begin{aligned} E[(S^2)^2] &= E\left(\frac{(n-1)^2}{n^4}\sum_i Y_i^4 + \frac{2(n-1)^2+4}{n^4}\sum_{i<j} Y_i^2 Y_j^2\right) \\ &= \frac{(n-1)^2}{n^3}\mu_4 + \frac{(n-1)^2+2}{n^3}(n-1)\sigma^4 \end{aligned} \tag{5.13}$$

where the last equality holds because $E(Y_i^4) = \mu_4$, $E(Y_i^2) = \sigma^2$ and there are $n(n-1)/2$ combinations of n r.v.s taken two at a time. Finally, since $\mathrm{Var}(S^2) = E[(S^2)^2] - E^2(S^2)$, we use eqs (5.13) and (5.12) to get

$$\mu_2(S^2) = \mathrm{Var}(S^2) = \frac{(n-1)^2}{n^3}\left(\mu_4 - \frac{n-3}{n-1}\sigma^4\right) \tag{5.14a}$$

So, in particular, if the original population is normal then the mean and variance of S^2 are given, respectively, by eq. (5.12) and by

$$\mathrm{Var}(S^2) = \frac{2(n-1)}{n^2}\sigma^4 \tag{5.14b}$$

where this last result follows from (5.14a) by taking eq. (2.42d) into account.

With rather cumbersome calculations one could then go on to obtain $\mu_3(S^2), \mu_4(S^2)$, etc. We do not do it here but limit ourselves to two further results worthy of mention: the first concerns the mean and variance of the sample moments A_k and their covariances. It is rather easy to determine

$$\begin{aligned} &E(A_k) = \alpha_k \\ &\mathrm{Var}(A_k) \equiv E\Big[(A_k - \alpha_k)^2\Big] = \frac{\alpha_{2k} - \alpha_k^2}{n} \\ &\mathrm{Cov}(A_k A_l) \equiv E\Big[(A_k - \alpha_k)(A_l - \alpha_l)\Big] = \frac{\alpha_{k+l} - \alpha_k\alpha_l}{n} \end{aligned} \tag{5.15}$$

where the first two equations are in agreement with the special cases (5.9) and (5.10) when one notes that $A_1 = M, \alpha_1 = \mu$ and $\alpha_2 = \sigma^2 + \mu^2$. For the order of magnitude of even and odd central moments of the A_k we have

$$\begin{aligned}\mu_{2m}(A_k) &\equiv E\left[(A_k - a_k)^{2m}\right] = O(n^{-m}), \\ \mu_{2m-1}(A_k) &\equiv E\left[(A_k - a_k)^{2m-1}\right] = O(n^{-m}),\end{aligned} \qquad m = 1, 2, \ldots \tag{5.16}$$

and it is easily seen that eq. (5.11b) are the special case $k = 1$ of (5.16).

The second result gives the covariance between the sample mean and the sample variance; it is left to the reader to show that

$$E\left[(M - \mu)\left(S^2 - \frac{n-1}{n}\sigma^2\right)\right] = E[(M - \mu)S^2] = \frac{n-1}{n^2}\mu_3 \tag{5.17}$$

which implies that for any symmetric distribution M and S^2 are uncorrelated. In fact, as it is probably known to the reader, μ_3 is a measure of skewness – or asymmetry or lopsidedness – of the distribution so that $\mu_3 = 0$ for any symmetric distribution. More specifically, the (adimensional) coefficient of skewness γ_1 is often used, where by definition

$$\gamma_1 = \frac{\mu_3}{\mu_2^{3/2}} = \frac{\mu_3}{\sigma^3} \tag{5.18}$$

With the above results at hand, one can determine the mean and variance of a number of (well-behaved) functions of sample moments by using the approximations given in Section 3.5.1. So, for example, if $k, l \geq 1$ are any two integers and $g(A_k, A_l)$ is a twice differentiable function in some neighbourhood of (α_k, α_l), then eq. (3.126) gives

$$E[g(A_k, A_l) \cong g(\alpha_k, \alpha_l) \tag{5.19a}$$

while eq. (3.128a) leads to

$$\begin{aligned}\mathrm{Var}[g(A_k, A_l)] \cong & \left(\frac{\partial g}{\partial A_k}\right)^2 \mathrm{Var}(A_k) + \left(\frac{\partial g}{\partial A_l}\right)^2 \mathrm{Var}(A_l) \\ & + 2\frac{\partial g}{\partial A_k}\frac{\partial g}{\partial A_l}\mathrm{Cov}(A_k A_l)\end{aligned} \tag{5.19b}$$

where it is understood that all derivatives are calculated at the point (α_k, α_l). Note, in particular, that eqs (5.19a) and (5.19b) can be used to approximate the mean and variance of the sample central moment C_k which, as shown by eq. (5.8), is a polynomial in $A_k, A_{k-1}, \ldots, A_1$.

Example 5.2(a) Consider the mean and variance of $C_2 = A_2 - A_1^2$. From eq. (5.19a) we get $E(C_2) = \alpha_2 - \mu^2 = \sigma^2$, which is the leading term in the exact result (5.12). On the other hand, eq. (5.19b) yields

$$\mathrm{Var}(C_2) \cong \frac{\alpha_4 - \alpha_2^2 + 8\mu^2\alpha_2 - 4\mu^4 - 4\mu\alpha_3}{n} = \frac{\mu_4 - \mu_2^2}{n}$$

which, as it should be expected, is the leading term of eq. (5.14) (the second equality is obtained by taking into account the relations between ordinary and central moments).

It is evident that the method leading to eqs (5.19a) and (5.19b) is essentially of analytical nature and, as such, it applies to all cases in which the assumptions of the relevant theorems are satisfied. These assumptions, in general, have do with the behaviour of the function g and, for this point, the reader is referred to books of mathematical analysis.

Example 5.2(b) As a second example, we calculate the variance of the so-called (sample) 'coefficient of variation' $V = S/M = \sqrt{C_2}/A_1$, provided that this quantity is bounded. We have

$$\frac{\partial V}{\partial C_2} = \frac{1}{2\mu\sqrt{\mu_2}}$$

$$\frac{\partial V}{\partial A_1} = -\frac{\sqrt{\mu_2}}{\mu^2}$$

so that using eq. (5.19b) and retaining only the leading terms in $\mathrm{Var}(C_2)$, $\mathrm{Var}(A_1)$ and $\mathrm{Cov}(C_2A_1)$ – see eqs (5.14), (5.10) and (5.17), respectively – we get

$$\mathrm{Var}(V) \cong \frac{\mu^2(\mu_4 - \mu_2^2) - 4\mu_3\mu_2\mu + 4\mu_2^3}{4n\mu_2\mu^4} \tag{5.20}$$

By similar calculations one could obtain, for instance, the approximate mean and variance of the sample counterpart of the coefficient of skewness (5.18).

5.3.1 *Asymptotic behaviour of sample characteristics*

The considerations of the preceding section readily extend to the multi-dimensional case and the reader is invited to work out the details. Here we turn our attention to another issue: the asymptotic behaviour of sample characteristics as n tends to infinity or, in practical applications, for large samples.

Starting with the sample mean M, we can use Markov's WLLN (Proposition 4.13) to determine that $M \to E(M)[P]$. In fact, since the X_i are iid r.v.s and $S_n = X_1 + \cdots + X_n = nM$, then $\mathrm{Var}(S_n) = n^2\mathrm{Var}(M) = n\sigma^2$

and $\mathrm{Var}(S_n)/n^2 \to 0$ as $n \to \infty$, showing that the assumptions of the theorem are all satisfied. Then, by virtue of eq. (5.9) we have $M \to \mu[P]$. Actually, by recalling the various form of SLLN given in Section 4.5, one can state the stronger result $M \to \mu[\text{a.s.}]$. More generally, as a consequence of Chebyshev's inequality (see also Proposition 4.12) and the first of eq. (5.15) we have $A_k(n) \to E(A_k) = \alpha_k[P]$ and, even more, by virtue of Proposition 4.19 $A_k(n) \to \alpha_k[\text{a.s.}]$ whenever α_k is finite. Note that here the n in parenthesis stresses the fact that the moments A_k depend on the sample size.

Clearly, similar statements are valid for the sample central moments and for any sample characteristic which is a continuous function of a finite number of the A_k. These convergence properties, in turn, imply that for large values of n the quantities calculated using the data of the experiment can be regarded as 'estimates' of the corresponding population characteristics. However, according to certain criteria used to evaluate the quality of the approximation, we will see in later sections that these may not always be the 'best' estimates one can find.

A second aspect to consider is the fact that the quantity $nA_k = \sum_i X_i^k$ is a sum of n independent variables – the X_i^k – which are independent by virtue of Proposition 3.3 and all have the same distribution. As a consequence, it follows that

Proposition 5.1(a) *As $n \to \infty$, the standardized variable*

$$\frac{nA_k - n\alpha_k}{\sqrt{n(\alpha_{2k} - \alpha_k^2)}} = \frac{\sqrt{n}(A_k - \alpha_k)}{\sqrt{\alpha_{2k} - \alpha_k^2}}$$

tends in distribution to the standard Gaussian r.v.

In fact, since eq. (5.15) imply $E(X_i^k) = \alpha_k$ and $\mathrm{Var}(X_i^k) = \alpha_{2k} - \alpha_k^2$ for all $i = 1, \ldots, n$, the result follows from Lindeberg–Levy CLT (Proposition 4.22). Also, note that Proposition 5.1 can be stated in different words by saying that A_k is asymptotically normal with mean α_k and variance $(\alpha_{2k} - \alpha_k^2)/n$ so that, in particular, the sample mean M is asymptotically normal with mean μ and variance σ^2/n (see also remark (c) after the proof of Proposition 4.22). In this regard, moreover, when sampling from a normal population we have the special result:

Proposition 5.1(b) *If the parent r.v. X is normal with mean μ and variance σ^2, M is exactly normal with mean μ and variance σ^2/n.*

The proof is immediate if we turn to CFs and note that

$$\begin{aligned}\varphi_M(u) &= E\{\exp[iu(X_1 + \cdots + X_n)/n]\}\\ &= E[iuX_1/n]\cdots E[iuX_n/n] = [\varphi_X(u/n)]^n\end{aligned}$$

Then, since φ_X has the form given in (2.52), φ_M is the CF of a Gaussian r.v. with mean μ and variance σ^2/n.

Continuing along the above line of reasoning, we can use Proposition 4.26 (the multi-dimensional CLT) to show that

Proposition 5.2 *The joint distribution of any finite number of sample moments is itself asymptotically normal.*

In fact, considering the two-dimensional case for simplicity, let $r, s \geq 1$ be any two integers; the vector $n(A_r, A_s)^T$ can be written as

$$n\begin{pmatrix}A_r\\ A_s\end{pmatrix} = \begin{pmatrix}\sum_i X_i^r\\ \sum_i X_i^s\end{pmatrix} = \begin{pmatrix}X_1^r\\ X_1^s\end{pmatrix} + \cdots + \begin{pmatrix}X_n^r\\ X_n^s\end{pmatrix}$$

where $\mathbf{X}_i = (X_i^r, X_i^s)^T$ are n iid two-dimensional vectors such that for all $i = 1, \ldots, n$ we have the mean $E(\mathbf{X}_i) = (\alpha_r, \alpha_s)^T$ and the covariance matrix

$$\mathbf{K} = \begin{pmatrix}\mathrm{Var}(X_i^r) & \mathrm{Cov}(X_i^r X_i^s)\\ \mathrm{Cov}(X_i^s X_i^r) & \mathrm{Var}(X_i^s)\end{pmatrix} = \begin{pmatrix}\alpha_{2r} - \alpha_r^2 & \alpha_{r+s} - \alpha_r\alpha_s\\ \alpha_{r+s} - \alpha_r\alpha_s & \alpha_{2s} - \alpha_s^2\end{pmatrix}$$

where the proof of the relation $\mathrm{Cov}(X_i^r X_i^s) = \alpha_{r+s} - \alpha_r\alpha_s$ is immediate. Then, Proposition 4.26 states that, as $n \to \infty$, the vector $\sqrt{n}(A_r - \alpha_r, A_s - \alpha_s)$ tends in distribution to a Gaussian two-dimensional vector with mean 0 and covariance matrix $\mathbf{K}$. The extension to a higher dimensional case is straightforward. Another important result is as follows:

Proposition 5.3 *Let $g(x, y)$ be a twice differentiable function in some neighbourhood of (α_r, α_s). Then, as $n \to \infty$, the r.v. $\sqrt{n}[g(A_r, A_s) - g(\alpha_r, \alpha_s)]$ tends in distribution to a normal variable with zero mean and variance* $\mathbf{D}^T\mathbf{K}\mathbf{D}$, *where* $\mathbf{D}$ *is the vector*

$$\mathrm{D} = \begin{pmatrix}\partial g/\partial A_r\\ \partial g/\partial A_s\end{pmatrix}$$

and it is understood that all derivatives are calculated at the point (α_r, α_s).

In order to sketch the proof, set $c_r = \partial g/\partial A_r$ and $c_s = \partial g/\partial A_s$. Since

$$g(A_r, A_s) - g(\alpha_r, \alpha_s) = c_r(A_r - \alpha_r) + c_s(A_s - \alpha_s) + \cdots$$

the variable $\sqrt{n}[g(A_r,A_s)-g(\alpha_r,\alpha_s)]$ is a sum of two r.v.s which, by virtue of Proposition 5.2, tend in distribution to a normal r.v. with zero mean, variances $c_r^2(\alpha_{2r}-\alpha_r)$ and $c_s^2(\alpha_{2s}-\alpha_s)$ and covariance $c_rc_s(\alpha_{r+s}-\alpha_r\alpha_s)$. Then, Proposition 5.3 follows from the fact that the sum of two dependent normal r.v.s A, B with means a, b and variances σ_A, σ_B is itself normal with mean $a+b$ and variance $\sigma_A^2+\sigma_B^2+2\text{Cov}(A,B)$ (see eq. (3.60)). Also note that one can equivalently state the theorem by saying that $g(A_r,A_s)$ is asymptotically normal with mean $g(\alpha_r,\alpha_s)$ and variance $n^{-1}\mathbf{D}^{\text{T}}\mathbf{KD}$ where $n^{-1}\mathbf{K}$ is the covariance matrix of the sample moments A_r, A_s. The extension to cases where g is a function of more than two moments is immediate and, by appropriately defining $\mathbf{D}$, the matrix notation $\mathbf{D}^{\text{T}}\mathbf{KD}$ still applies.

Example 5.3 The sample central moments C_k are functions of $A_1, A_2, \ldots, A_k$ and therefore Proposition 5.3 includes them as special cases. In fact, any C_k is asymptotically normal with mean μ_k and variance

$$\frac{1}{n}\left(\mu_{2k}-2k\mu_{k-1}\mu_{k+1}-\mu_k^2+k^2\mu_2\mu_{k-1}^2\right) \tag{5.21}$$

where eq. (5.21) can be obtained starting from eq. (5.8) and noting that the central moments do not depend on where we take the origin. Therefore, there is no loss of generality in assuming the origin at the population mean – that is, setting $\mu=0$ – so that $\alpha_k=\mu_k$ and all derivatives $\partial C_k/\partial A_j$ are zero except $\partial C_k/\partial A_k=1$ and $\partial C_k/\partial A_1=-k\mu_{k-1}$. Then, since

$$n^{-1}\mathbf{D}^{\text{T}}\mathbf{KD}=\text{Var}(A_k)+k^2\mu_{k-1}^2\text{Var}(A_1)-2k\mu_{k-1}\text{Cov}(A_kA_1)$$

the desired result follows by taking eq. (5.15) into account. So, for instance, the asymptotic variance of C_2 is $n^{-1}(\mu_4-\mu_2^2)$ which, as expected, coincides with the leading term of eq. (5.14a).

Returning to our main discussion, it is worth pointing out that the considerations above do not imply that asymptotic normality – although rather common – is a general rule. In order to give an example of sample characteristics which show a different behaviour in the limit of $n\to\infty$, we must first introduce the notion of 'order statistics'.

Suppose, for simplicity, that we are sampling from a continuous population; each realization of the sample $x_1,\ldots,x_n$ can be arranged in increasing order $x_{(1)}\le x_{(2)}\le\cdots\le x_{(n)}$ where, clearly, $x_{(1)}=\min(x_1,\ldots,x_n)$ and $x_{(n)}=\max(x_1,\ldots,x_n)$. Then, letting $X_{(k)}, k=1,\ldots,n$, denote the r.v. that has the value $x_{(k)}$ for each realization of the sample, we define a new sequence $X_{(1)},X_{(2)},\ldots,X_{(n)}$ of random variables satisfying $X_{(1)}\le\cdots\le X_{(n)}$. This new sequence is called the ordered series of the sample and $X_{(k)}$, in turn, is called the kth order statistic where, in particular, $X_{(1)}$ and $X_{(n)}$ are the extreme values of the sample. A first observation is that order statistics are not independent because information on one r.v. of the series provides information on other r.v.s: in fact, for example, if $X_{(k)} \geq x$ then we know that $X_{(k+1)}, \ldots, X_{(n)} \geq x$. A second observation is that sampling from an absolutely continuous populations prevents the possibility of two or more order statistics being equal since the probability of, say, $X_{(k)} = X_{(k+1)}$ is zero, thus justifying the expression 'for simplicity' at the beginning of this paragraph.

The PDF of $X_{(k)}$ can be obtained by noting that the event $X_{(k)} \leq x$ occurs whenever at least k out of the n independent r.v.s $X_1, \ldots, X_n$ are $\leq x$. Each one of these events has probability $F(x)$ – where $F(x)$ is the PDF of the parent r.v. X and $f(x) = F'(x)$ is its pdf. Therefore we have a binomial PDF given by

$$F_{(k)}(x) = \sum_{j=k}^{n} \binom{n}{j} [F(x)]^j [1 - F(x)]^{n-j} \tag{5.22a}$$

which is absolutely continuous if $F(x)$ is. Taking the derivative with respect to x leads to the pdf of $X_{(k)}$, that is,

$$\begin{aligned} F'_{(k)} &= \sum_{j=1}^{n} \binom{n}{j} jF^{j-1}(1-F)^{n-j} f - \sum_{j=k+1}^{n} \binom{n}{j-1} (n-j+1) F^j (1-F)^{n-j-1} f \\ &= \binom{n}{k} kF^{k-1}(1-F)^{n-k} f + \sum_{j=k+1}^{n} \left[\binom{n}{j} j - \binom{n}{j-1} (n-j+1) \right] \\ &\quad \times F^{j-1}(1-F)^{n-j} f \end{aligned}$$

and since the term within square brackets is zero we get

$$f_{(k)}(x) = \binom{n}{k} k [F(x)]^{k-1} [1 - F(x)]^{n-k} f(x) \tag{5.22b}$$

where it should be noted that in the extreme cases $k = 1$ and $k = n$, respectively, eq. (5.22a) reduces to eqs (2.73a) and (2.72a) while (5.22b) agrees with eqs (2.73b) and (2.72b).

In order to investigate the behaviour of order statistics as $n \to \infty$ we must distinguish between mid-terms and extremal terms of the ordered series. We call mid-terms the elements whose index is of the form $k = [pn]$ where p is any fixed number $0 < p < 1$ and the notation $[a]$ indicates the integer value of the number a. So, in these cases k depends on n and $k/n \to p$ as $n \to \infty$. On the other hand, we call extremal terms the elements of the series whose ordinal index is considered fixed throughout the limiting process and has either the form $k = r$ or $k = n - s + 1$, where r, s are any two fixed integers ≥ 1. Note that $k = n - s + 1$ always indicates the sth element from the top, irrespective of the sample size n which, in fact, is assumed to increase indefinitely.

Without entering into the details of the calculation, it can be shown (see, for example, [3] or [19]) that the mid-terms are asymptotically normal. However, the extremal terms are not. In fact, for instance, consider $X_{(r)}$ and define the new variable $\gamma = nF(x)$, denoting by $g_{(r)}(\gamma)$ its pdf. Since we must have $f_{(r)}(x)\,\mathrm{d}x = g_{(r)}(\gamma)\,\mathrm{d}\gamma$, we get from (5.22b)

$$g_{(r)}(\gamma) = \frac{r}{n}\binom{n}{r}\left(\frac{\gamma}{n}\right)^{r-1}\left(1-\frac{\gamma}{n}\right)^{n-r} = \frac{r}{\gamma}\binom{n}{r}\left(\frac{\gamma}{n}\right)^{r}\left(1-\frac{\gamma}{n}\right)^{n-r} \tag{5.23}$$

and – being $0 \le \gamma/n \le 1$ – we can use the limit (4.9) to obtain

$$\lim_{n\to\infty} g_{(r)}(\gamma) = \frac{\gamma^{r-1}}{(r-1)!}\mathrm{e}^{-\gamma} = \frac{\gamma^{r-1}}{\Gamma(r)}\mathrm{e}^{-\gamma} \tag{5.24}$$

where in the second expression we used the well-known 'gamma function' defined in Appendix C. Similarly, if for the sth statistic from the top $X_{(n-s+1)}$ one defines $g = n(1 - F(x))$ it is easy to determine that $g_{(s)}(\gamma)$ is again given by (5.23), the only difference being the index s in place of r. Therefore $\lim_{n\to\infty} g_{(s)}(\gamma) = \{\gamma^{s-1}/\Gamma(s)\}\mathrm{e}^{-\gamma}$.

The above limiting functions are gamma distributions – $\Gamma(\gamma; 1, r)$ and $\Gamma(\gamma; 1, s)$, respectively – which, however, represent the limit of a function of the relevant order statistics and not of the order statistics themselves. When $F(x)$ is given, it may sometimes be possible to obtain the explicit inverse relation, $x = F^{-1}(\gamma/n)$ or $x = F^{-1}(1 - \gamma/n)$ as appropriate, but these cases are rather rare. It is worth noting, nonetheless, that considerable work has been done in this direction and it has been found that the limiting distributions of (appropriately standardized) extreme statistics are only of three types, often denoted as Types I, II and III or EV1, EV2 and EV3 – where EV is the acronym for extreme values. Convergence to type I, II or III depends essentially on the 'tail' of the underlying distribution $F(x)$ and the rate of convergence is generally rather slow. For more details on this rich and interesting topic the interested reader can refer to [5, 10, 11, 22].

We close this section with two additional comments relevant to the above discussion. First, the sample counterpart of the p-quantile ζ_p – which is defined implicitly by the equation $F(\zeta_p) = p(0 < p < 1)$ – is a mid-term order statistic and therefore it is asymptotically normal. It can be shown that its mean and variance are, respectively, ζ_p and

$$\frac{p(1-p)}{nf^2(\zeta_p)} \tag{5.25}$$

In particular, since $\zeta_{1/2}$ defines the median of the population, its sample counterpart – $X_{[n/2]+1}$ if n is odd or any value between $X_{[n/2]}$ and $X_{[n/2]+1}$ if n is even – is asymptotically normal with mean $\zeta_{1/2}$ and variance $\{4nf^2(\zeta_{1/2})\}^{-1}$.

If, moreover, the parent r.v. is normal with parameters μ, σ^2, then the sample median is asymptotically normal with parameters $\mu, (2n)^{-1}\pi\sigma^2$.

The second comment refers to the extremal variables $X_{(r)}$ and $X_{(n-s+1)}$; if one considers their joint distribution it can be shown that they are asymptotically independent. Both results cited in these comments can be found in [3] or [19].

5.4 Point estimation

As stated at the beginning of this chapter, experimental data are a means to an end: to draw inferences on a population when, for whatever reason, it is not possible to examine the population in its entirety. Clearly, the type of inference – and therefore the desired final information – depends on the problem at hand. Nonetheless, some classes of problems are frequently encountered in practice and specific statistical methods have been devised to address them.

Here we consider the parametric model of eq. (5.1) with the aim of 'estimating' one or more unknown population parameters. This is one of the typical inference problems and we can choose to give our estimate in one of two distinct forms: (i) by assigning a specific value to the unknown parameter or (ii) by specifying an interval which – with a given level of confidence – includes the 'true' value of the parameter. One speaks of 'point estimation' in case (i) and of 'interval estimation' in case (ii) and it is understood that (i) and (ii) refer to each one of the unknown parameters when these are more than one. Point estimation is the subject of this and the following sections (Section 5.5 included).

Given a sample $X_1, \ldots, X_n$ we have considered in the previous sections a number of sample characteristics: each one has the form of a function $T(\mathbf{X}) = T(X_1, \ldots, X_n)$ and is a random variable which takes on the value $t = T(\mathbf{x}) = T(x_1, \ldots, x_n)$ after the experiment has been performed and we have obtained the realization $x_1, \ldots, x_n$. If, moreover, $T(\mathbf{X})$ contains no unknown quantities it is generally called a *statistic*. Intuitively, one would think of estimating an unknown population parameter by using the corresponding statistic so that, for instance, we could use M and S^2 as estimators of the population mean and variance μ and σ^2, respectively. As reasonable as this may sound, things are not always so clear-cut because other statistics can be used for the same purpose and, *a priori*, there seems to be no reason why M and S^2 should be preferred. In order to motivate our choice even in more complex cases, we must first try to evaluate the 'goodness' of estimators.

Let us call θ the unknown parameter to be estimated and let $T(\mathbf{X})$ – or, often, T_n or simply T, implicitly implying the dependence on the sample size – be the statistic used to estimate it. A first desirable property for T is that

$$E(T) = \theta \quad \text{for all } \theta \in \Theta \tag{5.26}$$

which, in words, is phrased by saying that T is an *unbiased* estimator (often we will write UE for short) of θ. Note that, strictly, one should write $E_\theta(T) = \theta$ because the Lebesgue–Stieltjies integral defining the expectation is an integral in $dF(x;\theta)$. This fact, however, is often tacitly assumed for parametric models of the type (5.1).

Defining the bias of T_n as $b = E(T_n) - \theta$ it is obvious that T_n is unbiased whenever $b = 0$. Note that eq. (5.26) does not imply $t = \theta$ for every realization of the sample; in fact some realizations will give $t - \theta > 0$ and some others will result in $t - \theta < 0$, however, on average, eq. (5.26) guarantees that there is no systematic error in the evaluation of θ. Also, if g is an arbitrary non-linear function, eq. (5.26) does not imply $E[g(T_n)] = g(\theta)$, this meaning, for example, that if T_n is an UE of σ^2 not necessarily $\sqrt{T_n}$ is an UE of the standard deviation σ.

Besides the bias, another measure of 'distance' from the true value is the *mean square error* (of T_n), defined as

$$\mathrm{Mse}(T_n) = E[(T_n - \theta)^2] \tag{5.27a}$$

which, using the identity $T - \theta = (T - E(T)) + (E(T) - \theta) = (T - E(T)) + b$, can be expressed as

$$\mathrm{Mse}(T) = \mathrm{Var}(T) + b^2 \tag{5.27b}$$

where $E[(T - E(T))^2] = \mathrm{Var}(T)$ by definition. Equation (5.27b), in addition, shows that the mean square error of an UE coincides with its variance. So, between any two estimators, say T, T', of the same parameter θ, it seems logical to prefer T if $\mathrm{Mse}(T) < \mathrm{Mse}(T')$ for all $\theta \in \Theta$. If, as it is often the case, we limit our choice to the class of UEs – let us denote this class by $u(\theta)$ – the 'best' estimator will be the one with minimum variance for all $\theta \in \Theta$. This minimum-variance-unbiased-estimator (MVUE) $\bar{T}$ is often called an *efficient* (or optimum) estimator of θ and satisfies the condition

$$\mathrm{Var}(\bar{T}) = \min_{T \in u(\theta)} \{\mathrm{Var}(T)\} \quad \text{for all } \theta \in \Theta. \tag{5.28}$$

although the concepts are sometimes distinguished because the estimator with minimum variance among all possible estimators (of a given parameter) may not be unbiased.

Clearly, one can also speak of *relative efficiency* and compare two estimators on the basis of the ratio of their variances by saying that – given $T, T' \in u(\theta)$ – T is more efficient than T' if $\mathrm{Var}(T) < \mathrm{Var}(T')$ for all $\theta \in \Theta$. In this regard, however, it should be noted that it may happen that $\mathrm{Var}(T) < \mathrm{Var}(T')$ for some values of θ but $\mathrm{Var}(T') < \mathrm{Var}(T)$ for other values of θ. Since the inequality must hold uniformly in θ – that is, for all $\theta \in \Theta$ – and θ is unknown, no efficiency comparison can be made in these cases. The same consideration applies to efficient estimators and (5.28) may hold for, say, T_1 for some θ and T_2 for some other θ. Then, efficiency is not enough to compare estimators. Within the class $u(\theta)$, the following results hold:

Proposition 5.4 *If $T_1, T_2 \in u(\theta)$ are two efficient estimators, then $T_1 = T_2$ where the equality $T_1 = T_2$ is understood in a probability sense, that is, if T_1 and T_2 satisfy eq. (5.28) then $P_\theta(\mathbf{X} \in \{\mathbf{x} : T_1(\mathbf{x}) \neq T_2(\mathbf{x})\}) = 0$ for all $\theta \in \Theta$.*

In other words, an efficient estimator, when it exists, is unique. The following proposition, on the other hand, states that efficiency is linear:

Proposition 5.5 *If T_1, T_2, respectively, are efficient estimators of θ_1, θ_2, then $a_1 T_1 + a_2 T_2$ is an efficient estimator of $a_1\theta_1 + a_2\theta_2$ for all $a_1, a_2 \in \mathbb{R}$.*

Both proofs can be found in Ref. [19].

A final remark is in order: in some cases an UE may not exist or, in other cases, a slightly biased estimator T_b can be preferred to an unbiased one T if $\text{Mse}(T_b) < \text{Mse}(T) = \text{Var}(T)$ for all $\theta \in \Theta$.

Other desirable properties of estimators consider their behaviour as $n \to \infty$ and not, as above, by regarding the sample size as fixed. These properties are called asymptotic and one says, for instance, that an estimator T is asymptotically unbiased if

$$\lim_{n\to\infty} E(T_n) = \theta \tag{5.29}$$

or equivalently $\lim_{n\to\infty} b_n = 0$, where we write b_n because the bias generally depends on the sample size. Clearly, an UE is asymptotically unbiased while the reverse, however, is not true in general.

Another asymptotic property is as follows: an estimator T_n of θ is *consistent* if $\lim_{n\to\infty} P(|T_n - \theta| < \varepsilon) = 1$ for all $\varepsilon > 0$, that is, if (see Section 4.3)

$$T_n \to \theta[P] \tag{5.30}$$

Some authors speak of weakly consistent estimator in this case and use the adjective 'strong' if $T_n \to \theta$ [a.s.] or, sometimes, if $T_n \to \theta$ $[M_2]$. In any case (see Propositions 4.6 and 4.8) strong consistency implies weak consistency and, in most cases, the definition of 'consistent' is understood in the sense of eq. (5.30). A useful sufficient condition to determine consistency is given by

Proposition 5.6 *T_n is a consistent estimator if (a) it is asymptotically unbiased and (b)* $\lim_{n\to\infty} \text{Var}(T_n) = 0$.

In fact, if (a) and (b) hold then eqs (5.27b) and (5.27a) imply $\lim_{n\to\infty} E[(T_n - \theta)^2] = 0$, that is, $T_n \to \theta[M_2]$ and therefore $T_n \to \theta[P]$. It is evident that

(b) only must hold if T_n is unbiased. Note also that requirements (a) and (b) are not necessary; in fact it can be shown that there are consistent estimators whose variance is not finite.

Owing to the properties of P-convergence we have also:

Proposition 5.7 *If T_n is a consistent estimator of θ and g is a continuous function, then $g(T_n)$ is a consistent estimator of $g(\theta)$.*

All the definitions and considerations above extend readily to the case of more than one unknown parameter $\theta_1, \theta_2, \ldots, \theta_k (k > 1)$ which, as noted in Section 5.2, can be considered as a k-dimensional vector.

To end this section, a final word of caution on asymptotic properties of estimators is not out of place. In practical cases, these properties provide valid criteria of judgement for large samples but lose their meaning for small samples. Unfortunately, the notions of 'small' or 'large' samples often depend on the problem at hand and cannot be made more precise without considering specific cases. A general rule of thumb requires $n > 30$ in order to be able to speak of 'large samples'; caution, however, must be exercised because the exceptions to this 'rule' are not rare.

Example 5.4(a) Equation (5.9) and the first of (5.15) show that the statistics M and A_k are UEs of the population parameters μ and α_k, respectively. Equation (5.12), however, shows that S^2 is a biased estimator of σ^2, the bias being $b = -\sigma^2/n$. Since $b \to 0$ as $n \to \infty$, S^2 is an asymptotically unbiased estimator of σ^2 (also, it is consistent because it satisfies the requirements of Proposition 5.6). For finite samples, nonetheless, the bias can be removed by considering the estimator

$$\bar{S}^2 = \frac{1}{n-1}\sum_{i=1}^{n}(X_i - M)^2 = \frac{n}{n-1}S^2 \tag{5.31a}$$

which satisfies $E(\bar{S}^2) = \sigma^2$ (note that some authors use the name 'sample variance' to denote the statistic $\bar{S}^2$). Also, for a normal population we have from eqs (5.14b) and (5.31a)

$$\text{Var}(\bar{S}^2) = \frac{2\sigma^4}{n-1} \tag{5.31b}$$

The procedure of bias removal shown above can be generalized to all cases in which $E(T) = c + d\theta$ – where c, d are two known constants – by defining $\bar{T} = (T - c)/d$. Then, the statistic $\bar{T}$ is an UE of θ. Another example of this type is the statistic C_3 as an estimator of μ_3 because $E(C_3) = n^{-2}(n-1)(n-2)\mu_3$ (the reader is invited to check this result).

In cases where the population mean μ is known the quantity

$$\hat{S}^2 = \frac{1}{n}\sum_{i=1}^{n}(X_i - \mu)^2 \tag{5.32}$$

is a statistic in its own right. It is left to the reader to show that (i) $\hat{S}^2$ is an UE of σ^2 and (ii) $\mathrm{Var}(\hat{S}^2) = n^{-1}(\mu_4 - \sigma^4) = 2n^{-1}\sigma^4$ where the last equality is due to the fact that, for a normal r.v., $\mu_4 = 3\sigma^4$ (see eq. (2.44d)).

Example 5.4(b) From the considerations above it is evident that, in general, there exist many unbiased estimators of a given parameter. As a further example of this, it is easy to show that any linear combination $\hat{T} = \sum_{i=1}^{n} c_i X_i$ such that $c_1 + c_2 + \cdots + c_n = 1$ is an UE of μ. Turning to its variance, however, we have $\mathrm{Var}(\hat{T}) = \sigma^2 \sum_i c_i^2$ and since

$$\sum_i c_i^2 = \sum_i \left(c_i - \frac{1}{n}\right)^2 + \frac{1}{n}$$

it follows that the minimum value of the sum $\sum_i c_i^2$ occurs when $c_i = 1/n$ for all $i = 1, \ldots, n$. Consequently, the sample mean M is the most efficient among all estimators (of μ) of the form $\hat{T}$. If, in particular, the sample comes from a normal population $N(\mu, \sigma^2)$, we noted at the end of the preceding section that the sample median – let us denote it by Z – is asymptotically normal with parameters μ and $(2n)^{-1}\pi\sigma^2$. For large samples, therefore, Z can be chosen as an estimator of μ but since (Proposition 5.1b) $\mathrm{Var}(M) < \mathrm{Var}(Z)$, the sample mean is more efficient than Z. We open here a short parenthesis: the fact that the sample median is less efficient than M should not lead the analyst to discard Z altogether as an estimator of μ. In fact, this statistic is much more robust than M and this quality is highly desirable in practice when the data may be contaminated by 'outliers'. We do not enter in any detail here but we only say that 'robust' in this context means that Z, as an estimator of the mean, is much less sensitive than M to the presence of outliers, where the term 'outlier' denotes an unexpectedly high or low value which, at first sight, does not seem to belong to the sample. As a matter of fact, this is often the case because outliers are generally due to recording, transmission or copying errors; in some cases, however, they may be true data of exceptional events. The interested reader can refer, for example, to Chapter 16 of [27].

Example 5.4(c) Turning briefly to asymptotic properties it is immediate to show, for instance, that M and A_k are consistent estimators of μ and α_k, respectively. In fact, they are unbiased and their variance – see eq. (5.10) and the second of (5.15) – satisfy condition (b) of Proposition 5.6. Also, having already noted that S^2 is an asymptotically unbiased estimator of σ^2 we can use eq. (5.14) to determine that – if μ_4 exists – then $\text{Var}(S^2) \to 0$ as $n \to \infty$ and therefore S^2 is a consistent estimator of σ^2 by virtue of Proposition 5.6.

Example 5.4(d) As a final example, let us suppose that the parent r.v. X of the sample is distributed according to a Cauchy pdf of the form

$$f(x;\theta) = \frac{1}{\pi[1+(x-\theta)^2]} \tag{5.33}$$

which represents our parametric model. Suppose further that we consider the sample mean M as an estimator of θ. Now, using characteristic functions it is not difficult to show that M has the same distribution as X and therefore the probability $P(|M-\theta| \geq \varepsilon)$ – being the same for all n – cannot tend to zero as $n \to \infty$. The conclusion is that M is not a consistent estimator of θ.

5.4.1 *Cramer–Rao inequality*

In the preceding section we defined the relative efficiency of estimators by restricting our attention to the class $u(\theta)$ of UEs. Within this class, the requirement of minimum variance – see eq. (5.28) – is the property of interest. Suppose, however, that somehow (we will have to say more about this later) we can find some unbiased estimators of a given parameter θ. Among these estimators, we can select the most efficient, but how do we know that there are no more efficient ones? In many cases, the Cramer–Rao inequality can answer this question. In fact, provided that some 'regularity conditions' are satisfied, it turns out that the variance of UEs is bounded from below; if we find an estimator whose variance equals this lower bound then we also know that this estimator – in the terms specified by Proposition 5.4 – is unique.

In order to keep things relatively simple, we consider the one-dimensional continuous case and denote by $f(x;\theta)$ the pdf of the parent r.v. X of the sample $(X_1,\ldots,X_n)$. Then the so-called likelihood function

$$L(\mathbf{x};\theta) = L(x_1,\ldots,x_n;\theta) = \prod_{i=1}^{n} f(x_i;\theta) \tag{5.34}$$

is the pdf of the sample. We assume the following regularity conditions:

(a) the set $\{x : f(x;\theta) > 0\}$ – that is, in mathematical terminology, the support of the pdfs $f(x;\theta)$ – does not depend on θ;
(b) the function $f(x;\theta)$ is differentiable with respect to θ;
(c) $\frac{\partial}{\partial\theta}\int f(x;\theta)\,dx = \int \frac{\partial}{\partial\theta} f(x;\theta)\,dx$
(d) $\frac{\partial}{\partial\theta}\int T(\mathbf{x})L(\mathbf{x};\theta)\,d\mathbf{x} = \int T(\mathbf{x})\frac{\partial}{\partial\theta}L(\mathbf{x};\theta)\,d\mathbf{x}$
where all (Lebesgue) integrals are on all space ($\mathbb{R}$ in (c) and $\mathbb{R}^n$ in (d))

(e) $E[U^2(\mathbf{X};\theta)] < \infty$ where the 'score' or 'contribution' function $U(\mathbf{X};\theta)$ is defined as

$$U(\mathbf{X};\theta) = \frac{\partial}{\partial\theta}\ln L(\mathbf{X};\theta) = \sum_{i=1}^{n}\frac{\partial}{\partial\theta}\ln f(X_i;\theta) \tag{5.35}$$

and the second equality descends from (5.34).

Proposition 5.8 (Cramer–Rao inequality) *Under the above regularity conditions, let $T = T(\mathbf{X}) \in u(\theta)$. Then*

$$\mathrm{Var}(T) \geq \frac{1}{E[U^2(\mathbf{X};\theta)]} = \frac{1}{I_n(\theta)} \tag{5.36}$$

where the function $E[U^2(\mathbf{X};\theta)]$, *being important in its own right, is denoted by* $I_n(\theta)$ *and called Fisher's information (on* θ*) contained in the sample* $\mathbf{X}$.

As a preliminary result, note that

$$E[U(\mathbf{X};\theta)] = 0 \tag{5.37}$$

In fact, since $f(X;\theta) = f(X_1;\theta) = \cdots = f(X_n;\theta)$, from (5.35) it follows

$$\begin{aligned} E[U(\mathbf{X};\theta)] &= \sum_i E\left(\frac{\partial}{\partial\theta}\ln f(X_i;\theta)\right) = nE\left(\frac{\partial}{\partial\theta}\ln f(X;\theta)\right) \\ &= n\int f(x;\theta)\frac{\partial}{\partial\theta}[\ln f(x;\theta)]\,\mathrm{d}x = n\int \frac{\partial}{\partial\theta}f(x;\theta)\,\mathrm{d}x \\ &= n\frac{\partial}{\partial\theta}\int f(x;\theta)\,\mathrm{d}x = 0 \end{aligned}$$

where we used the relation $\partial \ln f/\partial\theta = (1/f)(\partial f/\partial\theta)$ in the fourth equality, condition (c) in the fifth and the last equality holds because $\int f(x;\theta)\,\mathrm{d}x = 1$. Now, in order to prove eq. (5.36) we apply Cauchy–Schwarz inequality (eq. (3.21)) to the variables $(T(\mathbf{X}) - \theta)$ and $U(\mathbf{X};\theta)$

$$E^2[(T-\theta)U] \leq E[(T-\theta)^2]E(U^2) = \mathrm{Var}(T)E(U^2) \tag{5.38}$$

Since $E[(T-\theta)U] = E(TU) - \theta E(U) = E(TU)$, we use the relation $\partial \ln L/\partial\theta = (1/L)(\partial L/\partial\theta)$ and condition (d) to get

$$\begin{aligned} E(TU) &= \int T(\mathbf{x})L(\mathbf{x};\theta)\frac{\partial}{\partial\theta}[\ln L(\mathbf{x};\theta)]\,\mathrm{d}\mathbf{x} = \int T(\mathbf{x})\frac{\partial L(\mathbf{x};\theta)}{\partial\theta}\,\mathrm{d}\mathbf{x} \\ &= \frac{\partial}{\partial\theta}\int T(\mathbf{x})L(\mathbf{x};\theta)\,\mathrm{d}\mathbf{x} = \frac{\partial}{\partial\theta}E(T) = \frac{\partial}{\partial\theta}\theta = 1 \end{aligned}$$

so that the l.h.s of (5.38) is unity and Cramer–Rao inequality follows. Furthermore, by considering the explicit form of U of eq. (5.35) we get

$$
\begin{aligned}
I_n(\theta) = E(U^2) = E\left\{\left(\sum_i \frac{\partial}{\partial\theta}\ln f(X_i;\theta)\right)^2\right\} &= \sum_i E\left\{\left(\frac{\partial}{\partial\theta}\ln f(X_i;\theta)\right)^2\right\} \\
&+ \sum_{i\neq j} E\left\{\left(\frac{\partial}{\partial\theta}\ln f(X_i;\theta)\right)\left(\frac{\partial}{\partial\theta}\ln f(X_j;\theta)\right)\right\} \\
&= nE\left\{\left(\frac{\partial}{\partial\theta}\ln f(X;\theta)\right)^2\right\} = nI(\theta)
\end{aligned}
$$

where (i) the sum on $i \neq j$ is zero because of independence and of eq. (5.37) and (ii) $I(\theta) = E\{(\partial \ln f(X;\theta)/\partial\theta)^2\}$ is called Fisher's information and is the amount of information contained in one observation; the fact that $I_n(\theta) = nI(\theta)$ means that the information of the sample is proportional to the sample size.

In the light of these considerations we can rewrite (5.36) as

$$
\operatorname{Var}(T) \geq \frac{1}{nI(\theta)} \tag{5.39}
$$

If, in addition, f is twice θ-differentiable and we can interchange the signs of integration and derivative twice we have yet another form of the inequality. In fact, while proving eq. (5.37) we showed that $\int (\partial f/\partial\theta)\,\mathrm{d}x = 0$; under the additional assumptions, we can differentiate with respect to θ to get

$$
0 = \int \frac{\partial^2 f}{\partial\theta^2}\mathrm{d}x = \int \frac{1}{f}\left(\frac{\partial^2 f}{\partial\theta^2}\right) f\,\mathrm{d}x = E\left(\frac{1}{f}\frac{\partial^2 f}{\partial\theta^2}\right)
$$

and since we can use this last result to obtain

$$
\begin{aligned}
E\left(\frac{\partial^2 \ln f}{\partial\theta^2}\right) &= E\left(\frac{\partial}{\partial\theta}\left(\frac{1}{f}\frac{\partial f}{\partial\theta}\right)\right) = -E\left(\frac{1}{f^2}\left(\frac{\partial f}{\partial\theta}\right)^2\right) \\
&+ E\left(\frac{1}{f}\frac{\partial^2 f}{\partial\theta^2}\right) = -E\left(\left(\frac{\partial \ln f}{\partial\theta}\right)^2\right) = -I(\theta)
\end{aligned}
$$

Cramer–Rao inequality can be written as

$$
\operatorname{Var}(T) \geq -\left[nE\left(\frac{\partial^2 f(X;\theta)}{\partial\theta^2}\right)\right]^{-1} \tag{5.40}
$$

A few remarks are in order:

(1) the equal sign in (5.36) holds if and only if the two r.v. in Cauchy–Schwarz inequality are linearly related, that is, when

$$T(\mathbf{X}) - \theta = a(\theta)U(\mathbf{X};\theta) \tag{5.41}$$

where $a(\theta)$ is some function of θ.

(2) if T is an UE of a (differentiable) function $\tau(\theta)$ of θ, the numerator of Cramer–Rao inequality becomes $\{\tau'(\theta)\}^2$ and eq. (5.41) becomes $T - \tau(\theta) = a(\theta)U$. Whenever this relation holds, then $\text{Var}(T) = a(\theta)E(TU)$. Then, since the θ-derivative of $\tau(\theta) = E(T) = \int T(\mathbf{x})L(\mathbf{x};q)\,d\mathbf{x}$ is $\tau'(\theta) = E(TU)$, it follows that

$$\text{Var}(T) = a(\theta)\tau'(\theta) \tag{5.42}$$

which reduces to $\text{Var}(T) = a(\theta)$ if, as we considered above, $\tau(\theta) = \theta$.

(3) Cramer–Rao inequality establishes a lower bound for the variance of an UE; this does not imply that an estimator with such minimum variance exists (when this is not the case, one may use Bhattacharya's inequality; for more details see for instance Ref. [17] or [19]).

(4) the ratio between the lower bound and $\text{Var}(T)$ is called efficiency of the estimator and denoted by e_T, that is,

$$e_T = \frac{1}{I_n(\theta)\text{Var}(T)} = \frac{1}{nI(\theta)\text{Var}(T)} \tag{5.43}$$

where $0 \le e_T \le 1$ and $e_T = 1$ indicates a MVUE estimator.

(5) The discrete case, with only minor modifications is analogous to the continuous one.

Example 5.5(a) Consider a sample from a normal distribution with unknown mean $\theta = \mu$ and known variance σ^2. All regularity conditions are met and $f(x;\theta)$ is given by eq. (2.29a). Then $\partial \ln f(x;\mu)/\partial\mu = (x-\mu)/\sigma^2$ and Fisher's information is

$$I(\mu) = E\left(\left[\frac{\partial}{\partial\mu}\ln f(x;\mu)\right]^2\right) = \frac{1}{\sigma^4}E[(X-\mu)^2] = \frac{1}{\sigma^2}$$

which, as expected, implies that a smaller variance corresponds to a higher information. Now, considering M as an estimator of μ and knowing that (eq. (5.10)) $\text{Var}(M) = \sigma^2/n$ we get $e_M = 1$; therefore M is a MVUE estimator of μ. Also, eq. (5.41) must hold. In fact, using the expression of $f(X_i;\theta)$

pertinent to our case, we get from eq. (5.35)

$$U(\mathbf{X};\mu) = \frac{1}{\sigma^2}\sum_{i=1}^{n}(X_i - \mu) = \frac{n}{\sigma^2}(M - \mu)$$

which, in fact, is eq. (5.41) where $T - \theta = M - \mu$ and $a(\theta) = \sigma^2/n$. Note that, in agreement with eq. (5.42), $a(\theta) = \mathrm{Var}(M)$.

Example 5.5(b) Turning to a discrete case, consider a sample from a parent Poisson r.v. X with unknown parameter $\theta = \lambda$. From the pmf of eq. (4.1) we get the Fisher's information $I(\lambda) = \lambda^{-1}$ and since $\mathrm{Var}(X) = \lambda$ implies $\mathrm{Var}(M) = \lambda/n$, we have again $e_M = 1$.

(Incidentally, it is not out of place to point out that examples (a) and (b) must not lead to the (wrong) conclusion that M – although always unbiased – is always an efficient estimator of the mean.)

Example 5.5(c) Exponential Models. An important class of parametric models has the general form

$$f(x;\theta) = \exp\{A(\theta)B(x) + C(\theta) + D(x)\} \tag{5.44}$$

and is called exponential. Not all exponential models satisfy the regularity conditions, but for the ones that do the following considerations apply. Denoting by a prime the derivative with respect to θ, the score function is easily obtained as

$$U(\mathbf{X};\theta) = A'(\theta)\sum_{i=1}^{n} B(X_i) + nC'(\theta) = nA'(\theta)\left[\frac{1}{n}\sum_{i=1}^{n} B(X_i) + \frac{C'(\theta)}{A'(\theta)}\right]$$

which corresponds to eq. (5.41) once we set (see also remark (2))

$$\begin{aligned} T(\mathbf{X}) &= n^{-1}\sum_{i=1}^{n} B(X_i) \\ \tau(\theta) &= -C'(\theta)/A'(\theta) \\ a(\theta) &= [nA'(\theta)]^{-1} \end{aligned} \tag{5.45}$$

from which it follows that for the exponential class the statistic $T(\mathbf{X})$ is an efficient estimator of $\tau(\theta)$, where $T(\mathbf{X})$ and $\tau(\theta)$ are given by the first and second parts of eq. (5.45). This, by eq. (5.42), implies

$$\mathrm{Var}(T) = \frac{\tau'(\theta)}{nA'(\theta)} \tag{5.46}$$

Also, only a small effort is required to show that

$$I(\theta) = \tau'(\theta)A'(\theta) \tag{5.47}$$

By appropriately identifying the functions A, B, C and D, many practical models are, as a matter of fact, exponential. Examples (a) and (b), for instance, are two such cases. In fact, if we set $A(\theta) = \theta/\sigma^2$, $B(x) = x$, $C(\theta) = -\theta^2/2\sigma^2$ and $D(x) = -x^2/2\sigma^2$ we find example (a) and, as above, we determine that M is an efficient estimator of μ with $\mathrm{Var}(M) = \sigma^2/n$. The Poisson example of case (b), on the other hand, is obtained by setting $A(\theta) = \ln\theta$, $B(x) = x$, $C(\theta) = -\theta$ and $D(x) = -\ln x!$.

Example 5.5(d) As a further special case of exponential model, the reader is invited to consider a sample from a normal population with known mean μ and unknown variance $\theta^2 = \sigma^2$. By setting $A(\theta) = -1/2\theta^2$, $B(x) = (x-\mu)^2$ and $C(\theta) = -\ln(\theta\sqrt{2\pi})$ it turns out that $T(\mathbf{X}) = n^{-1}\sum(X_i - \mu)^2$ is an efficient estimator of $\tau(\theta) = \theta^2$. Also, the reader should check that eq. (5.41) for this case is $n^{-1}\sum_i(X_i-\mu)^2 - \theta^2 = n^{-1}\theta^3 U(\mathbf{X};\theta)$ and that $\mathrm{Var}(T) = 2\theta^4/n$, in agreement with result (ii) of Example 5.4(a).

The above examples show that for large samples the order of the variance of UEs is n^{-1}. This, as a matter of fact, is a general rule which applies to regular models. It is worth pointing out that in some cases of non-regular models it is possible to find UEs whose variance decreases more quickly than n^{-1} as n increases – that is, we can find UEs with variances smaller than the Cramer–Rao limit. Examples of these 'superefficient' estimators can be found, for instance, in Chapter 32 of [3] or in Chapter 2 of [19].

In closing this section, we briefly outline the case of more than one parameter, let us say k, so that $\mathbf{q} = (\theta_1, \theta_2, \ldots, \theta_k)^{\mathrm{T}}$ is a k-dimensional vector. Then, the score function is itself a vector $\mathbf{U} = (U_1, \ldots, U_k)^{\mathrm{T}}$ where $U_i(\mathbf{X};\mathbf{q}) = \partial \ln L(\mathbf{X};\mathbf{q})/\partial\theta_i$ and one can form the $k \times k$ information matrix of the sample as

$$\mathbf{I}_n(\mathbf{q}) = E(\mathbf{U}\mathbf{U}^{\mathrm{T}}) = n\mathbf{I}(\mathbf{q}) \tag{5.48}$$

(the second equality is the vector counterpart of the one-dimensional relation $I_n(\theta) = nI(\theta)$, valid in our experiment of repeated independent trials). The ijth element $I_{ij}(\mathbf{q})$ of $\mathbf{I}(\mathbf{q})$ – which, in turn, is the information matrix of one observation – is given by $(i,j = 1,2,\ldots,k)$

$$I_{ij}(\mathbf{q}) = E\left(\frac{\partial \ln f(X;\mathbf{q})}{\partial \theta_i}\frac{\partial \ln f(X;\mathbf{q})}{\partial \theta_j}\right) = -E\left(\frac{\partial^2 \ln f(X;\mathbf{q})}{\partial \theta_i \partial \theta_j}\right) \tag{5.49}$$

and the last relation holds if $f(x;\mathbf{q})$ is twice differentiable with respect to the parameters $\theta_1,\ldots,\theta_k$. Clearly, both $\mathbf{I}_n$ and $\mathbf{I}$ are symmetric, that is, $\mathbf{I}_n = \mathbf{I}_n^{\mathrm{T}}$ and $\mathbf{I} = \mathbf{I}^{\mathrm{T}}$. Given these preliminary notions, let $T(\mathbf{X})$ be an unbiased estimator of some function $\tau(\mathbf{q}) = \tau(\theta_1,\ldots,\theta_k)$ of the unknown parameters; then the Cramer–Rao inequality is now written

$$\mathrm{Var}(T) \geq \mathbf{d}^{\mathrm{T}}\mathbf{I}_n^{-1}\mathbf{d} = \frac{1}{n}\mathbf{d}^{\mathrm{T}}\mathbf{I}^{-1}\mathbf{d} \tag{5.50}$$

where $\mathbf{d} = \mathbf{d}(\mathbf{q})$ is the vector of derivatives $\mathbf{d}(\mathbf{q}) = (\partial\tau/\partial\theta_1,\ldots,\partial\tau/\partial\theta_k)^{\mathrm{T}}$ and, similarly to eq. (5.41), the equality holds if and only if

$$T(\mathbf{X}) - \tau(\mathbf{q}) = [\mathbf{a}(\mathbf{q})]^{\mathrm{T}}\mathbf{U}(\mathbf{X};\mathbf{q}) \tag{5.51}$$

for some vector function $\mathbf{a} = (a_1,\ldots,a_k)^{\mathrm{T}}$ of the parameters (note that, in general, $a_i = a_i(\mathbf{q})$ for all $i = 1,\ldots,k$). Moreover, as in the one-dimensional case, one calls efficient an estimator of $\tau(\mathbf{q})$ whose variance coincides with the r.h.s. of (5.50). Finally, since it is evident that eq. (5.50) holds only if $\mathbf{I}_n(\mathbf{q})$ (and therefore $\mathbf{I}(\mathbf{q})$) is non-singular for all $\mathbf{q} \in \Theta$, this assumption is generally added to the other defining conditions of regularity.

5.4.2 *Sufficiency and completeness of estimators*

In order to evaluate the 'goodness' of an estimator, another desirable property – besides the ones considered so far – is sufficiency. The definition is: given an unknown parameter θ, an estimator $T(\mathbf{X})$ of θ is sufficient (or exhaustive for some authors) if the conditional likelihood $L(\mathbf{x}|T = t;\theta)$ does not depend on θ. Equivalently, T is sufficient if the conditional probability $P_\theta(\mathbf{X} \in A|T = t)$ does not depend on θ for any event $A \subset \Xi$.

This definition is not self-evident and some further comments may help. In essence, sufficiency requires that the values $t = T(x_1,\ldots,x_n)$ taken on by the statistic T must contain all the information we can get on θ. In other words, suppose that two realizations of the sample $\mathbf{x}$ and $\mathbf{x}'$ both lead to the value $t = T(\mathbf{x}) = T(\mathbf{x}')$. If the function $L(\mathbf{x}|T = t;\theta)$ depended on θ then we would have, say, $L(\mathbf{x}|T = t) > L(\mathbf{x}'|T = t)$ for $\theta \in \Theta_1$ and $L(\mathbf{x}|T = t) < L(\mathbf{x}'|T = t)$ for $\theta \in \Theta_2$, where $\Theta_1 \cup \Theta_2 = \Theta$ and $\Theta_1 \cap \Theta_2 = \emptyset$ (i.e. the sets Θ_1, Θ_2 form a partition of the parameter space Θ). Therefore, knowing which one of the two realization has occurred provides more information than just the fact of knowing that $T = t$. So, for instance, if $\mathbf{x}$ has occurred, we would tend to think that, preferably, $\theta \in \Theta_1$. If, on the other hand $L(\mathbf{x}|T = t) = L(\mathbf{x}'|T = t)$ for all $\theta \in \Theta$ then the specific realization of the sample leading to $T = t$ is irrelevant and – for a fixed sample size n – the equality $T = t$ summarizes all that we can know in order to estimate θ. This is why T is a 'sufficient' estimator of θ.

In practice, it may be difficult to determine sufficiency just by using the definition above. Often, an easier way to do it is to use Neyman's theorem (which some authors give as the definition of sufficiency)

Proposition 5.9(a) (Neyman's factorization theorem) *A statistic $T(\mathbf{X})$ is sufficient for θ if and only if the likelihood function can be factorized into the product of two functions $g(T(\mathbf{x});\theta)$ and $h(\mathbf{x})$, that is,*

$$L(\mathbf{x};\theta) = g(T(\mathbf{x});\theta)h(\mathbf{x}) \tag{5.52}$$

(where it should be noted that the factor g depends on $\mathbf{x}$ *only through $T(\mathbf{x})$).*

In fact, since

$$L(\mathbf{x}|t;\theta) = \frac{P_\theta(\mathbf{X} = \mathbf{x} \cap T = t)}{P_\theta(T = t)} = \frac{L(\mathbf{x};\theta)}{\sum L(\mathbf{x}';\theta)}$$

(the sum at the denominator is over all realizations $\mathbf{x}'$ giving $T = t$) if we assume that the factorization (5.52) holds we get $L(\mathbf{x}|t;\theta) = h(\mathbf{x})/\sum h(\mathbf{x}')$ and therefore, according to the definition above, T is sufficient (if $\mathbf{x}$ is such that $T(\mathbf{x}) \neq t$ then $L(\mathbf{x}|t;\theta) = 0$; consequently $L(\mathbf{x}|t;\theta)$ does not depend on θ for any realization of the sample). The proof of the reverse statement – that is, if $L(\mathbf{x}|t;\theta)$ does not depend on θ then eq. (5.52) holds – is left to the reader.

Example 5.6(a) Let $\mathbf{X}$ be a sample from a Poisson variable (see eq. (4.1)) of unknown parameter θ. Then

$$L(\mathbf{x};\theta) = \prod_{i=1}^{n} \mathrm{e}^{-\theta}\frac{\theta^{x_i}}{x_i!} = \mathrm{e}^{-n\theta}\frac{\theta^{x_1+\cdots+x_n}}{x_1!\cdots x_n!} \tag{5.53}$$

and eq. (5.52) holds with $g = \mathrm{e}^{-n\theta}\theta^{x_1+\cdots+x_n}$ and $h = (x_1!\cdots x_n!)^{-1}$. It follows that the statistic $T(\mathbf{X}) = X_1 + \cdots + X_n$ is a sufficient estimator of θ. Alternatively, in this case we could also use the definition by noting (Section 4.2) that T is itself a Poisson variable of parameter $n\theta$. So, $P_\theta(T = t) = \{\mathrm{e}^{-n\theta}(n\theta)^t\}/t!$ and $L(\mathbf{x}|t;\theta)$ is independent on θ because

$$L(\mathbf{x}|t;\theta) = \frac{\mathrm{e}^{-n\theta}\theta^t}{\{P_\theta(T = t)\}x_1!\cdots x_n!} = \frac{t!}{n^t(x_1!\cdots x_n!)}$$

Example 5.6(b) Let $\mathbf{X}$ be a sample from a Gaussian variable with unknown mean $\mu = \theta$ and known variance σ^2. Then, defining $T(\mathbf{x}) = x_1 + \cdots + x_n$ we have

$$\begin{aligned} L(x;\theta) &= \prod_{i=1}^{n} \frac{1}{\sqrt{2\pi}\sigma} \exp\left(-\frac{(x_i-\theta)^2}{2\sigma^2}\right) \\ &= \left(\frac{1}{\sqrt{2\pi}\sigma}\right)^n \exp\left(-\frac{1}{2\sigma^2}\left(\sum_i x_i^2 - 2\theta T(\mathbf{x}) + n\theta^2\right)\right) \end{aligned} \tag{5.54}$$

and since Neyman's theorem holds by choosing

$$g(T(\mathbf{x});\theta) = \exp\left[-\frac{1}{2\sigma^2}(2\theta T(\mathbf{x}) + n\theta^2)\right]$$

$$h(\mathbf{x}) = (\sqrt{2\pi}\sigma)^{-n} \exp\left(-\frac{1}{2\sigma^2}\sum_i x_i^2\right)$$

the statistic $T(\mathbf{X}) = X_1 + \cdots + X_n$ is a sufficient estimator of θ.

A corollary to Proposition 5.9(a) is

Proposition 5.10 *(i) If the function z is one-to-one and T is sufficient for θ, then $Z = z(T)$ is also a sufficient estimator of θ. Moreover, (ii) $Z = z(T)$ is a sufficient estimator of $\hat{\theta} = z(\theta)$.*

In fact, the relation $L(\mathbf{x};\theta) = g(z^{-1}(Z);\theta)h(\mathbf{x}) = g_1(Z;\theta)h(\mathbf{x})$ proves part (i) while part (ii) follows easily by also considering the relation $\theta = z^{-1}(\hat{\theta})$. An immediate consequence of the corollary is that if $T(\mathbf{X}) = X_1 + \cdots + X_n$ is a sufficient estimator for the mean μ of a population, so is $M = T/n$.

At this point, an important observation is that Neyman's factorization (5.52) implies eq. (5.41) which – as we have determined – characterizes efficient (MVUE) estimators. In other words, this means that the class of sufficient statistics (for the parameter θ) includes the MVUE of θ when this estimator exists (note, however, that sufficient statistics may exist even when there is no MVUE). Moreover, Rao–Blackwell theorem states that the following:

Proposition 5.11 (Rao–Blackwell) *The MVUE, when it exists, is a function of a sufficient statistic.*

In fact, let $\mathbf{X}$ be a sample from a population with an unknown parameter θ, $T(\mathbf{X})$ a sufficient statistic for θ and $T_1(\mathbf{X})$ an arbitrary UE of θ. Then $E(T_1|T)$ (note that in strict symbolism we should write $E_\theta(T_1|T)$) is a function of the form $H(T)$ which takes on the value $H(t) = E(T_1|t)$ when $T = t$. Since

T, T_1 are random variables in their own right, we can use eq. (3.89a) to get $E[H(T)] = E[E(T_1|T)] = E(T_1) = \theta$ where the last equality holds because T_1 is unbiased. The consequence is that $H(T)$ is itself an UE of θ. In addition to this, eq. (3.91) shows that $\text{Var}(T_1) = E[\text{Var}(T_1|T)] + \text{Var}[H(T)]$ which – since $E[\text{Var}(T_1|T)] \geq 0$ – implies

$$\text{Var}[H(T)] \leq \text{Var}(T_1) \tag{5.55}$$

(the equal sign holds if and only if $E[\text{Var}(T_1|T)] = E\{[T_1 - E(T_1|T)]^2\} = 0$, that is, whenever $T_1 = H(T)$ – or, more precisely, when $P\{T_1 = H(T)\} = 1$). At this point one could conclude that $H(T)$ is (i) an UE of θ and (ii) more efficient than T_1. Before doing this, however, one must show that $H(T)$ is a statistic, that is, does not depend on θ. By recalling eq. (3.88) we can write

$$H(t) = E(T_1|t) = \int T_1(\mathbf{x}) L(\mathbf{x}|t;\theta)\, d\mathbf{x}$$

and note that both $L(\mathbf{x}|t;\theta)$ and $T_1(\mathbf{x})$ do not depend on θ because, respectively, T is sufficient and T_1 is a statistic. So, $H(t)$ does not depend on θ; moreover, as t varies the r.v. $H(T)$ takes on the values $H(t)$ with a density $f_T(t)$ which is itself independent on θ (T is a statistic). Consequently, as desired, $H(T)$ is a statistic.

Despite its intrinsic importance, Rao–Blackwell theorem is of little help in explicitly finding the MVUE (assuming that it exists). In fact, given a sufficient and an unbiased estimator, T and T_1 respectively, we can construct the UE $H(T)$ which – although more efficient than T_1 – may not be the MVUE of θ. In principle, by using $H(T)$ and another sufficient statistic, we expect to be able to find an even more efficient (than $H(T)$) UE. However, if the original sufficient statistic is complete (see definition below), it turns out that $H(T)$ is the MVUE of the parameter θ. This is stated in the following proposition:

Proposition 5.12 (Lehmann–Scheffé theorem) *Let $T(\mathbf{X})$ be a sufficient and complete statistic for θ and $T_1(\mathbf{X})$ an UE of θ. Then $H(T) = E(T_1|T)$ is the efficient estimator of θ.*

Before showing why this is so, we give the definition of completeness: a sufficient statistic T is complete if for any (bounded) function $\varphi(T)$ the relation

$$E_\theta[\varphi(T)] = 0 \quad \text{for all } \theta \in \Theta$$

implies $\varphi(t) = 0$ for almost all values $t = T(\mathbf{x})$ (the term 'almost all' refers to the measure P_θ and indicates that $P_\theta\{\varphi(T(\mathbf{x})) = 0\} = 1$ for all $\theta \in \Theta$).

Returning to Proposition 5.12, assume that there exists another UE $K(T)$ depending on T. Defining $L(T) = H(T) - K(T)$, we have $E[L(T)] = \theta - \theta = 0$ for all θ and this, by completeness, implies $H(t) = K(t)$ a.e. which, in turn, shows that $H(T)$ is the unique UE depending on T. Let now $\tilde{T}$ be an arbitrary UE. By virtue of the considerations above $J(T) = E(\tilde{T}|T)$ is unbiased, $\text{Var}[J(T)] \le \text{Var}(\tilde{T})$ and the equality holds iff $\tilde{T} = J(T)$. Since $H(T)$ is the only UE depending on T, then we must have $J(T) = H(T)$ and this proves the theorem.

At this point, two closing remarks on sufficiency are worthy of mention.

First we outline the generalization to the case of k unknown parameters. In this case the following definition applies: the vector $\mathbf{T} = (T_1, \ldots, T_k)$ is called a (jointly) sufficient statistic for $\mathbf{q} = (\theta_1, \ldots, \theta_k)$ if the function $L(\mathbf{x}|t_1, \ldots, t_k; \mathbf{q})$ does not depend on $\mathbf{q}$. Neyman's theorem, on the other hand, becomes:

Proposition 5.9(b) *The k-dimensional statistic* $\mathbf{T}(\mathbf{X}) = (T_1(\mathbf{X}), \ldots, T_k(\mathbf{X}))$ *is (jointly) sufficient for* $\mathbf{q} = (\theta_1, \ldots, \theta_k)$ *if and only if the likelihood function can be expressed as the product*

$$L(\mathbf{x}; \mathbf{q}) = g(T_1(\mathbf{x}), \ldots, T_k(\mathbf{x}); \mathbf{q})h(\mathbf{x}) \tag{5.56}$$

So, for example, it is easy to show that $\mathbf{T} = (T_1, T_2)$ – where $T_1 = \sum_i X_i$ and $T_2 = \sum_i X_i^2$ – is a sufficient statistic for the two-dimensional Gaussian model with unknown mean and variance. Using the sufficient statistic $\mathbf{T}$ we can then construct the well-known estimators $M = n^{-1}T_1$ and $\bar{S}^2 = (n-1)^{-1}[T_2 - n^{-1}T_1^2]$ (see eq. (5.31)) of μ and σ^2.

The second and final remark may appear rather obvious at first glance but – we believe – deserves to be stated explicitly: sufficiency depends on the adopted statistical model. In other words, if the model is changed, a given sufficient statistic may no longer be sufficient in the new model. As a consequence, we should never discard the raw data and replace them with sufficient statistics. In fact, although the main advantage of sufficient statistics is to reduce the dimensionality of the sample without losing any information on the unknown parameter(s), it should also be kept in mind that the sample itself $\mathbf{X} = (X_1, \ldots, X_n)$ is always a sufficient statistic irrespective of the adopted statistical model. Consequently – since the model may always be changed in the light of new evidence or of new assumptions – it is always good practice to preserve the original data. As an example, consider a sequence of binomial trials with unknown probability of success $\theta = p$. The order of successes and failures is clearly unimportant in a model of independent trials and the sufficient statistic $T = X_1 + \cdots + X_n$ is equivalent to the sample as far as the estimation of θ is concerned. However, it can be shown [8] that it is not so if a new model of dependent trials is postulated.

(Incidentally, under the assumption of binomial independent – that is, Bernoulli – trials, the reader is invited to show that $T = X_1 + \cdots + X_n$ is, indeed, a sufficient statistic.)

5.5 Maximum likelihood estimates and some remarks on other estimation methods

In regard to point estimation, not much has been said so far on the way in which we can find 'good' estimators although, in the preceding section, we have implicitly given a method of finding the MVUE of a parameter θ by using an UE T_1, a sufficient complete statistic T and calculating the conditional expectation $E(T_1|T)$. This procedure, however, often involves computational difficulties and is seldom used in practice. Other methods, in fact, have been devised and the most popular by far is the so-called 'method of maximum likelihood', introduced by Fisher in 1912 (although the definition of likelihood, also due to Fisher, appeared later). Before considering this, however, it is worth spending a few words on other methods with the main intention of simply illustrating – without any claim of completeness – other approaches to the problem.

One of the oldest estimation procedures is Pearson's 'methods of moments' and consists in equating an appropriate number of sample moments to the corresponding population moments which, in turn, depend on the unknown parameters. By considering as many moments as there are parameters, say k, one solves the resulting equations for $\theta_1, \ldots, \theta_k$ thus obtaining the desired estimates. In mathematical terms, if $j = 1, \ldots, k$ and $a_j = A_j(\mathbf{x})$ are the sample moments of the observed realization $\mathbf{x} = (x_1, \ldots, x_n)$, one must solve the set of equations

$$\alpha_j(\theta_1, \ldots, \theta_k) = a_j, \quad j = 1, 2, \ldots, k \tag{5.57a}$$

whose result is in the form

$$\theta_j = t_j(a_1, \ldots, a_k), \quad j = 1, 2, \ldots, k \tag{5.57b}$$

where the t_j's – that is, the values taken on by the estimators T_j's at $\mathbf{X} = \mathbf{x}$ – are obtained as functions of the sample moments. Recalling the developments of Section 5.3.1, this last observation on the T_j implies, under fairly general conditions, two desirable properties: for large samples the T_j are (i) consistent and (ii) asymptotically normal. Often, however, they are biased and their efficiency, as Fisher himself has pointed out [9] may be rather poor. For small samples, moreover, it should be kept in mind that sample moments may significantly differ from their population counterparts, thus leading to poor estimates. This is especially true if higher-order moments must be used because in these cases $n < 100$ is generally considered a small sample.

As an example of the method, consider a sample $\mathbf{X}$ from a population with unknown mean $\mu = \alpha_1$ and variance $\sigma^2 = \alpha_2 - \alpha_1^2$. Equations (5.57a) and (5.57b) are simply $\alpha_j = a_j$ $(j = 1, 2)$, and since $a_1 = m = n^{-1} \sum_i x_i$ and $a_2 = n^{-1} \sum_i x_i^2$, we get $t_1 = m$ and $t_2 = a_2 - m^2 = n^{-1} \sum_i (x_i - m)^2$. The desired estimators are therefore

$$T_1 = M$$

$$T_2 = A_2 - M^2 = \frac{1}{n} \sum_i (X_i - M)^2$$

where we already know (eq. (5.12)) that $T_2 = S^2$ is a biased estimator of σ^2. In all, however, the method has the advantage of simplicity and the 'moments-estimates' can be used as a first approximation in view of a more refined analysis.

A second method of estimation is based on Bayes' formula (eq. (3.79) in the continuous case). If we consider the unknown parameter θ as a value taken on by a r.v. Q with pdf $f_Q(\theta)$ – which, somehow, must be known by some prior information and for this reason is called '*a priori*' density – Bayes' formula yields (taking eq. (3.80b) into account)

$$f(\theta|\mathbf{x}) = \frac{f(\mathbf{x}|\theta) f_Q(\theta)}{\int_{-\infty}^{\infty} f(\mathbf{x}|\theta) f_Q(\theta)\, d\theta} \tag{5.58}$$

Then, by defining Bayes' estimator (of θ) as $T_B \equiv E(Q|X)$, its value t_B corresponding to the realization $\mathbf{x}$ is taken as the estimate of θ, that is,

$$t_B = E(Q|\mathbf{X} = \mathbf{x}) = \int_{-\infty}^{\infty} \theta f(\theta|\mathbf{x})\, d\theta \tag{5.59}$$

A few additional comments on this method are worthy of mention. First, the function $f(\mathbf{x}|\theta)$ at the numerator of (5.58) is just the pdf $f(\mathbf{x};\theta)$ that specifies the statistical model. However, the point of view is different; instead of seeing θ as a deterministic quantity and postulating the existence of a 'true' value θ_0 which – were it known – would provide the 'exact' probabilistic description by means of $f(\mathbf{x};\theta_0)$, the Bayesian approach considers θ as a random variable and writes $f(\mathbf{x}|\theta)$ to mean that the realization $\mathbf{x}$ is conditioned by the event $Q = \theta$. In this light, the 'a posteriori' density $f(\theta|\mathbf{x})$ provides information on θ after the realization $\mathbf{x}$ has been obtained and consequently we can use it to calculate the quantity $E(Q|\mathbf{X} = \mathbf{x})$ which, in turn – being the mean value of Q given that the event $\mathbf{X} = \mathbf{x}$ has occurred – is a good candidate as an estimate of θ. Nonetheless, a key point of the method is how well we know $f_Q(\theta)$. This, clearly, depends on the specific case under study although it has been argued that a uniform distribution for Q may be used in cases of no or very little prior information (a form of the so-called 'principle of indifference'). We do not enter into the details of this debated issue, which is outside the scope of the book, and pass to the main subject of this section: the method of maximum likelihood.

Consider the statistical model (5.1) with k unknown parameters $\mathbf{q} = (\theta_1, \ldots, \theta_k)$. Once a realization of the sample $\mathbf{x}$ has been obtained, the likelihood $L(\mathbf{x}; \mathbf{q})$ is a function of $\mathbf{q}$ only; consequently, we can write $L(\mathbf{q})$ and note that this function expresses the probability (density) of obtaining the result that, in fact, has been obtained, that is, $\mathbf{x}$. In this light it is reasonable to assume as 'good' estimates of the unknown parameters the values $\hat{\mathbf{q}} = (\hat{\theta}_1, \ldots, \hat{\theta}_k)$ that maximize $L(\mathbf{q})$, that is,

$$L(\hat{\mathbf{q}}) = \operatorname*{Max}_{\mathbf{q} \in \Theta} L(\mathbf{q}) \tag{5.60}$$

where it should be noted that the maximum is taken on the parameter space and not on all the possible values that make mathematical sense for $L(\mathbf{q})$. Owing to (5.60), $\hat{\theta}_1, \ldots, \hat{\theta}_k$ are called maximum likelihood (ML) estimates of $\theta_1, \ldots, \theta_k$. As $\mathbf{x}$ varies, we will obtain different values of $\hat{q}$ and this correspondence leads to the definition of 'maximum likelihood estimators' (MLE) as those statistics $\hat{T}_1(\mathbf{X}), \ldots, \hat{T}_k(\mathbf{X})$ which, respectively, take on the values $\hat{\theta}_1, \ldots, \hat{\theta}_k$ when $\mathbf{X} = \mathbf{x}$. In practice, the ML estimates are obtained by finding the maximum of the log-likelihood function $l(\mathbf{q}) = \ln L(\mathbf{q})$ (which is equivalent to maximizing $L(\mathbf{q})$), that is, by first solving the likelihood equations

$$\frac{\partial l(\theta_1, \ldots, \theta_k)}{\partial \theta_j} = 0, \quad j = 1, 2, \ldots, k \tag{5.61}$$

and then checking which solution is an absolute maximum (in fact, the solutions of eq. (5.61) – if there are any – determine the stationary points of $l(\mathbf{q})$, which can be minima, maxima or saddle points). The whole procedure is generally rather easy if we have one (or two) unknown parameter(s) but it is evident that computational difficulties may arise for higher values of k. In these cases one must resort to numerical techniques of solution of eq. (5.61) and the Newton–Raphson iteration method is frequently used for this task. The subject, however, is beyond our scope and the reader interested in computational aspects may refer, for instance, to [29] (Incidentally, in regard to the determination of the maximum among the solutions of (5.61), it may be worth recalling a theorem of analysis which states the following: If $l(\mathbf{q})$ is twice differentiable and Θ is an open set of $\mathbb{R}^k$, a maximum is attained at $\hat{\mathbf{q}}$ if the quadratic form $(\mathbf{q} - \hat{\mathbf{q}})^T \mathbf{H}(\hat{\mathbf{q}})(\mathbf{q} - \hat{\mathbf{q}})$ defined by the Hessian matrix $\mathbf{H}(\mathbf{q}) = [\partial^2 l / \partial \theta_i \partial \theta_j]$ $(i, j = 1, \ldots, k)$ is negative definite).

Example 5.7(a) Considering a sequence of n Bernoulli trials, the statistical model is clearly given by (5.2). Then, the ML estimate of the parameter $\theta = p$ is easily obtained by ignoring the terms with the factorials (which do not involve θ) and writing $l(\theta) = x \ln \theta + (n - x) \ln(1 - \theta)$. Taking the derivative

$$\frac{\partial l}{\partial \theta} = \frac{x}{\theta} - \frac{n - x}{1 - \theta} = 0$$

we get the solution $\hat{\theta} = x/n$, which is a maximum because $\partial^2 l/\partial\theta^2$ is negative at the point $\theta = \hat{\theta}$. Also note that the ML estimate coincides with the observed frequency of success. This specific example is one among many others that, *a posteriori*, justifies the relative frequency approach to probability discussed in Chapter 1.

Example 5.7(b) In the case of a sample from a normal population with unknown mean $\mu = \theta_1$ and variance $\sigma^2 = \theta_2$, the reader is invited to determine that the ML estimates are $\hat{\theta}_1 = n^{-1} \sum_i x_i = m$ and $\hat{\theta}_2 = n^{-1} \sum_i (x_i - m)^2 = s^2$ so that the MLE are M and S^2, respectively.

The examples above do not do justice to the ML method because the reader can easily check that the method of moments yields the same estimators. In general, however, this is not so and the reason why the ML method is so widely adopted lies in the good properties of MLEs. The first can be called the 'covariance' property with respect to parameter transformations; in fact, referring to eq. (5.4) we have

Proposition 5.13 *If* $\hat{\mathbf{q}} = (\hat{\theta}_1, \ldots, \hat{\theta}_k)$ *is the MLE of* $\mathbf{q} = (\theta_1, \ldots, \theta_k)$ *and* h *a one-to-one mapping from* Θ *to* Ψ $(\Theta, \Psi \subset R^k)$, $\hat{\mathbf{r}} = h(\hat{\mathbf{q}})$ *is the MLE of* $h(\mathbf{q})$.

The proof is immediate because the function $h^{-1} : \Psi \to \Theta$ exists and

$$\operatorname*{Max}_{\mathbf{q}\in\Theta} L(\mathbf{q}) = \operatorname*{Max}_{\mathbf{r}\in\Psi} L(h^{-1}(\mathbf{r})) \equiv \operatorname*{Max}_{\mathbf{r}\in\Psi} L_{\mathbf{r}}(\mathbf{r})$$

(we note in passing that the explicit form of $L_{\mathbf{r}}$ is obtained by simply setting $h^{-1}(\mathbf{r})$ in the original likelihood function L; the differential elements must not be included because we transform the parameters and not the variables).

So, for instance, the fact that S^2 is the MLE of σ^2 in a normal model with known mean and unknown variance tells us that $S = \sqrt{\{n^{-1} \sum (X_i - \mu)^2\}}$ is the MLE of the standard deviation σ. A useful consequence of Proposition 5.13, moreover, is that some problems can be cast in a simpler form by an appropriate change of parameters; in these cases we can solve the simpler problem – thus finding the ML estimates $\hat{\mathbf{r}} = (\hat{r}_1, \ldots, \hat{r}_k)$ – and then determine $\mathbf{q} = (\theta_1, \ldots, \theta_k)$ by means of h^{-1}. A nice example of this is given in Ref. [19]

(Chapter 2, Example 2.22) where a bivariate normal model (see eqs (3.61a) and (3.61b)) is considered and the ML estimates of $\sigma^2 = \theta_1$ and $\rho = \theta_2$ are determined by introducing the new parameters $r_1 = -[2\sigma^2(1-\rho^2)]^{-1}$ and $r_2 = \rho[\sigma^2(1-\rho^2)]^{-1}$. Then, the desired result

$$\begin{aligned}\hat{\sigma}^2 &= (2n)^{-1}\sum_i \left(x_i^2 + y_i^2\right) \\ \hat{\rho} &= 2\sum_i x_i y_i \Big/ \sum_i \left(x_i^2 + y_i^2\right)\end{aligned} \tag{5.62}$$

is obtained with a noteworthy simplification of the calculations.

A final remark on Proposition 5.13: some authors speak of 'invariance' property. This term, however, would imply that the MLEs remain unchanged; since, in fact, they do change according to the transformation law h, we think that the term 'covariance' should be preferred.

Other properties concern the relation between MLEs, efficient estimators and sufficient statistics, stated by the following two results, respectively.

Proposition 5.14 *If a MVUE $T(\mathbf{X})$ of θ exists, then $T(\mathbf{X}) = \hat{T}(\mathbf{X})$.*

For regular problems, in fact, if a MVUE of θ exists it satisfies eq. (5.41). This, together with the likelihood equation (5.61) yields the desired result.

Proposition 5.15 *If $T(\mathbf{X})$ is a sufficient statistic for θ and the MLE $\hat{T}(\mathbf{X})$ of θ exists and is unique, then $\hat{T}$ is a function of T.*

The proof is almost immediate: since T is sufficient, Neyman's factorization (5.52) holds and maximizing L is equivalent to maximizing g which, in turn, depends on T. Consequently, the MLE itself will be a function of T.

Before turning to the asymptotic properties of MLEs – which will be the subject of the next section – we point out two facts and state without proof an interesting result worthy of mention. First, MLEs, although asymptotically unbiased (see the following section), are often biased. Second, the ML method can be used in cases more general than the one considered here, that is, independent drawings from a fixed distribution. For instance, the example (taken from Ref. [19]) on parameter transformation and mentioned above is a case in which, in fact, independence does not hold.

Finally, the following proposition [30] provides an interesting characterization of some probability distributions based on a ML estimate:

Proposition 5.16 *For $n \geq 3$, let $\mathbf{X}$ be a sample from a continuous population with pdf of the form $f(x-\theta)$ and let $X_{(1)} \leq X_{(2)} \leq \cdots \leq X_{(n)}$ be the corresponding order statistics. If $\hat{T} = \sum_i a_i X_{(i)}$ with $a_i \geq 0$ and $a_1 + \cdots + a_n = 1$ is the MLE of θ, then*

(a) *if* $a_1 = \cdots = a_n = 1/n$, *then* f *is a normal density;*
(b) *if* $a_1 + a_n = 1; a_1 a_n > 0$, *then* f *is a uniform density;*
(c) *if* $a_j + a_{j+1} = 1; a_j a_{j+1} > 0$ *with* $j \in \{1, 2, \ldots, n-1\}$, *then* f *is a Laplace density.*

In regard to point (c), we call a Laplace r.v. a continuous r.v. X whose pdf is

$$f_X(x) = \frac{1}{2\beta} \exp\left(-\frac{|x-\alpha|}{\beta}\right) \tag{5.63}$$

where $x \in \mathbb{R}$ and the two parameters are such that $\alpha \in \mathbb{R}; \beta > 0$. Its CF is

$$\varphi_X(u) = \frac{e^{i\alpha u}}{1 + \beta^2 u^2} \tag{5.64}$$

while its mean and variance are, respectively

$$\begin{aligned} E(X) &= \alpha \\ \mathrm{Var}(X) &= 2\beta^2 \end{aligned} \tag{5.65}$$

5.5.1 *Asymptotic properties of ML estimators*

As a matter of fact, some important properties of MLEs are asymptotic in nature. Since their proofs, however, are generally rather lengthy, this section is limited to the statement of the main results. For details, the interested reader can refer to more specialized literature (see, for instance, [3, 17, 19, 26, 28]).

Assuming, as it is often the case, that we are dealing with a regular problem (Section 5.4.1) and that the likelihood function L_n attains its maximum at an interior point of Θ for all n (this, in other words, means that the MLE exists for all n), then:

(1) $\hat{T}_n \to \theta[P]$, that is, the MLE is (weakly) consistent;
(2) the r.v. $\sqrt{n}(\hat{T}_n - \theta)$ converges in distribution to a normal r.v. with zero mean and variance $1/I(\theta)$ or, equivalently, the MLE $\hat{T}_n$ is asymptotically normal with mean θ and variance given by the Cramer–Rao limit $\{nI(\theta)\}^{-1}$ (eq. (5.39)).

Although we do not provide the proofs of the above statements, some comments are not out of place. First, it should be noted that result (1) can be strengthened and strong consistency (in the sense of a.s. convergence) can be proven (see, for instance, [1]). Second, we have noted in the preceding section that the ML method does not always lead to unbiased estimators;

they are, however, asymptotically unbiased because the bias – which, anyway, can generally be removed for finite values of n – tends to zero as n^{-1} when we let $n \to \infty$. Besides the minor inconvenience of bias for finite n, a more important property is given in point (2) in regard to the variance of MLEs. In fact, if we introduce the notion of asymptotic efficiency $\bar{e}_T$ of an estimator T as $e_T = \lim_{n\to\infty} e_T$, then $e_{\hat{T}} = 1$, meaning that MLEs are asymptotically efficient.

Now, this fact does not imply that MLEs are the only asymptotically normal and asymptotically efficient estimators but it has been shown that, in general, MLEs have better efficiency properties for large values of n (Refs. [23, 24]). In regard to this last observation, we note in passing that (2) does not generally imply that $\mathrm{Var}(\hat{T}_n) \to \{nI(\theta)\}^{-1}$ as $n \to \infty$ (D-convergence does not imply convergence of the moments); however, for a large class of asymptotically normal estimators the variance can be expressed as

$$\mathrm{Var}(T) = \frac{1}{nI(\theta)} + \frac{a_2(\theta)}{n^2} + \cdots$$

and the estimator with the minimum $a_2(\theta)$ is to be preferred (second-order efficiency). Quite often it turns out that MLEs are such estimators.

Another remark on result (2) is that cases where the asymptotic variance depends on the unknown parameter are rather common. An appropriate parameter transformation can fix the problem by maintaining, at the same time, asymptotic normality. In fact, if h is a differentiable function with $h' \neq 0$ then it can be shown that the variable $\sqrt{n}\{h(\hat{T}_n) - h(\theta)\}$ is asymptotically normal with zero mean and variance $[h'(\theta)]^2/I(\theta)$. Enforcing the condition that this new variance equals a constant – say b^2 – we get $h'(\theta) = b\sqrt{I(\theta)}$ and therefore

$$h(\theta) = a + b\int \sqrt{I(\theta)}\,\mathrm{d}\theta \tag{5.66}$$

where both constants a, b can be chosen so that $h(\theta)$ is in simple form.

Example 5.8(a) Consider a sample from a Poisson variable (eq. (4.1)) with unknown parameter $\lambda = \theta$. Since $\partial^2 f/\partial\theta^2 = -x/\theta^2$ then

$$I(\theta) = E(x/\theta^2) = \frac{1}{\theta^2}\sum_x x\frac{\theta^x \mathrm{e}^{-\theta}}{x!} = \frac{1}{\theta} \tag{5.67}$$

because the sum – being the mean of the parent r.v. X – equals θ. It follows from (5.67) that the Cramer–Rao limit is θ/n. On the other hand, the MLE of θ is obtained by taking the logarithm of eq. (5.53) and equating its derivative to zero; the reader can easily check that the result is

$$\hat{T}_n = M = \frac{1}{n}\sum_i X_i \tag{5.68}$$

whose variance is θ/n (eq. (5.10), taking into account that $\mathrm{Var}(X) = \theta$). So, as expected, the MLE is consistent and in this case it is also efficient because (eq. (5.43)) $e_{\hat{T}} = 1$. Moreover, from result (2) we know that $\sqrt{n}(\hat{T}_n - \theta)$ is asymptotically normal with zero mean and a variance which depends on the parameter, that is, $1/I(\theta) = \theta$. Setting $a = 0$ and $b = 1$ in eq. (5.66) we get $h(\theta) = 2\sqrt{\theta}$ so that the new variable $2\sqrt{n}(\sqrt{\hat{T}_n} - \sqrt{\theta})$ is asymptotically standard-normal, that is, with zero mean and unit variance. Alternatively, setting $a = 0$ and $b = 1/2$ we have that $Y_n = \sqrt{n}(\sqrt{\hat{T}_n} - \sqrt{\theta})$ is asymptotically normal with zero mean and $\mathrm{Var}(Y_n) = 1/4$.

Example 5.8(b) When the model is non-regular, asymptotic normality may not hold. As an example, in the uniform model $f(x;\theta) = 1/\theta$ for $0 \le x \le \theta$ (and zero otherwise) the likelihood function is

$$L(\mathbf{x};\theta) = \begin{cases} 1/\theta^n, & x_{(n)} \equiv \max\limits_{1\le i\le n} x_i \le \theta \\ 0, & \text{otherwise} \end{cases}$$

and $T = X_{(n)}$ – where $X_{(n)}$ is the nth order statistic – is a sufficient statistic for θ. Also, the likelihood function is monotone decreasing for $\theta \ge x_{(n)}$ and therefore it attains its maximum at $\theta = x_{(n)}$ where, however, there is a discontinuity. So, even if we can call $T = X_{(n)}$ the MLE of θ, this is not a solution of the likelihood equation (5.61) and we may not expect property (2) to hold. In fact, we already know from Section 5.3.1 that the extreme value of the sample $X_{(n)}$ is not asymptotically normal.

The above results still hold in the case of several parameters. Explicitly, referring to the considerations at the end of Section 5.4.1, property (2) becomes

(2′) the r.v. $\sqrt{n}(\hat{\mathbf{T}}_n - \mathbf{q})$ is asymptotically normal with zero mean and variance $\{\mathbf{I}(\mathbf{q})\}^{-1}$
or, in case we are estimating a scalar function $\tau(\mathbf{q}) = \tau(\theta_1, \ldots, \theta_k)$ of the unknown parameters:

(2″) the r.v. $\sqrt{n}\{\hat{T}_n - \tau(\mathbf{q})\}$ is asymptotically normal with zero mean and variance $\mathbf{d}^T\mathbf{I}^{-1}\mathbf{d}$, where $\mathbf{d}(\mathbf{q}) = (\partial\tau/\partial\theta_1, \ldots, \partial\tau/\partial\theta_k)^{\mathrm{T}}$.

5.6 Interval estimation

Within the framework of the statistical model (5.1), we have discussed in the preceding sections the subject of 'point estimation' which, in essence, consists in (a) finding a 'good' estimator $T(\mathbf{X})$ of the unknown parameter θ and (b) using the data from the experiment – that is, the realization of the sample $\mathbf{x}$ – to calculate the numerical value $t = T(\mathbf{x})$. Then, on the basis of a number of considerations on what is meant by 'good', we expect t to be a reliable estimate of θ (broadly speaking, we could call it our educated 'best-guess' on the true value of θ).

The procedure above is well justified if the main question of the estimation problem is 'what value should I use for θ?'. If, however, one is more interested in specifying a range of values within which he/she can confidently expect θ to lie, then the method of 'interval estimation' provides a better way to tackle the problem. In perspective, moreover, one should consider that a point estimate is almost meaningless without a statement of its 'reliability'.

So, still keeping the model (5.1) as our starting point, we now wish to determine an interval which contains the true value of θ – though unknown – at a specified 'confidence level' (CL for short) $\gamma = 1-\alpha$ $(0 < \gamma < 1)$. This, in other words, means that we have to find two statistics T_1, T_2, with $T_1 < T_2$, such that

$$P_\theta\{T_1(\mathbf{X}) < \theta < T_2(\mathbf{X})\} = \gamma \tag{5.69a}$$

for all $\theta \in \Theta$. In this case we call (T_1, T_2) a γ-confidence interval (often γ-CI) for θ and T_1, T_2, respectively, the lower and upper confidence limits. Note that eq. (5.69a) defines a random interval which, on the one hand, depends on the sample $\mathbf{X}$ but, on the other hand, does not depend on θ (because both limits are statistics).

By carrying out an experiment we obtain a realization of the sample $\mathbf{x}$ and, accordingly, the values $t_1 = T_1(\mathbf{x})$ and $t_2 = T_2(\mathbf{x})$ for the two statistics; the interval (t_1, t_2) is then an estimate of the γ-CI. At this point, one could be tempted to say that θ belongs to (t_1, t_2) with a probability γ. This statement, however, is wrong because (t_1, t_2) is not a random interval and therefore the true value of θ either belongs to it or it does not. The correct interpretation must be given in terms of relative frequency of success: if the experiment is repeated many times – thus obtaining many estimates of (T_1, T_2) – the resulting estimated intervals will contain the true value of θ in $100\gamma\%$ of the cases. Conversely, in the long run we will be wrong in $100\alpha\%$ of the cases. This, in essence, is the meaning of the term 'confidence' in this context.

Now, before showing how to determine the confidence limits, some additional remarks on eq. (5.69a) are in order:

(i) If the population under study is discrete it may not be possible to meet condition (5.69a) exactly; in this case we call γ-CL the smallest interval such that

$$P_\theta\{T_1(\mathbf{X}) < \theta < T_2(\mathbf{X})\} \geq \gamma \tag{5.69b}$$

for all $\theta \in \Theta$.

(ii) The statistic $D_\gamma(\mathbf{X}) = T_2 - T_1$ is the length of the CI. This quantity can be considered as a measure of precision of our estimate: given, say, two methods of interval estimation and a CL γ, the method leading to the smaller D_γ is to be preferred. Whichever the adopted method, however, it is reasonable to expect that there must be a relation between D and γ because – for a fixed sample size n – a higher confidence level (or, equivalently, a lower α) is paid at the price of a larger interval. In fact, choosing an unreasonably high value of γ generally leads to a CI which is too large to be of any practical use (and consequently to almost no information on θ). If we want a high CL and an interval of acceptable length we can, of course, increase the sample size. Since this operation is generally costly, it is evident that any procedure of interval estimation implicitly implies a compromise between confidence level, interval length and sample size.

(iii) Equation (5.69) defines a two-sided interval but in some applications one-sided intervals are required; these intervals have the form $(-\infty, T_2)$ or (T_1, ∞).

(iv) In case of several unknown parameters, the CI for an individual component, say θ_i, is still given by (5.69) and the same applies in case of a scalar function $\tau(\mathbf{q})$ of the unknown parameter(s). Clearly, θ is replaced by θ_i in the former case and by $\tau(\mathbf{q})$ in the latter. More specifically, a γ-confidence region for the vector parameter $\mathbf{q} = (\theta_1, \ldots, \theta_k)$ is a random subset $C_\gamma(\mathbf{X}) \subset \Theta$ such that for all $\mathbf{q} \in \Theta$ we have

$$P_\mathbf{q}\{\mathbf{q} \in C_\gamma(\mathbf{X})\} \geq \gamma \tag{5.69c}$$

The general technique used to determine confidence intervals is based on the search of a so-called pivot quantity. This is a r.v. of the form $G(\mathbf{X};\theta)$ – that is, it depends on the sample and on the unknown parameter and therefore it is not a statistic – such that (1) its distribution f_G does not depend on θ and (2) for every $\mathbf{x}$ the function $G(\mathbf{x};\theta)$ is continuous and strictly monotone in θ.

Then, given $\gamma \in (0,1)$ there are many ways in which we can choose $g_1 < g_2$ so that the relation

$$P_\theta\{g_1 < G(\mathbf{X};\theta) < g_2\} = \int_{g_1}^{g_2} f_G(g)\,\mathrm{d}g = \gamma \tag{5.70}$$

holds. If, for every $\mathbf{x}$, we define $T_1(\mathbf{x})$ and $T_2(\mathbf{x})$ – with $T_1 < T_2$ – as the solutions (with respect to θ) of the equations $G(\mathbf{x};\theta) = g_1$ and $G(\mathbf{x};\theta) = g_2$, respectively, eq. (5.70) is equivalent to eq. (5.69). Note that T_1, T_2 are well-defined because they are obtained by means of the inverse (with respect to θ) function G^{-1}, which, in turn, is well-defined by virtue of condition (2). So, if G is monotonically increasing then $T_1(\mathbf{x}) = G^{-1}(\mathbf{x};g_1)$ and $T_2(\mathbf{x}) = G^{-1}(\mathbf{x};g_2)$ while, on the other hand, $T_1(\mathbf{x}) = G^{-1}(\mathbf{x};g_2)$ and $T_2(\mathbf{x}) = G^{-1}(\mathbf{x};g_1)$ if G is monotonically decreasing.

The question at this point is how to construct a pivot quantity. A number of useful results given in Appendix C will be of help in this task (see also the following examples) but here we outline a general procedure. Suppose we are dealing with an absolutely continuous model; it can be shown that if the parent r.v. X has a PDF $F_X(x;\theta)$ which is continuous and strictly monotone in θ then

$$G(\mathbf{X};\theta) = -\sum_{i=1}^{n} \ln F(X_i;\theta) \tag{5.71}$$

is a pivot quantity for the interval estimation of θ. The proof, which we only outline here, is based on the fact if X has a continuous and monotonically increasing PDF $F(x)$ then the chain of equalities

$$F_Y(y) = P(Y \le y) = P\{F(X) \le y\} = P\{X \le F^{-1}(y)\} = F[F^{-1}(y)] = y$$

shows that the r.v. $Y \equiv F(X)$ has a uniform distribution on the interval $(0,1)$. Consequently, each r.v. $F(X_i;\theta)$ in (5.71) is uniformly distributed on $(0,1)$, $-\ln F(X_i;\theta)$ has a $\Gamma(1,1)$ distribution and $G(\mathbf{X};\theta)$ has a $\Gamma(1,n)$ pdf, that is,

$$f_G(g) = \frac{g^{n-1}e^{-g}}{\Gamma(n)} \tag{5.72}$$

which does not depend on θ. Since $G(\mathbf{X};\theta)$ is evidently continuous and monotone in θ, it follows that it is a pivot quantity. So, by taking (5.72) into account and choosing g_1, g_2 such that eq. (5.70) holds, the solutions of the equations $-\sum \ln F(x_i;\theta) = g_1$ and $-\sum \ln F(x_i;\theta) = g_2$ give the desired confidence interval. This last step, in practice, is often the most difficult part.

Before giving some examples, we mention the following useful result (whose proof is immediate):

Proposition 5.17 *If (T_1, T_2) is a γ-CI for θ and h is a strictly monotone function, then $h(T_1)$ and $h(T_2)$ are the limits of the γ-CI for $h(\theta)$. The interval is $(h(T_1), h(T_2))$ if h is monotonically increasing and $(h(T_2), h(T_1))$ if h is monotonically decreasing.*

Example 5.9(a) Let **X** be a sample from a normal population with unknown mean $\mu = \theta$ and known variance. In this case only a small effort is required to see that the r.v. $G = \sqrt{n}(M - \theta)/\sigma$ is a pivot quantity (condition (1) above follows from the fact that $G \approx N(0,1)$ – see Section 5.3.1, Proposition 5.1(b) – and therefore its pdf does not depend on θ). Consequently, our γ-CI has the form

$$(T_1, T_2) = \left(M - \frac{g_2\sigma}{\sqrt{n}}, M - \frac{g_1\sigma}{\sqrt{n}}\right) \tag{5.73}$$

where g_1, g_2 are any two numbers such that $g_1 < g_2$ and $\Phi(g_2) - \Phi(g_1) = \gamma$ (where $\Phi(x) = (\sqrt{2\pi})^{-1}\int_{-\infty}^{x} e^{-t^2/2}\,dt$ is the PDF of a standard normal r.v.). The shortest interval can be obtained by minimizing the function

$$D_\gamma(g_1, g_2) = \frac{\sigma}{\sqrt{n}}(g_2 - g_1) \tag{5.74}$$

under the constraint $\Phi(g_2) - \Phi(g_1) = \gamma$ (we note in passing that this is a rather rare case where D_γ does not depend on **X**). Using the well-known method of Lagrange undeterminate multipliers and taking into account that the standard normal pdf is an even function we get $g_1 = -g_2$. Then, since $\Phi(-x) = 1 - \Phi(x)$ it follows that $\Phi(g_1) = (1-\gamma)/2 = 1 - \Phi(g_2)$ and $\Phi(g_2) = (1+\gamma)/2$. By calling $c_{(1+\gamma)/2}$ the $(1+\gamma)/2$-quantile of the standard normal distribution, that is, $c_{(1+\gamma)/2} = \Phi^{-1}[(1+\gamma)/2]$ (this, in other words, is that particular value of g_2 that minimizes the interval length) the desired γ-CI for the mean is

$$(T_1, T_2) = \left(M - c_{(1+\gamma)/2}\frac{\sigma}{\sqrt{n}}, M + c_{(1+\gamma)/2}\frac{\sigma}{\sqrt{n}}\right) \tag{5.75a}$$

where the values of $c_{(1+\gamma)/2}$ can be found in statistical tables. The interval length is in this case

$$D_\gamma = 2c_{(1+\gamma)/2}\frac{\sigma}{\sqrt{n}} \tag{5.75b}$$

So, for instance, if $\gamma = 0.95$ then $(1+\gamma)/2 = 0.975$ and we find $c_{0.975} = 1.960$ while at a higher confidence level, say $\gamma = 0.99$, we get $(1+\gamma)/2 = 0.995$ and $c_{0.995} = 2.576$. As noted in point (ii) eq. (5.75b) shows that a higher CL, for a given sample size n, is paid at the price of a longer interval; for a given confidence level, on the other hand, the interval length can only be reduced by increasing n.

Suppose now that we had used the median Z instead of M. We have pointed out at the end of Section 5.3.1 that Z is asymptotically normal with mean μ and standard deviation $\sigma\sqrt{\pi/2n}$, that is, $r = \sqrt{\pi/2}$ times the standard deviation of M. If, just for the sake of the argument, we suppose that the error of the approximation can be neglected (in other words, we pretend that the distribution of Z is exactly normal) we get the γ-CI $(Z \pm c_{(1+\gamma)/2} r\sigma/2\sqrt{n})$, which is longer than (5.75) although the risk of error is the same.

Example 5.9(b) Consider now the (more frequent) case in which the variance is not known. Since $\bar{S}^2$ (eq. (5.31)) is an unbiased estimator of σ^2 we may think of using $G = \sqrt{n}(M - \theta)/\bar{S}$ as a pivot quantity. In this case, however, it can be shown that $G \approx St(n-1)$ and therefore the quantiles of the Student distribution (with $n-1$ degrees of freedom) will have to be used in specifying our confidence interval for the mean. The symmetry of the distribution suggests that we can parallel the considerations above on g_1, g_2 and arrive at the CI

$$(T_1, T_2) = \left(M - t_{(1+\gamma)/2;n-1} \frac{\bar{S}}{\sqrt{n}}, M + t_{(1+\gamma)/2;n-1} \frac{\bar{S}}{\sqrt{n}} \right) \tag{5.76}$$

where, denoting by $S_{(n-1)}$ the Student PDF with $n-1$ degrees of freedom, we have $t_{(1+\gamma)/2;n-1} = S_{(n-1)}^{-1}[(1+\gamma)/2]$. The values of these quantiles are also easily found on statistical tables for ν (the number of degrees of freedom) up to 40–50. Tables for higher values of ν are not given because $St(\nu) \to N(0,1)$ as $\nu \to \infty$ and the normal approximation is already rather good for $\nu \geq 30$.

Note that now D_γ depends on the sample (through $\bar{S}$) and therefore the interval length is a r.v. which can only be determined after we have carried out our experiment. Nonetheless, also in this case we expect the considerations of point (ii) to hold.

As a numerical example of cases (a) and (b) suppose that we test 20 similar products and obtain an average weight of $M = 100.2$ g. If we know that the population standard deviation is, say, $\sigma = 4$ g, the 95%-CI for M is (eq. (5.75))

$$\left(100.2 - 1.96 \frac{4}{\sqrt{20}}, 100.2 + 1.96 \frac{4}{\sqrt{20}} \right) = (98.45, 101.95)$$

If, on the other hand, we make no assumptions on the variance and calculate it from the data obtaining, say, $\bar{s} = 3.80$ g, we use eq. (5.76) to get

$$\left(100.2 - 2.093 \frac{3.8}{\sqrt{20}}, 100.2 + 2.093 \frac{3.8}{\sqrt{20}} \right) = (98.42, 101.98)$$

because for $\gamma = 0.95$, $(1+\gamma)/2 = 0.975$ and we find from the tables (for $\nu = 19$) the quantile $t_{0.975;19} = 2.093$. Note that the second interval is larger than the first even if the estimated standard deviation is smaller than the true σ. This situation may occur in practice because in the second case the uncertainty on the standard deviation also plays a part. Moreover, if we carried out another experiment on other 20 items giving, by chance, $M = 100.2$, the first interval would not change while the second will because of the new estimate of $\bar{S}$.

A further consideration on example (a) is that eq. (5.75b) gives us the possibility to determine the minimum sample size needed to achieve a specified 'precision' of our estimate at a given CL. In fact, if the 'precision' is measured by D_γ, there may be cases in which we do not want our CI to exceed a given length L. This condition is expressed by the relation $2c_{(1+\gamma)/2}\sigma/\sqrt{n} \le L$ which can be solved for n to give

$$n \ge \left(\frac{2c_{(1+\gamma)/2}\sigma}{L}\right)^2 \tag{5.77}$$

Example 5.10(a) Suppose that we are still dealing with a normal model; now, however, we know the mean μ and the variance is unknown. Setting $\theta = \sigma$, this means that we are looking for a CI for the function $\tau(\theta) = \theta^2$. It is not difficult to see that

$$G(\mathbf{X};\theta) = \frac{1}{\theta^2}\sum_{i=1}^{n}(X_i - \mu)^2 \tag{5.78}$$

is a pivot quantity. Now, since $(X_i - \mu)/\theta \approx N(0,1)$ it is known (Appendix C) that $(X_i - \mu)^2/\theta^2 \approx \chi^2(1)$ from which it follows that $G(\mathbf{X};\theta) \approx \chi^2(n)$ by the reproducibility property of the χ^2 distribution. Solving the equations $G(\mathbf{x};\theta) = g_1$ and $G(\mathbf{x};\theta) = g_2$ we get a CI of the form

$$(T_1(\mathbf{X}), T_2(\mathbf{X})) = \left(g_2^{-1}\sum_i (X_i - \mu)^2, g_1^{-1}\sum_i (X_i - \mu)^2\right) \tag{5.79}$$

where – denoting by $K_n(x)$ the PDF of the distribution $\chi^2(n)$ – g_1, g_2 must satisfy the condition $K_n(g_2) - K_n(g_1) = \gamma$. A common choice is to select a so-called 'central' interval, that is, to choose g_1, g_2 as the $(1 \mp \gamma)/2$ quantiles of $\chi^2(n)$, respectively. This gives

$$\begin{aligned} g_1 &= K_n^{-1}[((1-\gamma)/2] = \chi^2_{(1-\gamma)/2;n} \\ g_2 &= K_n^{-1}[((1+\gamma)/2] = \chi^2_{(1+\gamma)/2;n} \end{aligned} \tag{5.80}$$

so that the CI (5.79) is explicitly written as

$$(T_1, T_2) = \left(\frac{nS^2}{\chi^2_{(1+\gamma)/2;n}}, \frac{nS^2}{\chi^2_{(1-\gamma)/2;n}}\right) \tag{5.81}$$

and the values of the quantiles can be found on statistical tables. So, for instance, if we are looking for a 95%-CI and $n = 20$, then $(1-\gamma)/2 = 0.025$ and $(1+\gamma)/2 = 0.975$. Since on tables of χ^2 quantiles we find $\chi^2_{0.025;20} = 9.59$ and $\chi^2_{0.975;20} = 34.17$, our interval is $(0.59S^2, 2.09S^2)$.

Two remarks on this example: first, it is a direct consequence of Proposition 5.17 that the interval $(\sqrt{T_1}, \sqrt{T_2})$ – where T_1, T_2 are as in (5.81) – is a γ-CI for the standard deviation σ. Second, using Lagrange's method one can determine that the estimated interval (5.81) is not optimal, that is, is not the shortest one. A quantitative evaluation, however, is not immediate and requires a numerical solution. For a 95%-CI it can be shown that the shortest interval involves two quantities α_1, α_2 such that $\alpha_1 + \alpha_2 = 1-\gamma$ and the corresponding quantiles are 9.96 and 35.23 (instead of 9.59 and 34.17).

Example 5.10(b) If, as it often happens, also the mean of the population is not known, a pivot quantity is given by (5.78) by simply substituting M in place of μ, that is, $G(\mathbf{X};\theta) = (n-1)\bar{S}^2/\theta^2 = (n-1)\bar{S}^2/\tau$ (where, as above, $\tau(\theta) = \theta^2$). In this case $G(\mathbf{X};\theta) \approx \chi^2(n-1)$ and we get the CI for the variance

$$(T_1, T_2) = \left(\frac{n-1}{\chi^2_{(1+\gamma)/2;n-1}} \bar{S}^2, \frac{n-1}{\chi^2_{(1-\gamma)/2;n-1}} \bar{S}^2 \right) \tag{5.82}$$

so that, for instance, for $n = 20$ and $\gamma = 0.95$ we find in tables the two quantiles $\chi^2_{(1+\gamma)/2;n-1} = \chi^2_{0.975;19} = 32.85$ and $\chi^2_{(1-\gamma)/2;n} = \chi^2_{0.025;19} = 8.907$. As above, the central CI (5.81) is not the shortest interval but it is the most frequently used in practice. If, at this point we also want a CI for the mean, we proceed exactly as in Example 5.9(b) thus obtaining the interval (5.76) which – owing to the symmetry of the Student distribution – is the shortest among all intervals of the form $(M - a_1\bar{S}, M + a_2\bar{S})$.

Example 5.11(a) From the preceding examples it appears that the determination of CIs for the (unknown) mean of a normal population involves (i) standardized normal quantiles if the variance is known or (ii) Student quantiles – with the appropriate number of degrees of freedom – if the variance is not known. Provided that collective independence of the r.v.s involved in the estimation problem applies, this is a general fact. Suppose in fact, that we want to find a CI for the difference $\mu_1 - \mu_2$ where $\mu_1 = \theta_1, \mu_2 = \theta_2$ are the means of two normal populations with variances σ_1^2, σ_2^2, respectively. Also, let $\mathbf{X} = (X_1, \ldots, X_n)$ and $\mathbf{Y} = (Y_1, \ldots, Y_m)$ be the samples taken from the two populations and M_1, M_2 the two sample means.

If the variances are known then we can exploit the fact that

$$G = \frac{M_1 - M_2 - (\theta_1 - \theta_2)}{\sqrt{n^{-1}\sigma_1^2 + m^{-1}\sigma_2^2}} \approx N(0,1) \tag{5.83}$$

and therefore G is a pivot quantity. Proceeding exactly as in Example 5.9(a) we obtain the CI

$$\left(M_1 - M_2 \pm c_{(1+\gamma)/2}\sqrt{\frac{\sigma_1^2}{n} + \frac{\sigma_2^2}{m}} \right) \tag{5.84}$$

If the variances are not known we use the estimators $\bar{S}_1^2, \bar{S}_2^2$ (or S_1^2, S_2^2) instead of the population variances. Using these estimators, it is convenient to introduce the 'pooled' variance

$$S_p^2 = \frac{(n-1)\bar{S}_1^2 + (m-1)\bar{S}_2^2}{n+m-2} = \frac{nS_1^2 + mS_2^2}{n+m-2} \tag{5.85}$$

because it can be shown (Appendix C) that the r.v.

$$G = \frac{M_1 - M_2 - (\theta_1 - \theta_2)}{S_p\sqrt{n^{-1} + m^{-1}}} \tag{5.86}$$

is distributed as a Student variable with $n+m-2$ degrees of freedom. This is our pivot quantity for the case at hand and we can parallel Example 5.9(b) to get the CI

$$\begin{aligned} &\left(M_1 - M_2 \pm t_{(1+\gamma)/2;n+m-2} S_p \sqrt{n^{-1} + m^{-1}} \right) \\ &\quad = \left(M_1 - M_2 \pm t_{(1+\gamma)/2;n+m-2} \sqrt{\frac{m+n}{mn(m+n-2)}\left(nS_1^2 + mS_2^2\right)} \right) \end{aligned} \tag{5.87}$$

where the second expression has been written in terms of the sample variances S_1^2, S_2^2.

Example 5.11(b) As above, let $\mathbf{X} = (X_1, \ldots, X_n)$ and $\mathbf{Y} = (Y_1, \ldots, Y_m)$ be independent samples from normal populations with unknown variances $\sigma_1^2 = \theta_1^2, \sigma_2^2 = \theta_2^2$, respectively. Now we wish to determine a CI for the ratio $\tau(\theta_1, \theta_2) = \theta_1^2/\theta_2^2$. The pivot quantity for this problem is obtained by noting that (Appendix C) $Z_1 = (n-1)\bar{S}_1^2/\sigma_1^2 \approx \chi^2(n-1)$ and $Z_2 = (m-1)\bar{S}_2^2/\sigma_2^2 \approx \chi^2(m-1)$ so that the r.v.

$$G = \frac{\bar{S}_1^2/\theta_1^2}{\bar{S}_2^2/\theta_2^2} = \frac{1}{\tau}\left(\frac{\bar{S}_1^2}{\bar{S}_2^2}\right) \tag{5.88}$$

has a Fisher distribution with $n-1$ and $m-1$ degrees of freedom. Solving (5.88) for τ we get an interval of the form

$$\left(g_2^{-1}\frac{\bar{S}_1^2}{\bar{S}_2^2}, g_1^{-1}\frac{\bar{S}_1^2}{\bar{S}_2^2}\right) \tag{5.89a}$$

so that denoting by $F_{(1-\gamma)/2;n-1,m-1}$ and $F_{(1+\gamma)/2;n-1,m-1}$, respectively, the $(1-\gamma)/2$ and $(1+\gamma)/2$ quantiles of the distribution $\mathrm{Fsh}(n-1,m-1)$ the desired CI for the variance ratio is

$$\begin{aligned}&\left(\frac{\bar{S}_1^2/\bar{S}_2^2}{F_{(1+\gamma)/2;n-1,m-1}}, \frac{\bar{S}_1^2/\bar{S}_2^2}{F_{(1-\gamma)/2;n-1,m-1}}\right)\\ &\quad=\left(\frac{\bar{S}_1^2/\bar{S}_2^2}{F_{(1+\gamma)/2;n-1,m-1}}, F_{(1+\gamma)/2;m-1,n-1}\frac{\bar{S}_1^2}{\bar{S}_2^2}\right)\end{aligned} \tag{5.89b}$$

where in the second expression we took into account the property $F_{(1-\gamma)/2;n-1,m-1} = \{F_{(1+\gamma)/2;m-1,n-1}\}^{-1}$. So, for instance, if $\gamma = 0.90$, $n = 20$ and $m = 15$ we find $F_{0.95;19,14} = 2.40$ and $F_{0.95;14,19} = 2.26$ and our interval is $(0.417\bar{S}_1^2/\bar{S}_2^2, 2.26/\bar{S}_1^2/\bar{S}_2^2)$.

Example 5.11(c) As an example of a non-normal model, consider a sample taken from an exponential population with unknown mean (i.e. the statistical model is expressed in terms of the pdfs $f(x;\theta) = \theta^{-1}\mathrm{e}^{-x/\theta}$). Now, since $X_i \approx \mathrm{Exp}(\theta)$ it follows that (Appendix C) $2X_i/\theta \approx \mathrm{Exp}(2) = \chi^2(2)$ and therefore $G = 2\theta^{-1}\sum_i X_i \approx \chi^2(2n)$. It is left to the reader to fill in the easy details and arrive at the central CI

$$\left(\frac{2\sum X_i}{\chi^2_{(1+\gamma)/2;2n}}, \frac{2\sum X_i}{\chi^2_{(1-\gamma)/2;2n}}\right) = \left(\frac{2nM}{\chi^2_{(1+\gamma)/2;2n}}, \frac{2nM}{\chi^2_{(1-\gamma)/2;2n}}\right) \tag{5.90}$$

As a numerical example, let $\gamma = 0.90$ and $n = 10$. We find $\chi^2_{(1+\gamma)/2;2n} = \chi^2_{0.95;20} = 31.41$ and $\chi^2_{(1-\gamma)/2;2n} = \chi^2_{0.05;20} = 10.85$; consequently $(0.64M, 1.84M)$.

At this point a remark on notation is in order: whenever we have spoken of quantiles we meant lower quantiles. Some statistical tables report lower quantiles, but some other tables do not. In other words, if F_G is the PDF under consideration (Gaussian, Student, χ^2, Fisher, or else, depending on the problem) and f_G its density, our convention so far is that

$$\begin{aligned} F_G(g_1) &= \int_{-\infty}^{g_1} f_G(g)\,dg = (1-\gamma)/2 = \alpha/2 \\ F_G(g_2) &= \int_{-\infty}^{g_2} f_G(g)\,dg = (1+\gamma)/2 = 1-\alpha/2 \end{aligned} \tag{5.91}$$

(we recall that $\gamma = 1 - \alpha$ by definition) so that the area under the pdf to the left of g_1 equals $\alpha/2$ and we can say, equivalently, that g_1 is the $\alpha/2$-lower quantile or, as we did, the $(1 - \gamma)/2$-lower quantile. Similarly, g_2 is the $(1 - \alpha/2)$-lower quantile or, equivalently, the $(1 + \gamma)/2$-lower quantile. In fact, for instance, one often finds – e.g. see [25] – the interval (5.81) written as $(nS^2/\chi^2_{1-\alpha/2;n}, nS^2/\chi^2_{\alpha/2;n})$.

From the first of eq. (5.91), however, it follows that the area to the right of g_1 is $1 - \alpha/2$, that is, $P(G > g_1) = \int_{g_1}^{\infty} f_G\,dg = 1 - \alpha/2$. Since the value of the area to the right of a given point is used to define the so-called 'upper quantile' of a distribution, the other convention sees g_1 is the upper $(1 - \alpha/2)$-upper quantile. By the same token, g_2 is the upper $\alpha/2$-upper quantile. Obviously, nothing changes for the degrees of freedom.

So, for instance, one can find eq. (5.81) written in terms of upper quantiles as $(nS^2/\chi_{\alpha/2;n}, nS^2/\chi_{1-\alpha/2;n})$ and now, if we look for a 95%-CI with, say, $n = 20$, we find (see, for instance, Table 4 on [4] or Table C in Appendix II of [7]) $\chi^2_{\alpha/2;n} = \chi^2_{0.025;20} = 34.17$ and $\chi^2_{1-\alpha/2;n} = \chi^2_{0.975;20} = 9.59$. Obviously, the resulting interval is the same as above. In the following, in order to avoid confusion, we will explicitly state which type of quantile we are using; it must be the analyst's care to check the tables at his/her disposal.

Besides this observation on symbolism, it may also be worth spending a few words on some other interesting aspects of interval estimation. We start with the vector parameter case, which was briefly mentioned in remark (iv) at the beginning of this section.

The general technique used to construct confidence regions is based on the fact that eq. (5.69c) is equivalent to

$$P_{\mathbf{q}}\{\mathbf{X} \in H(\mathbf{q})\} \ge \gamma \tag{5.92}$$

where, for every $\mathbf{q} \in \Theta$, the set $H(\mathbf{q})$ is the subset of the sample space Ξ containing all those realizations $\mathbf{x}$ (i.e. all those values taken on by $\mathbf{X}$) such that the confidence region constructed with these $\mathbf{x}$ will include $\mathbf{q}$. So, the desired confidence region is found by determining the sets $H(\mathbf{q})$ satisfying inequality (5.92). Since, for a given CL, the sets $H(\mathbf{q})$ can be chosen in many ways, the confidence region thus constructed is not unique and the problem remains of finding a 'minimal' confidence region. In practice, one generally finds the sets $H(\mathbf{q})$ with the help of some vector statistic $\mathbf{T}(\mathbf{X})$ with known distribution. As an example, we can reconsider Example 5.10(b) – normal model with unknown mean and variance – where we determined separate CIs for the mean and the variance. If, however, one considers the two-dimensional vector parameter $\mathbf{q} = (\mu, \sigma^2) = (\theta_1, \tau)$, it is wrong to deduce that the rectangle delimited by the intervals (5.76) and (5.82) is a γ-confidence region for $\mathbf{q}$. This is because the pivot quantities used to construct the CIs are related. Since it can be shown [19] that for a normal population the components of the two-dimensional statistic $\mathbf{T} = (M, S^2)$ are independent, we can use the results (5.76) and (5.82) to obtain the set

$$H(\mathbf{q}) = \left\{ \mathbf{x} : \sqrt{n/\tau}\,|m - \theta_1| < a; b' < ns^2/\tau < b'' \right\} \tag{5.93}$$

where $a = t_{(1+\gamma_1)/2;n-1}$, $b' = \chi^2_{(1-\gamma_2)/2;n-1}$ and $b'' = \chi^2_{(1+\gamma_2)/2;n-1}$. Moreover, the quantities γ_1, γ_2 – owing to the independence of M and S^2 – must satisfy the condition $\gamma_1\gamma_2 = \gamma$ in order to have a γ-confidence region. Solving the inequalities which define $H(\mathbf{q})$ we find $\tau > n(m - \theta_1)^2/a$ and $ns^2/b'' < \tau < ns^2/b'$. In the (θ_1, τ)-plane, therefore, the confidence region is the part of the plane bounded by the parabola $\tau = n(m - \theta_1)^2/a$ and the two straight lines $\tau = ns^2/b''$ and $\tau = ns^2/b'$.

Returning now to the one-dimensional case, a second consideration is the answer to the question: given a point estimator $T(\mathbf{X})$ (of θ) with known distribution $F_T(t;\theta)$, can we construct a CI for θ? Intuitively, the answer is yes and, in fact, it is so. Let us assume that $F_T(t;\theta)$ is continuous and monotone in θ. Then, for every value of $\theta \in \Theta$ it is possible to define two numbers $t_1, t_2 (t_1 < t_2)$ such that

$$P_\theta\{t_1 < T(\mathbf{X}) < t_2\} = F_T(t_2;\theta) - F_T(t_1;\theta) = \gamma \tag{5.94}$$

Although they are not random quantities (because they are two realizations of $T(\mathbf{X})$) , t_1, t_2 will be different for different values of θ; consequently, we can write $t_1(\theta), t_2(\theta)$ and note that these two functions will generally be monotonically increasing in θ (if t is any sort of reasonable estimate of θ, it should increase as θ increases). Moreover, in order to uniquely define t_1, t_2 one generally seeks a central interval by choosing them so that

$$\begin{aligned} F_T(t_1;\theta) &= (1-\gamma)/2 \\ F_T(t_2;\theta) &= (1+\gamma)/2 \end{aligned} \tag{5.95}$$

In the (θ, t)-plane we will therefore be able to identify a region bounded by the two functions $t_1(\theta), t_2(\theta)$. This region, by construction, is such that eq. (5.94) holds for any fixed value of $\theta \in \Theta$; but the important point is that for any fixed value of t it defines two values $\theta_1(t), \theta_2(t)$ – that is, the intersection of the horizontal line t with the curves $t_1(\theta), t_2(\theta)$ – such that the interval (θ_1, θ_2), in the long run, will bracket θ in γ% of the cases. This is precisely the notion of confidence interval for θ and therefore $(T_1(\mathbf{X}), T_2(\mathbf{X}))$, where $T_i(\mathbf{X}) = \theta_i(T(\mathbf{X}))$ for $i = 1, 2$, is the desired γ-CI. So, under the assumptions above, we can in practice proceed as follows: given $T(\mathbf{X})$ we obtain the realization $\mathbf{x}$ and consequently the estimate $t = T(\mathbf{x})$; then, solving for θ the equations $F_T(t;\theta) = (1-\gamma)/2$ and $F_T(t;\theta) = (1+\gamma)/2$ we determine the extremes θ_1 and θ_2 of the γ-interval. By so doing, in the long run, we will be wrong $(1-\gamma)\%$ of the times.

We close this section with a final observation on the examples above where, as the reader has probably noticed, we often assume a normal population as the starting statistical model. Although, clearly, the assumption of normality is not always justified in practice, we just point out two facts in its favour: (a) it has been shown that moderate and, sometimes, even significant departures from normality lead to acceptable results in many cases and (b) if we suspect serious departures from normality, there is always the possibility of trying a transformation of the parent r.v. X (see, for instance, Ref. [2]) because $\log(X)$, $\sqrt{X}$ or some other function of it are often more nearly normal.

Nonetheless, it goes without saying that in practical cases it is always advisable to check the basic assumption itself by carrying out a preliminary normality test on the data (this aspect is delayed to Chapter 6).

5.6.1 *Asymptotic confidence intervals*

Consider a point estimator $T_n(\mathbf{X})$ of the unknown parameter θ such that the r.v. $\sqrt{n}(T_n - \theta)$ is asymptotically normal with zero mean and variance $\sigma^2(\theta)$. If $\sigma^2(\theta)$ is a continuous function then it can be shown [19] that $\sqrt{n}(T_n - \theta)/\sigma(T_n) \to N(0,1)$ [D] as $n \to \infty$. Consequently, for all θ we have

$$P_\theta\left(\frac{\sqrt{n}|T_n - \theta|}{\sigma(T_n)} < c\right) \to \Phi(c) - \Phi(-c) = 2\Phi(c) - 1 = \gamma \tag{5.96a}$$

where $c \equiv c_{(1+\gamma)/2}$ is the $(1+\gamma)/2$-quantile of the standard normal distribution introduced in Example 5.9(a) and $\sigma(T_n)$ is the standard deviation of T_n. Since the relation above can be rewritten as

$$P_\theta\left(T_n - c_{(1+\gamma)/2}\frac{\sigma(T_n)}{\sqrt{n}} < \theta < T_n + c_{(1+\gamma)/2}\frac{\sigma(T_n)}{\sqrt{n}}\right) \to \gamma \tag{5.96b}$$

it follows that $(T_n \pm c_{(1+\gamma)/2}\sigma(T_n)/\sqrt{n})$ is an asymptotic γ-CI for θ, where it is evident that the smaller is $\sigma(T_n)$ the shorter is the interval. As a consequence, asymptotically efficient estimators will give the asymptotically shortest interval.

If we recall from Section 5.5.1 that for regular models maximum-likelihood estimators are (i) asymptotically normal and (ii) asymptotically efficient with variance $1/nI(\theta) = 1/I_n(\theta)$ – that is, the Cramer–Rao limit – then the interval

$$\left(\hat{T}_n \pm \frac{c_{(1+\gamma)/2}}{\sqrt{nI(\theta)}}\right) \tag{5.97a}$$

(where $\hat{T}_n$ is the ML estimator of θ) is the asymptotically shortest γ-CI for θ. Then, in order to 'stablize' the variance – that is, make it independent on θ – one may proceed as in Section 5.5.1 (eq. (5.66) and Example 5.8(a)) to obtain the confidence interval for $h(\theta)$

$$\left(h(\hat{T}_n) \pm c_{(1+\gamma)/2} b/\sqrt{n}\right) \tag{5.97b}$$

where, for simplicity, we chose $a = 0$ in eq. (5.66). If h is a monotone function we can then solve the resulting inequalities for θ to get the desired asymptotic γ-CI for the parameter θ.

Owing to their nature, asymptotic CIs are exact only in the limit of $n \to \infty$ but in common practice they are often used as approximate confidence intervals when the sample is large – with the obvious understanding that the larger is the sample, the better is the approximation. As it should be expected, however, the notion of 'large' sample depends on the problem at hand because the rate of convergence to the normal distribution is not the same for all estimators. Nonetheless, it is a widely adopted rule of thumb that $n > 30$ can be considered a large sample when estimating confidence intervals for means while $n > 100$ is the 'dividing line' between small and large samples when estimating confidence intervals for variances.

Example 5.12(a) In Example 5.8(a), we determined that the sample mean M is the ML estimator of the parameter θ of a Poisson model. Also, we found $I(\theta) = 1/\theta$ and noted that – choosing $a = 0$ and $b = 1/2$ in eq. (5.66) – the r.v. $\sqrt{n}(\sqrt{M} - \sqrt{\theta})$ is asymptotically normal with zero mean and variance 1/4. Then, it follows from eq. (5.97b) that $(\sqrt{M} \pm c_{(1+\gamma)/2}/2\sqrt{n})$ is, for large samples, an approximate γ-CI for $\sqrt{\theta}$; consequently

$$\left(\left(\sqrt{M} - c_{(1+\gamma)/2}/2\sqrt{n}\right)^2, \left(\sqrt{M} + c_{(1+\gamma)/2}/2\sqrt{n}\right)^2\right) \tag{5.98}$$

is the approximate γ-CI for θ.

Example 5.12(b) For a sequence of n Bernoulli trials we have seen in Example 5.7(a) that the ML estimate of the parameter $\theta = p$ is the observed frequency of success x/n (which coincides with the sample mean M if 1 counts as a success and 0 counts as a failure). It is left to the reader to show that

$$I(\theta) = \frac{1}{\theta(1-\theta)} \tag{5.99}$$

and therefore the approximate CI for θ is

$$\left(M \pm \frac{c_{(1+\gamma)/2}}{\sqrt{n}}\sqrt{\theta(1-\theta)}\right) \tag{5.100}$$

The stabilizing transformation can be obtained from eq. (5.66) which, setting $a = 0$ and $b = 1/2$, yields

$$h(\theta) = \frac{1}{2}\int \frac{d\theta}{\sqrt{\theta(1-\theta)}} = \arcsin(\sqrt{\theta}) \tag{5.101}$$

so that $(\arcsin(\sqrt{M}) \pm c_{(1+\gamma)/2}/2\sqrt{n})$ is the approximate CI for $\arcsin\sqrt{\theta}$.

5.7 A few notes on other types of statistical intervals

The somewhat detailed discussion of the preceding sections on confidence intervals should not lead one to think that they are the only statistical intervals used in practice. Besides CIs, in fact, it is rather common in many applications to consider 'tolerance intervals' (TI) or 'prediction intervals' (PI), where the choice between the three types is dictated by the final scope of the analysis. So, referring for the most part to Chapter 5 of [27], this section is simply meant to outline the main ideas behind these different concepts of statistical intervals. Before we do this, however, it is worth recalling that (a) the basic assumption is to draw a random sample from some population and (b) the statistical inferences are only valid for the population from which the sample was selected. In general, moreover, the assumption of normality is often made even if it may not be strictly met in practice. In this regard, the considerations at the end of Section 5.6 apply and in case of strong evidence of non-normality, one may always consider the possibility of using distribution-free methods (see, for instance, Ref. [13]).

Tolerance intervals are needed when we are interested in an interval which will contain a certain percentage of the population. In this case, therefore, we will have two percentages: the percentage of population included in the interval and the confidence level – often, as for CIs, 90, 95 or 99% – associated to the interval. This second percentage is usually included in the name and one speaks of 90%, 95% or 99%-TI, respectively.

Assuming a sample from a normal population, tolerance intervals are generally given in the form $(M \pm c_{T,R}(n)S)$ and the values of $c_{T,R}$ – where the subscript T is for 'tolerance' and R indicates the percentage of population contained in the interval – can be found in statistical tables for different values of the sample size n. So, for instance, for $n = 15$ and a 95%-TI, we find the values $c_{T,90} = 2.48$, $c_{T,95} = 2.95$ and $c_{T,99} = 3.88$.

As the name itself implies, prediction intervals have to do with future observations. More specifically, a PI is needed when we are interested in an interval which will contain a specified number k of future observations from the population under study. So, for instance, given the population of daily flights from, say, New York to Chicago, a pilot may not be interested in the average delay of these flights, but in the delay of the next flight in which he/she will be flying. Similarly, a customer purchasing a small number of units of a given product is not interested in the long-run performance of the process from which his/her units are a sample, but in the quality of those particular units that he is buying.

As for the other types of intervals, we associate to a PI a confidence level but now the second defining number is k, the number of future observations to be included in the interval. Again, the interval is given in the form $(M \pm c_{P,k}(n)\bar{S})$ where the subscript P is for 'prediction' and the values of $c_{P,k}$ can be found in statistical tables. As a numerical example, suppose that we have $n = 10$ observations from a normal population and we are interested in the values of $k = 2$ further randomly selected observations from that population. For $n = 10$, at a 95% CI we find the value $c_{P,2} = 2.79$ so that our 95%-PI is $(M \pm 2.79\bar{S})$, where M and $\bar{S}$ are the mean and (unbiased) standard deviation calculated on the basis of the ten observations at our disposal. An important difference between the types of intervals is that CIs become smaller and smaller as the sample size increases while it is not so for TIs and PIs.

Finally, it is worth noting that there exist other types of prediction intervals such as, for instance, the PI to contain – at a given confidence level – the mean of k future observations or the standard deviation of k future observations. For more detailed information the interested reader can refer to [13 and 14].

5.8 Summary and comments

The theory of Probability is an elegant and elaborate construction well worthy of study in its own right. Statistics, broadly speaking, is the other face of the coin because it provides the methods and techniques by which – on the basis of a limited number of observed data – we can make (inductive) inferences and/or draw conclusions on specific real-word problems where randomness is involved. In other words, one can safely say that Statistics 'sees these problems from a different angle', although it is evident that it must necessarily rely on Probability theory in order to be effective. The approach of Statistics is explained in Section 5.2, where the concept of statistical model is introduced together with the definitions of 'sample', 'realization of the sample' and some notes on the important aspect of data collection.

With Section 5.3 we turn to more practical considerations by noting that one of the first step in every analysis is to use the experimental data to calculate the so-called 'sample characteristics' where, by analogy, each one of them is generally the counterpart of a well-defined probabilistic quantity. In this light, therefore, one speaks of sample mean, sample variance, kth order (ordinary and central) sample moment, etc., and of their realizations which, in turn, may change from experiment to experiment because the realization of the sample, as a matter of fact, does change from experiment to experiment. Being random variables themselves, moreover, it makes sense to speak of mean, variance, etc. – and, more generally, of the probability distribution – of sample characteristics. All these aspects are discussed in Section 5.3 by implicitly assuming the sample size n as fixed. This, however, is not the whole story because another important issue is considered in Section 5.3.1: the behaviour of sample characteristics as the sample size increases indefinitely – that is, mathematically speaking, as $n \to \infty$. In the limit, in fact, some important properties of both theoretical and practical interest show up: theoretical because an infinite sample is an evident impossibility and consequently these asymptotic properties can never be realized in full, but practical because it can often be assumed that they are, to a certain extent, satisfied by large samples, thereby providing useful working approximations in many cases.

Having introduced the concept of sample characteristic and, in particular, of statistic – that is, a sample characteristic containing no unknown quantities – both Sections 5.4 and 5.5 and all their subsections are dedicated to the subject of point estimation. In essence, the problem consists in estimating one or more unknown parameters of a supposedly known type of distribution by means of an appropriate statistic. The type of distribution provides the underlying statistical model while the observed data are used to calculate the 'appropriate' statistic which, we hope, will estimate the unknown parameter(s) within an acceptable degree of accuracy.

Since this kind of problem is fundamental in almost all statistical applications, the first step is to specify some criteria by which we may be able to decide whether a given statistic can qualify as a 'good' – or even, if and when possible, as the 'best' – estimator for the parameter under investigation. In this respect, in fact, it is not sufficient to rely solely on analogy – that is, using the sample mean to estimate the mean, the sample variance for the variance, etc. – because it can be shown that this intuitive approach, although useful in some cases, may even be misleading in some other cases. Among the most important criteria to judge an estimator, Section 5.4 considers unbiasedness, asymptotic unbiasedness, efficiency and consistency. Then, in regard to efficiency, Section 5.4.1 deals with a fundamental result applying to the so-called regular problems: this is the Cramer–Rao inequality which, by establishing a lower limit for the variance of an estimator, can indicate the best estimator – when it exists – in terms of efficiency. In the process, the definition of Fisher's information is given and all the concepts above are generalized to the case of a k-dimensional (vector) parameter and to a scalar function of a vector parameter.

Another desirable property of estimators is sufficiency. The definition is not self-evident and, often, is also of little practical use for the purpose of identifying sufficient estimators. The required explanations are given in Section 5.4.2, where it is also shown that Neyman's factorization theorem provides an easier way to assess sufficiency and that – Rao–Blacwell theorem – the so-called MVUE (minimum variance unbiased estimator), when it exists, is a function of a sufficient statistic. The property of completeness, moreover, is introduced in order to state Lehmann–Scheffé theorem which, in turn, specifies the general form of a MVUE as a function of a sufficient and complete statistic and an unbiased estimator (for the unknown parameter under study).

Finally, the way in which we can find estimators with the above properties – or at least some of them – is explained in Section 5.5. Although not the only one, the most popular technique for this purpose is the so-called ML method. The name itself is self-explanatory and consists in maximizing the likelihood function (or, more often, its natural logarithm) with respect to the unknown parameter(s). The 'method of moments' and 'Bayes' method' are also briefly considered in Section 5.5 but it is noted that, in general, ML estimators have a number of desirable properties and here, probably, lies the reason for the method's popularity. Particularly worthy of mention are the asymptotic properties of ML estimators considered in Section 5.5.1.

The most appropriate solution to many problems is not in the form of a point estimate because the main concern is often a range of values within which we can confidently hope to find the true value of the unknown parameter. This is a so-called problem of interval estimation and is the subject of Section 5.6. So, by first specifying a confidence level γ, our goal is to determine two statistics T_1, T_2 such that eq. (5.69) holds; these statistics, once we find them, are the lower and upper limit of the CI, respectively. At this point we use the experimental data to calculate their realizations t_1, t_2 and say that (t_1, t_2) is the desired γ-CI.

The general technique by which the task of finding T_1, T_2 is accomplished is explained in Section 5.6 and the many worked-out examples show that confidence intervals are always specified in terms of quantiles of an appropriate distribution where, on the one hand, the 'appropriate' distribution (frequently the Gaussian, the χ^2 or the Fisher distribution) depend on the parameter under study while, on the other hand, the quantiles to be used in actually calculating the interval depend on the confidence level. In any case, however, it is pointed out that we cannot say that the true value θ of the parameter lies in the interval (t_1, t_2) with probability θ. This is because (t_1, t_2) is a 'deterministic' interval with nothing random in it and therefore θ either belongs to it or it does not. The correct statement is given in terms of the long-run interpretation of confidence intervals: by repeating the estimation procedure many times – thus obtaining many confidence intervals – θ will fall in these intervals in $100\gamma\%$ of the cases. Also, another general fact is that the procedure of interval estimation must be based on a compromise between sample size and confidence level. For a given sample size, in fact, a higher confidence level corresponds to a longer interval and therefore an unreasonably high value of γ will lead to an interval which may be too large to be of any practical use. The interval length, on the other hand, can be decreased by either choosing a lower confidence level or by increasing the sample size, or both. Increasing the sample size, however, is generally costly and, in some cases, may not even be practicable. So, a correct balance of these quantities must be agreed upon at the planning stage and, clearly, it is the analyst's responsibility – depending on the importance of the problem at hand – to suggest a viable solution.

Finally, it is noted that cases in which finding a confidence interval turns out to be a very difficult task are not rare. For large samples, however, a practical solution is the use of asymptotic confidence intervals and this is the subject of Section 5.6.1. In Section 5.7, moreover, we briefly introduce the concepts of 'tolerance intervals' and 'prediction intervals' by also giving a number of specific references for the reader interested in more details on these further aspects of interval estimation.

References and further reading

[1] Azzalini, A., *'Inferenza Statistica: una Presentazione Basata sul Concetto di Verosimiglianza'*, 2nd edn., Springer-Verlag Italia, Milano (2001).

[2] Bartlett M.S., *'The Use of Transformations'*, Biometrics, 3, pp. 39–52 (1947).

[3] Cramér, H., *'Mathematical Methods of Statistics'*, 19th edn., Princeton Univ. Press, Princeton (1999).

[4] Crow, E.L., Davis, F.A., Maxfield, M.W., *'Statistics Manual'*, Dover, New York (1960).

[5] de Haan, L., *'Sample extremes: an Elementary Introduction'*, Stat. Neerlandica, 30, 161–172 (1976).

[6] Di Crescenzo, A., Ricciardi, L.M., *'Elementi di Statistica'*, Liguori Editore, Napoli (2000).

[7] Duncan, A.J., *'Quality Control and Industrial Statistics'*, 5th edn., Irwin, Homewood, Illinois (1986).

[8] Edwards, A.W.F., *'Likelihood'*, The Johns Hopkins University Press, Baltimore (1992).

[9] Fisher, R.A., *'On the Mathematical Foundations of Theoretical Statistics'*, PTRS, 222 (1921).

[10] Galambos, J., *'The Asymptotic Theory of Extreme Order Statistics'*, 2nd edn., Krieger, Malabar (1987).

[11] Gnedenko, B.V., *'Sur la Distribution Limite du Terme Maximum d'une Série Aléatoire'*, *Ann. Math.*, 44, 423–453 (1943).

[12] Green, J.R., Margerison, D., *'Statistical Treatment of Experimental Data'*, Elsevier, Amsterdam (1977).

[13] Hahn, G.J., Meeker, W.Q., *'Statistical Intervals: a Guide for Praticioners'*, Wiley, New York (1990).

[14] Hahn, G.J., *'Statistical Intervals for a Normal Population. Part I. Tables, Examples and Applications'*, *Journal of Quality Technology*, July, 115–125

(1970); *'Part II. Formulas, Assumptions, some Derivations', Journal of Quality Technology*, October, 195–206 (1970).

[15] Huff, D., *'How to Lie With Statistics'*, W. W. Norton & Company, New York (1954).

[16] Keeping, E.S., *'Introduction to Statistical Inference'*, Dover, New York (1995).

[17] Klimov, G., *'Probability Theory and Mathematical Statistics'* Mir Publishers, Mosow (1986).

[18] Kottegoda, N.T., Rosso, R., *'Statistics, Probability and Reliability for Civil and Environmental Engineers'*, McGraw-Hill, New York (1998).

[19] Ivchenko, G., Medvedev, Yu., *'Mathematical Statistics'*, Mir Publishers, Moscow (1990).

[20] Mandel, J., *'The Statistical Analysis of Experimental Data'*, Dover, New York, (1984).

[21] Mendenhall, W., Wackerly, D.D., Scheaffer, R.L., *'Mathematical Statistics with Applications'*, 4th end., PWS-KENT Publishing Company, Boston (1990).

[22] Nasri-Roudsari, D., Cramer, E., *'On the Convergence Rates of Extreme Generalized Order Statistics'* www.math.uni-oldenburg.de/preprints/get/source/Rates.pdf

[23] Pace, L., Salvan, A., *'Teoria della Statistica'*, CEDAM, Padova (1996)

[24] Rao, C.R., *'Asymptotic Efficiency and Information', Proc. 4th Berkeley Symp. Math. Stat. Prob.* **1**, 531–545 (1961).

[25] Rinne, H., *'Taschenbuch der Statistik'*, Verlag Harri Deutsch, Frankfurt am Main (2003).

[26] Serfling, R.J., *'Approximation Theorems in Mathematical Statistics'*, John Wiley & Sons, New York (1980).

[27] Wadsworth, H.M. (editor), *'Handbook of Statistical Methods for Engineers and Scientists'*, McGraw-Hill, New York (1990).

[28] Zacks, S., *'The Theory of Statistical Inference'*, John Wiley & Sons, New York (1971).

[29] Thisted, R.A., *'Elements of Statistical Computing'*, Chapman & Hall, London (1988).

[30] Buczolich Z., Székely G.J., *Adv. Appl. Math.*, **10**, 439–256 (1992).

6 The test of statistical hypotheses

6.1 Introduction

Broadly speaking, any assumption on the distribution of one or more random variables observed in an experiment is a *statistical hypothesis*. The hypothesis may be based on theoretical considerations, on the analysis of other (similar) experiments or it may just be an educated guess suggested by reasonableness or common sense, whatever these terms mean. In any case, it must be checked by actually performing the experiment and by devising some method which – in the light of the acquired data – gives us the possibility to decide whether to accept it or reject it. This, it should be clear from the outset, does not imply that our decision will be right because, as in any procedure of statistical inference, the best we can do (unless we examine the entire population) is to reduce the probability of being wrong to an acceptable level, where the term 'acceptable' generally depends on the problem at hand, the seriousness of the consequences of being wrong and, last but not least, the cost of the experiment. Consequently, we will not state our conclusions by saying 'our hypothesis is true (false)' but 'the observed data are in favour (against) our hypothesis', and we will continue our work behaving *as if* the hypothesis were true (false).

The methods by means of which we make our decision are called *statistical tests* and are the subject of this chapter. We will first illustrate the main ideas from a general point of view and then turn to typical classes of problems and specific examples.

6.2 General principles of hypotheses testing

Let us start with some definitions. The hypothesis to be tested, generally denoted by H_0, is called the *null hypothesis* and it is tested against an *alternative hypothesis* H_1. The two hypotheses are regarded as mutually exclusive and exhaustive. This is to say that if we accept H_0 then we reject H_1 and conversely, but it does not mean that – given H_0 – the hypothesis H_1 is the one and only alternative to H_0. As a matter of fact, it is often possible to conceive of several alternatives to H_0, say H_1', H_1'' and so on, but the point is that once H_0 and H_1 have been formulated, the test leading to the acceptance/rejection of H_0 necessarily leads to the rejection/acceptance of H_1. Clearly, it is the analyst's responsibility to select the most appropriate pair of hypotheses for the problem at hand.

So, given H_0 and H_1, the experimental data form the evidence on the basis of which we decide to accept or reject H_0. Due to the intrinsic uncertainty of any statistical inference, our decision may be right or wrong; however, we may be wrong in two ways:

(a) by rejecting H_0 when in fact it is true; or
(b) by accepting H_0 when in fact it is false.

The common terminology defines (a) a *type I error* (or rejecting error) and (b) a *type II error* (or acceptance error). Ideally, one would like both possibilities of error to be as small as possible, but since it turns out that, for a fixed sample size n, it is generally not possible to decrease one type without increasing the other, some sort of compromising strategy must be adopted. We will come to this point shortly.

In essence, any statistical test is a rule by which a realization $\mathbf{x} = (x_1, \ldots, x_n)$ of the sample $\mathbf{X} = (X_1, \ldots, X_n)$ is used to make a decision about the assumption H_0. More specifically, this is done by dividing the sample space Ξ into two disjoint sets Ξ_0, Ξ_1 – called the *acceptance region* and the *rejection* (or *critical*) *region*, respectively – such that $\Xi_0 \cup \Xi_1 = \Xi$. As the names themselves imply, Ξ_0 contains all $\mathbf{x}$ which lead to the acceptance of H_0 while Ξ_1 contains all $\mathbf{x}$ which lead to the rejection of the null hypothesis. In this light, the basic formulation of a statistical test is as follows:

Let $\mathbf{x}$ be a realization of the sample $\mathbf{X}$. If $\mathbf{x} \in \Xi_0$ we accept the null hypothesis H_0; if, on the other hand, $\mathbf{x} \in \Xi_1$ we reject H_0 (and therefore accept H_1). Then, the two possibilities of error correspond to the cases: (a) $\mathbf{x} \in \Xi_1$ when H_0 is true and (b) $\mathbf{x} \in \Xi_0$ when H_0 is false.

The selection of the acceptance and rejection regions is strictly related to two other aspects: the test chosen for a given null hypothesis and the 'goodness' of the test. In fact, since it is reasonable to expect that a given null hypothesis H_0 can be tested by different methods and that each method will define its acceptance and rejection regions, the problem arises of which test to choose among all possible tests on H_0. The choice, we will see, depends also on the alternative hypothesis H_1 but for the moment we assume both H_0 and H_1 as given. Now, an intuitive solution to this problem is, for a specified sample size n, to call 'best' the test which makes the possibility of error as small as possible and choose this one. This aspect, however, deserves further consideration because – keeping in mind that we do not know if H_0 is true or not – the two types of error must be considered simultaneously.

If, as it is customary, we denote by α and β the probabilities of committing a type I and type II error respectively, it turns out that we cannot simultaneously make them as small as we wish. This fact is evident if we examine the two extreme cases. If we choose $\alpha = 0$ we will never make a type I error and this, in turn, means that $\Xi = \Xi_0$ because we will always accept H_0 regardless of the observed realization $\mathbf{x}$. This is a correct decision if H_0 is true (which we do not know); if, however, H_0 is false, our choice of accepting it no matter what – since $\Xi_1 = \Xi_0^C = \emptyset$ – implies $\beta = 1$. Conversely, choosing $\beta = 0$ means that $\Xi = \Xi_1$ and $\Xi_0 = \emptyset$; therefore we will always reject H_0, a circumstance which prevents us from committing a type II error if H_0 is false, but implies $\alpha = 1$ if H_0 is true. Between the two extremes there are many possible intermediate cases corresponding to different choices of Ξ_0 and Ξ_1 but it is a general fact that reducing α tends to increase β and viceversa.

The usually adopted strategy to overcome this difficulty is due to Neyman and Pearson and is based on the consideration that in most cases one type of error has more serious consequences than the other. Consequently, we fix a value for the probability of the worst error and, among all possible tests, we choose the one that minimizes the probability of the other error. Since the problem is often formulated in such a way that the type I error is the worst, the strategy consists in specifying a value for α and – if and when possible – choosing the test with the smallest value of β (or, equivalently, the maximum value of $1 - \beta$) compatible with the prescribed value of α. This specified value of α – which, clearly, depends on practical considerations about the problem at hand – defines the *significance level* of the test.

Before turning to other general aspects of hypothesis testing, we open a short parenthesis on notation. Often one denotes the probabilities of type I and type II errors by $P(H_1|H_0)$ and $P(H_0|H_1)$, respectively. This symbolism does not mean that we are dealing with conditional probabilities in the strict sense, but it is just a convenient way of indicating – in the two cases – the accepted hypothesis (in the first 'slot' within parenthesis) and the true hypothesis (in the second 'slot').

Returning to the main discussion, an observation of practical nature is that the critical (rejection) region is frequently defined by means of a so-called *test function* $T(\mathbf{X})$, where $T(\mathbf{X})$ is a statistic which must be appropriately chosen for the problem at hand. Having chosen a test statistic, the critical region will then be expressed in one of the following forms

$$\Xi_1 = \begin{cases} \{\mathbf{x} : T(\mathbf{x}) \geq c\} \\ \{\mathbf{x} : T(\mathbf{x}) \leq c\} \\ \{\mathbf{x} : |T(\mathbf{x})| \geq c\} \end{cases} \tag{6.1}$$

where c is a real number which depends on the significance level α. This, in other words, means that for every α the set $\mathrm{T} = \{t : t = T(\mathbf{x}), \mathbf{x} \in \Xi\}$ of all possible values of T is divided into two subsets T_0, T_1, where T_1 will include all those realizations $t = T(\mathbf{x})$ which lead to the rejection of H_0.

A second comment worthy of mention is that, for a fixed sample size n, some problems of hypothesis testing do not lend themselves easily to a solution. Things, however, often get better if we adopt an asymptotic approach by letting the sample size tend to infinity. This is a frequently adopted strategy but it should be kept in mind that the final results are then valid only for large samples. For moderate sample sizes, however, they can often be considered as useful working approximations.

Having outlined the general philosophy of the statistical testing, it can be of help at this point to have an idea of some typical of types of hypotheses encountered in practice. The following is a short list:

(1) *Hypothesis on the form of distribution*: In this case we make n independent observations of a r.v. X with unknown distribution $F_X(x)$ and use the acquired data $\mathbf{x}$ to check if the distribution of X is, as we assume, $F(x)$. The null hypothesis is then written $H_0 : F_X(x) = F(x)$. The function $F(x)$, in turn, may be (a) completely defined or (b) may belong to a certain class – for example, normal, Poisson, or else – the uncertainty being on one (or more) parameter(s) θ of the distribution. An example of this latter type can be $H_0 : F_X(x) = N(\mu, \theta)$, meaning that we want to test the hypothesis that X has a normal distribution with known mean μ and unknown variance θ.

(2) *Hypothesis of independence*: In this case we have, for example, a two-dimensional r.v. $\mathbf{X} = (X, Y)$ with unknown PDF $F_{\mathbf{X}}(x, y)$ and we have reasons to believe that X and Y are independent. Then, the null hypothesis is symbolically expressed as $H_0 : F_{\mathbf{X}}(x, y) = F_X(x)F_Y(y)$.

(3) *Hypothesis of homogeneity*: We carry out a series of m independent experiments – each experiment consisting of n trials – obtaining the results $(x_{1i}, \ldots, x_{ni})$, where $i = 1, \ldots, m$. Our basic assumption in this case is that these data are homogeneous, that is, they are all observations of the same random variable. Then, since the null hypothesis is that the distribution law is the same for all the experiments, we symbolically express the problem as $H_0 : F_1(x) = F_2(x) = \cdots = F_m(x)$, where we denoted by $F_i(x)$ the (unknown) distribution of the ith experiment.

Clearly, the types of hypothesis considered above do not cover all the possibilities because the list has been given mainly for illustrative purposes. Other specific cases will be examined in due time if and whenever needed in the course of future discussions.

6.3 Parametric hypotheses

If the hypothesis to be tested concerns one or more unknown parameters of a supposedly known type of probability distribution, one speaks of parametric hypotheses. The basic procedure is similar to what has been done in Chapter 5 – that is, we start from the statistical model (5.1) and, on the basis of the acquired data, draw inferences on θ – but the details differ. In Chapter 5, in fact, we did not formulate any hypothesis whatsoever on θ and our main concern was simply to determine a reliable estimate of it, either in the form of a numerical value or a confidence interval. Now we do formulate an hypothesis – the null hypothesis H_0 – and the scope is to accept it or reject it depending on whether H_0 is reasonably consistent with the observed data or not. This kind of approach is generally more convenient when, following the experiment, we must make a 'yes or no' decision and take action accordingly. For the moment we ignore the fact that the two problems – parametric hypothesis testing and confidence interval estimation – are, in fact, related and delay the discussion of this aspect to Section 6.3.4.

Denoting, as in Chapter 5, the parameter space by Θ, the general form of the null and alternative hypotheses is

$$\begin{aligned} &H_0 : \theta \in \Theta_0 \\ &H_1 : \theta \in \Theta_1 \end{aligned} \tag{6.2}$$

where Θ_0, Θ_1 are two subsets of Θ such that $\Theta_0 \cap \Theta_1 = \emptyset$ and $\Theta_0 \cup \Theta_1 = \Theta$. More specifically, we call *simple* any hypothesis which specifies the probability distribution completely, otherwise we speak of *composite* (or compound) hypothesis. So, for instance, $H_0 : \theta = \theta_0$ and $H_1 : \theta = \theta_1$ (where θ_0 and θ_1 are given numerical values) are simple hypotheses while $H_0 : \theta \geq \theta_0$, $H_1 : \theta \neq \theta_0$ or, say, $H_1 : \theta < \theta_0$ are composite hypotheses. Depending on the problem at hand, we may have any one of the three possibilities (i) both the null and alternative hypotheses are simple, (ii) one is simple and the other is composite and (iii) both hypotheses are composite.

Before examining the various cases, we must return for a moment to the discussion of Section 6.2 on how to select a 'good' test, a choice which – we recall – requires a closer look at the two types of error. In case of parametric hypotheses, they generally depend on θ and can be written as

$$\begin{aligned} &\alpha(\theta) = P_\theta(\mathbf{X} \in \Xi_1 | \theta \in \Theta_0) \\ &\beta(\theta) = P_\theta(\mathbf{X} \in \Xi_0 | \theta \in \Theta_1) \end{aligned} \tag{6.3}$$

If we define the so-called power function $W(\theta)$ as

$$W(\theta) = \begin{cases} P_\theta(X \in \Xi_1 | \theta \in \Theta_0) = \alpha(\theta) \\ P_\theta(X \in \Xi_1 | \theta \in \Theta_1) = 1 - \beta(\theta) \end{cases} \tag{6.4}$$

we recognize $1 - \beta$ as the probability of not making a type II error. Since an ideal test will result in $W(\theta) = 0$ if H_0 is true (i.e. $\theta \in \Theta_0$) and $W(\theta) = 1$ if H_0 is false (i.e. $\theta \in \Theta_1$), the function W can be used to compare different tests on a given pair of hypothesis H_0, H_1. In this light, in fact, we have the following definitions:

(i) we call *size* of a test the quantity

$$\alpha = \sup_{\theta \in \Theta_0} W(\theta) \tag{6.5}$$

(note that for parametric hypotheses the terms 'size' and 'significance level' are interchangeable).

(ii) given a test T on a pair of hypotheses H_0, H_1, let α be its size and $\beta(\theta)$ its probability of a type II error. Then T is called the ***uniformly most powerful*** test if, for any other test T^* (on H_0, H_1) of size $\alpha^* \le \alpha$, we have $\beta^*(\theta) \ge \beta(\theta)$ for all $\theta \in \Theta_1$. This, in other words, means that the uniformly most powerful test T – denoting by $W(\theta)$ its power function – satisfies the inequality

$$W(\theta) \ge W^*(\theta) \quad \text{for all } \theta \in \Theta_1 \tag{6.6}$$

Also, a desirable property for a test is unbiasedness. A test T is called unbiased if

$$W(\theta) \ge \alpha \quad \text{for all } \theta \in \Theta_1 \tag{6.7}$$

so that we have a higher probability of rejecting H_0 when it is false than rejecting it when it is true.

At this point, another word of caution is in order because mistakes and misunderstandings are rather frequent: the power function considers the probabilities of rejecting H_0 when it is true and when it is false. This is, in essence, the main idea of hypothesis testing and nothing can be said about the probability of H_0 being true or false. So, if H_0 is accepted at, say, the 5% significance level it does not mean that the probability of H_0 being true is 95%. This distinction, as a matter of fact, is fundamental and should always be kept in mind when reporting the results.

6.3.1 *Simple hypotheses: Neyman–Pearson's lemma*

A uniform more powerful test does not always exist because uniformity, that is, the condition 'for all $\theta \in \Theta_1$', is rather strong and we may have cases in which two tests, say T_1, T_2, cannot be compared because $W_1(\theta) < W_2(\theta)$ for some values of θ in Θ_1 while $W_1(\theta) > W_2(\theta)$ for some other values of θ in Θ_1. A most powerful test, however, always exists when we are dealing with a pair of simple hypothesis, that is, the case in which eq. (6.2) have the form

$$\begin{aligned} H_0 &: \theta = \theta_0 \\ H_1 &: \theta = \theta_1 \end{aligned} \tag{6.8}$$

where θ_0, θ_1 are two specific numerical values for the unknown parameter. Equation (6.8), in other words, imply that the parameter space consists of only two points – that is, $\Theta = \{\theta_0, \theta_1\}$ – and that the distribution of the r.v. X is either $F_0(x) = F(x; \theta_0)$ or $F_1(x) = F(x; \theta_1)$, where F is a known type of PDF (normal, exponential, Poisson or else). Assuming that F_0 and F_1 are both absolutely continuous with densities $f_0(x)$ and $f_1(x)$, respectively (with $f_0, f_1 > 0$), the following theorem – known as Neyman–Pearson's lemma – holds.

Proposition 6.1 (Neyman–Pearson's lemma) *Let (6.8) be the null and alternative hypotheses and* $\mathbf{x} = (x_1, \ldots, x_n)$ *be a realization of the sample* $\mathbf{X} = (X_1, \ldots, X_n)$. *The most powerful test of size* α *is specified by the critical region*

$$\Xi_1 = \{\mathbf{x} : l(\mathbf{x}) \le c\} \tag{6.9}$$

where c $(c \ge 0)$ *is such that* $P_{\theta_0}[l(\mathbf{X}) \le c] = \alpha$ *and* $l(\mathbf{X})$ *is a statistic called the 'likelihood-ratio' and defined as*

$$l(\mathbf{X}) \equiv \frac{L(\mathbf{X}; \theta_0)}{L(\mathbf{X}; \theta_1)} = \frac{\prod_i f_0(X_i)}{\prod_i f_1(X_i)} \tag{6.10}$$

In order to simplify the notation in the proof of the theorem let us call A the rejection region (6.9) and let B be the rejection region of another test of size α (on the hypotheses (6.8)). Then

$$\int_A L(\mathbf{x}; \theta_0)\, d\mathbf{x} = \int_B L(\mathbf{x}; \theta_0)\, d\mathbf{x} = \alpha$$

because both tests have size α. Noting that both A and B can be expressed as the union of two disjoint sets by writing $A = (A \cap B) \cup (A \cap B^C)$ and $B = (A \cap B) \cup (B \cap A^C)$ respectively, the equality above implies

$$\int_{A \cap B^C} L(\mathbf{x}; \theta_0)\, d\mathbf{x} = \int_{B \cap A^C} L(\mathbf{x}; \theta_0)\, d\mathbf{x} \tag{6.11}$$

By the definition of A, moreover, it follows that $L(\mathbf{x}; \theta_1) \ge L(\mathbf{x}; \theta_0)/c$ for $\mathbf{x} \in A$ and, clearly, $L(\mathbf{x}; \theta_1) < L(\mathbf{x}; \theta_0)/c$ for $\mathbf{x} \in A^C$. Using these inequalities, eq. (6.11) leads to the chain of relations

$$\int_{A\cap B^C} L(\mathbf{x};\theta_1)\,\mathrm{d}\mathbf{x} \ge \int_{A\cap B^C} \frac{L(\mathbf{x};\theta_0)}{c}\,\mathrm{d}\mathbf{x}$$

$$= \int_{B\cap A^C} \frac{L(\mathbf{x};\theta_0)}{c}\,\mathrm{d}\mathbf{x} > \int_{B\cap A^C} L(\mathbf{x};\theta_1)\,\mathrm{d}\mathbf{x}$$

which, in turn, are used to get

$$\int_A L(\mathbf{x};\theta_1)\,\mathrm{d}\mathbf{x} = \int_{A\cap B} L(\mathbf{x};\theta_1)\,\mathrm{d}\mathbf{x} + \int_{A\cap B^C} L(\mathbf{x};\theta_1)\,\mathrm{d}\mathbf{x}$$

$$> \int_{A\cap B} L(\mathbf{x};\theta_1)\,\mathrm{d}\mathbf{x} + \int_{B\cap A^C} L(\mathbf{x};\theta_1)\,\mathrm{d}\mathbf{x}$$

$$= \int_B L(\mathbf{x};\theta_1)\,\mathrm{d}\mathbf{x} \tag{6.12}$$

meaning that the probability $1-\beta$ is higher for the test with rejection region $A = \Xi_1$. In fact, since the quantity $1-\beta$ (i.e. the probability of rejecting H_0 when H_0 is false or, equivalently, of accepting H_1 when H_1 is true) of a given test is obtained by integrating $L(\mathbf{x};\ \theta_1)$ over its rejection region, eq. (6.12) proves the theorem because the test corresponding to B is any test of size α on the hypotheses (6.8).

In addition, we can show that the test is always unbiased. In fact, in the rejection region $A = \Xi_1$ (eq. (6.9)) we have $L(\mathbf{x};\theta_0) \le cL(\mathbf{x};\theta_1)$ which, if $c \le 1$, implies $L(\mathbf{x};\theta_0) \le L(\mathbf{x};\theta_1)$ and therefore

$$\alpha = \int_A L(\mathbf{x};\theta_0)\,\mathrm{d}\mathbf{x} \le \int_A L(\mathbf{x};\theta_1)\,\mathrm{d}\mathbf{x} = 1-\beta = W(\theta_1)$$

On the other hand, in the acceptance region A^C we have $L(\mathbf{x};\theta_0) > cL(\mathbf{x};\theta_1)$ and therefore $L(\mathbf{x};\theta_0) > L(\mathbf{x};\theta_1)$ whenever $c > 1$. Consequently

$$1-\alpha = \int_{A^C} L(\mathbf{x};\theta_0)\,\mathrm{d}\mathbf{x} > \int_{A^C} L(\mathbf{x};\theta_1)\,\mathrm{d}\mathbf{x} = \beta = 1 - W(\theta_1)$$

thus showing that condition (6.7) holds in any case.

Example 6.1(a) As an application of Neyman–Pearson's lemma, consider a normal r.v. with known variance σ^2. On the basis of the random sample $\mathbf{X}$

and the observed data $\mathbf{x}$ we want to test the pair of simple hypotheses (6.8) on the unknown mean $\mu = \theta$ where, for definiteness, we assume $\theta_1 > \theta_0$.

We have

$$l(\mathbf{x}) = \exp\left(-\frac{1}{2\sigma^2}\sum_{i=1}^{n}[(x_i - \theta_0)^2 - (x_i - \theta_1)^2]\right)$$
$$= \exp\left(\frac{n}{2\sigma^2}\left(\theta_1^2 - \theta_0^2\right) - \frac{nm}{\sigma^2}(\theta_1 - \theta_0)\right)$$

where $m = \sum x_i$ is the realization of the sample mean calculated from the data. The inequality defining the rejection region (6.9) holds if

$$m \geq \frac{(\theta_1 + \theta_0)}{2} - \frac{\sigma^2 \log c}{n(\theta_1 - \theta_0)} \tag{6.13a}$$

or, equivalently, if

$$\frac{\sqrt{n}(m - \theta_0)}{\sigma} \geq \frac{\sqrt{n}(\theta_1 - \theta_0)}{2\sigma} - \frac{\sigma \log c}{\sqrt{n}(\theta_1 - \theta_0)} \equiv t(c) \tag{6.13b}$$

Now, noting that (Proposition 5.1(b)) the r.v. $Z = \sqrt{n}(M - \theta_0)/\sigma$ is standard normal if H_0 is true, we have $P_{\theta_0}[l(\mathbf{X}) \leq c] = P_{\theta_0}[Z \geq t(c)] = \alpha$ and therefore $t(c)$ is the α-upper quantile of the standard normal distribution. This quantity is found on statistical tables and is frequently denoted by the special symbol z_α (we find, for instance, for $\alpha = 0.05; 0.025; 0.01$ – the most commonly adopted values of α – the upper quantiles $z_{0.05} = 1.645$, $z_{0.025} = 1.960$ and $z_{0.01} = 2.326$, respectively). Then, in agreement with Neyman–Pearson's lemma, it follows that the most powerful test for our hypotheses is defined by the critical region

$$\Xi_1 = \left\{\mathbf{x} : m \geq \theta_0 + z_\alpha \frac{\sigma}{\sqrt{n}}\right\} \tag{6.14a}$$

and its power is $W(\theta_1) = 1 - \beta = P_{\theta_1}\{M \geq \theta_0 + z_\alpha \sigma/\sqrt{n}\}$. This quantity can be obtained by noting that under the alternative hypothesis the r.v. $\sqrt{n}(M - \theta_1)/\sigma$ is standard normal. Consequently, $W(\theta_1)$ equals the r-upper quantile of the standard normal distribution, where $r = z_\alpha - \sqrt{n}(\theta_1 - \theta_0)/\sigma$. Since $r < z_\alpha$ the area (under the standard normal pdf) to the right of r is greater than the area to the right of z_α – which, by definition, equals α. This shows that, as expected, the test is unbiased.

As a numerical example, suppose that we fix a significance level $\alpha = 0.025$ and we wish to test the simple hypotheses $H_0 : \theta = 15.0; H_1 : \theta = 17.0$ knowing that the standard deviation of the underlying normal population is

$\sigma = 2.0$. Suppose further that we carry out $n = 20$ measurements leading to $m = 16.2$. Since the rejection region in this case is

$$\Xi_1 = \left\{ m \geq 15.0 + 1.96 \frac{2.0}{\sqrt{20}} \right\} = \{m \geq 15.88\}$$

and $m = 16.2$ falls in it, we reject the null hypothesis and accept H_1. Moreover, we can calculate the power of the test noting that $r = -2.51$ so that the corresponding upper quantile is $0.994 = W(\theta_1)$ and the probability of a type II error is $\beta = 1 - 0.994 = 0.006$.

Following the same line of reasoning as above, it is easy to determine that the rejection region in the case $\theta_1 < \theta_0$ is

$$\Xi_1 = \left\{ \mathbf{x} : m \leq \theta_0 - z_\alpha \frac{\sigma}{\sqrt{n}} \right\} \tag{6.14b}$$

because we get the condition $P_{\theta_0}[l(\mathbf{X}) \leq c] = P_{\theta_0}[Z \leq t(c)] = \alpha$ thus implying that now $t(c)$ – where $t(c)$ is as in eq. (6.13b) – is the α-lower quantile of the standard normal distribution. Owing to the symmetry of the distribution, this lower quantile is $-z_\alpha$ and therefore eq. (6.14b) follows.

As a further development of the exercise, consider the following problem: in the case $\theta_1 > \theta_0$ we have fixed the probability of a type I error to a value α, what (minimum) sample size do we need to obtain a probability of type II error smaller than a given value β? The probability of a type II error is

$$P_{\theta_1}(M < \theta_0 + z_\alpha \sigma / \sqrt{n}) = P_{\theta_1}\{Z < z_\alpha - \sqrt{n}(\theta_1 - \theta_0)/\sigma\}$$

where Z is the standard normal r.v. $Z = \sqrt{n}(M - \theta_1)/\sigma$. The desired upper limit β is obtained when $z_\alpha - \sqrt{n}(\theta_1 - \theta_0)/\sigma$ equals the β-lower quantile of the standard normal distribution. If we denote this lower quantile by q_β we get

$$n = \frac{\sigma^2 (z_\alpha - q_\beta)^2}{(\theta_1 - \theta_0)^2} \tag{6.15}$$

and consequently $\tilde{n} = [n] + 1$ (the square brackets denote the integer part of the number) is the minimum required sample size. So, for instance, taking the same numerical values as above for $\alpha, \theta_0, \theta_1, \sigma$, suppose we want a maximum probability of type II error $\beta = 0.001$. Then, the minimum sample size is $\tilde{n} = 26$ (because $z_{0.025} = 1.96$, $q_{0.001} = -3.09$ and eq. (6.15) gives $n = 25.5$).

Example 6.1(b) Consider now a normal population with known mean μ and unknown variance $\sigma^2 = \theta^2$. Somehow we know that the variance is either θ_0^2 or θ_1^2 (with $\theta_1^2 > \theta_0^2$) and the scope of the analysis is to test the pair of simple hypotheses

$$\begin{aligned} H_0 &: \theta^2 = \theta_0^2 \\ H_1 &: \theta^2 = \theta_1^2 \end{aligned} \tag{6.16}$$

Again, we use Neyman–Pearson's lemma to obtain the most powerful test for the case at hand. Since

$$\begin{aligned} l(\mathbf{x}) &= \left(\frac{\theta_1}{\theta_0}\right)^n \exp\left(-\frac{1}{2}\left(\frac{1}{\theta_0^2} - \frac{1}{\theta_1^2}\right)\sum_{i=1}^n (x_i - \mu)^2\right) \\ &= \left(\frac{\theta_1}{\theta_0}\right)^n \exp\left(-\frac{\theta_1^2 - \theta_0^2}{2\theta_1^2}\sum_{i=1}^n \left(\frac{x_i - \mu}{\theta_0}\right)^2\right) \end{aligned}$$

the inequality defining the rejection region (6.9) holds if

$$\sum_{i=1}^n \left(\frac{x_i - \mu}{\theta_0}\right)^2 \geq \frac{2\theta_1^2}{\theta_1^2 - \theta_0^2}[n \log(\theta_1/\theta_0) - \log c] \equiv t(c) \tag{6.17}$$

Under the hypothesis H_0, each one of the n independent r.v.s $Y_i = [(X_i - \mu)/\theta_0]^2$ is distributed according to the χ^2 probability law with one degree of freedom. Consequently, the sum $Y = \sum Y_i$ has a χ^2 distribution with n degrees of freedom and the relation $P_{\theta_0}[l(\mathbf{X}) \leq c] = P_{\theta_0}[Y \geq t(c)] = \alpha$ means that $t(c)$ must be the α-upper quantile of this distribution. Then, denoting this quantile by the symbol $\chi^2_{\alpha;n}$, the rejection region for the test is

$$\Xi_1 = \left\{\mathbf{x} : \sum_{i=1}^n (x_i - \mu)^2 \geq \theta_0^2 \chi^2_{\alpha;n}\right\} \tag{6.18a}$$

As a numerical example, suppose we fix a significance level $\alpha = 0.05$ and we wish to test the hypotheses $H_0 : \theta^2 = 3.0; H_1 : \theta^2 = 3.7$ for a normal population with mean $\mu = 18$. If we carry out an experiment consisting of, say, 15 measurements $\mathbf{x} = (x_1, \ldots, x_{15})$, our rejection region will be

$$\Xi_1 = \left\{\mathbf{x} : \sum_{i=1}^{15} (x_i - 18)^2 \geq 3.0\chi^2_{\alpha;n}\right\} = \left\{\mathbf{x} : \sum_{i=1}^{15} (x_i - 18)^2 \geq 74.988\right\}$$

because from statistical tables we get the upper quantile $\chi^2_{0.05;15} = 24.996$. It is left to the reader to show that in the case $\theta_1^2 < \theta_0^2$ the rejection region is

$$\Xi_1 = \left\{ \mathbf{x} : \sum_{i=1}^{n} (x_i - \mu)^2 \leq \theta_0^2 \chi^2_{1-\alpha;n} \right\} \tag{6.18b}$$

where $\chi^2_{1-\alpha;n}$ is the $(1 - \alpha)$-upper quantile (or, equivalently, the α-lower quantile) of the χ^2 distribution with n degrees of freedom.

So, the basic idea of Proposition 6.1 is rather intuitive: since the likelihood-ratio statistics (6.10) can be considered as a relative measure of the 'weight' of the two hypotheses , $l(\mathbf{x}) > 1$ suggests that the observed data support the null hypothesis while the relation $l(\mathbf{x}) < 1$ tends to imply the opposite conclusion and the specific value of $l(\mathbf{x})$ – that is c in eq. (6.9) – below which we reject H_0 depends on α, that is, the risk we are willing to take of making a type I error. In this light, therefore, it is evident that nothing would change if, as some authors do, one defined the likelihood-ratio as $l(\mathbf{X}) = L(\mathbf{X}; \theta_1)/L(\mathbf{X}; \theta_0)$ and considered the rejection region $\Xi_1 = \{l(\mathbf{x}) \geq c\}$ with $P_{\theta_0}[l(\mathbf{X}) \geq c] = \alpha$.

When the probability distributions are discrete the same line of reasoning leads to the most powerful test for the simple hypotheses (6.8). Discreteness, however, often introduces one minor inconvenience. In fact, since the likelihood-ratio statistic takes on only discrete values, say $l_1, l_2, \ldots, l_k, \ldots$, it may not be possible to satisfy the condition $P_{\theta_0}[l(\mathbf{X}) \leq c] = \alpha$ exactly. The following example will clarify this situation.

Example 6.2 At the significance level α, suppose that we want to test the simple hypotheses (6.8) (with $\theta_1 > \theta_0$) on the unknown parameter $p = \theta$ of a binomial model. Defining $y = \sum_{i=1}^{n} x_i$ we have

$$l(\mathbf{x}) = \left(\frac{\theta_0}{\theta_1} \right)^y \left(\frac{1 - \theta_0}{1 - \theta_1} \right)^{n-y}$$

and $l(\mathbf{x}) \leq c$ if

$$y \geq \left(\log \frac{\theta_1}{\theta_0} + \log \frac{1 - \theta_0}{1 - \theta_1} \right)^{-1} \left(n \log \frac{1 - \theta_0}{1 - \theta_1} - \log c \right) \equiv t(c)$$

Under the null hypothesis, the r.v. $Y = X_1 + \cdots + X_n$ – being the sum of n binomial r.v.s – is itself binomially distributed with parameter θ_0, and in order to meet the condition $P_{\theta_0}[l(\mathbf{X}) \leq c] = P_{\theta_0}[Y \geq t(c)] = \alpha$ exactly there should exist an (integer) index $k = k(\alpha)$ such that

$$\sum_{m=k(\alpha)}^{n} \binom{n}{m} \theta_0^m (1 - \theta_0)^{n-m} = \alpha \tag{6.19}$$

If such an index does exist – a rather rare occurrence indeed – the test attains the desired significance level and the rejection region is

$$\Xi_1 = \left\{ \mathbf{x} : y = \sum_{i=1}^{n} x_i \geq k(\alpha) \right\} \tag{6.20}$$

However, the most common situation by far is the case in which eq. (6.19) is not satisfied exactly but we can find an index $r = r(\alpha)$ such that

$$\alpha' \equiv \sum_{m=r(\alpha)}^{n} \binom{n}{m} \theta_0^m (1-\theta_0)^{n-m} < \alpha < \sum_{m=r(\alpha)-1}^{n} \binom{n}{m} \theta_0^m (1-\theta_0)^{n-m} \equiv \alpha'' \tag{6.21}$$

At this point we can define the rejection region as (a) $\Xi_1 = \{\mathbf{x} : y \geq r(\alpha)\}$ or as (b) $\Xi_1 = \{\mathbf{x} : y \geq r(\alpha) - 1\}$, knowing that in both cases we do not attain the desired level α but we are reasonably close to it. In case (a), in fact, the actual significance level α' is slightly lower than α ($r(\alpha)$ is the minimum index satisfying the left-hand side inequality of (6.21)) while in case (b) the actual significance level α'' is slightly greater than α ($r(\alpha) - 1$ is the maximum index satisfying the right-hand side inequality of (6.21)). Also, in terms of power we have

$$1 - \beta' = \sum_{m=r(\alpha)}^{n} \binom{n}{m} \theta_1^m (1-\theta_1)^{n-m} < \sum_{m=r(\alpha)-1}^{n} \binom{n}{m} \theta_1^m (1-\theta_1)^{n-m} = 1 - \beta''$$

and, as expected, $\beta' > \beta''$. In the two cases, respectively, Proposition 6.1 guarantees that these are the most powerful tests at levels α' and α''.

Besides the cases (a) and (b) – which in most applications will do – a third possibility called 'randomization' allows the experimenter to attain the desired level α exactly. Suppose that we choose the rejection region (b) associated to a level $\alpha'' > \alpha$ and defined in terms of the index $s(\alpha) \equiv r(\alpha) - 1$. Under the null hypothesis, let us call P_0 the probability of the event $Y = s(\alpha)$, that is,

$$P_0 \equiv P_{\theta_0}\{Y = s(\alpha)\} = \binom{n}{s(\alpha)} \theta_0^{s(\alpha)} (1-\theta_0)^{n-s(\alpha)}$$

(which, on the graph of the PDF $F_0(x)$, is the jump $F_0(r) - F_0(s)$) and let us introduce the 'critical (or rejection) function' $g(\mathbf{x})$ defined as

$$g(\mathbf{x}) = \begin{cases} 1, & y > s(\alpha) \\ (P_0 + \alpha - \alpha'')/P_0, & y = s(\alpha) \\ 0, & y < s(\alpha) \end{cases} \tag{6.22}$$

Then we reject H_0 if $y > s(\alpha)$, we accept it if $y < s(\alpha)$ and, if $y = s(\alpha)$, we reject it with a probability $(P_0 + \alpha - \alpha'')/P_0$ – or, equivalently, accept it with the complementary probability $(\alpha'' - \alpha)/P_0$. This means that if $y = s(\alpha)$ we have to set up another experiment with two possible outcomes, one with probability $(P_0 + \alpha - \alpha'')/P_0$ and the other with probability $(\alpha'' - \alpha)/P_0$; we reject H_0 if the first outcome turns out, otherwise we accept it. The probability of type I error of this randomized test (i.e. its significance level) is obtained by taking the expectation of the critical function and we get, as expected

$$\begin{aligned} P(H_1|H_0) &= E_{\theta_0}[g(\mathbf{x})] = P_{\theta_0}\{y > s(\alpha)\} \\ &\quad + \frac{P_0 + \alpha - \alpha''}{P_0} P_{\theta_0}\{y = s(\alpha)\} \\ &= \alpha'' - P_0 + \frac{P_0 + \alpha - \alpha''}{P_0} P_0 = \alpha \end{aligned}$$

The reader is invited to:

(a) show that the case $\theta_0 > \theta_1$ leads to the rejection region $\Xi_1 = \{\mathbf{x} : y \leq r(\alpha)\}$ where, taking $r(\alpha)$ as the maximum index satisfying the inequality

$$\alpha' \equiv \sum_{m=0}^{r(\alpha)} \binom{n}{m} \theta_0^m (1 - \theta_0)^{n-m} \leq \alpha \tag{6.23}$$

the attained significance level α' is slightly lower than α (unless we are so lucky to have the equal sign in (6.23));

(b) work out the details of randomization for this case.

So, in the light of Example 6.2 we can make the following general considerations on discrete cases:

(i) Carrying out a single experiment, discreteness generally precludes the possibility of attaining the specified significance level α exactly. Nonetheless we can find a most powerful test at a level $\alpha' < \alpha$ or at a level $\alpha'' > \alpha$.

(ii) At this point, we can either be content of α' (or α'', whichever is our choice) or – if the experiment leads to a likelihood-ratio value on the border between the acceptance and rejection regions – we can 'randomize' the test in order to attain α. In the first case we lack the probability $\alpha - \alpha'$ while we have a probability $\alpha'' - \alpha$ in excess in the second case. Broadly speaking, randomization compensates for this part by adding a second experimental stage.

(iii) This second stage can generally be carried out by looking up a table of random numbers. So, referring to Example 6.2, suppose that we get $(P_0 + \alpha - \alpha'')/P_0 = 0.65$ and *before* the experiment we have arbitrarily selected a certain position (say, 12th from the top) of a certain column at a certain page of a two-digit random numbers table. If that number lies between 00 and 64 we reject H_0 and accept it otherwise. By so doing, we have performed the most powerful test of size α on the simple hypotheses (6.8).

6.3.2 *A few notes on sequential analysis*

So far we have considered the sample size n as a number fixed in advance. Even at the end of Example 6.1(a), once we have chosen the desired values of α and β, eq. (6.15) shows that n can be determined before the experiment is carried out. A different approach due to Abraham Wald and called 'sequential analysis' leads to a decision on the null hypothesis without fixing the sample size in advance but by considering it as a random variable which depends on the experiment's outcomes.

It should be pointed out that sequential analysis is a rather broad subject worthy of study in its own right (see, for instance, Wald's book [22]) but here we limit ourselves to some general comments relevant to our present discussion.

As in the preceding section, suppose that we wish to test the two simple hypotheses (6.8). For $k = 0, 1$ let

$$L_{km} \equiv L(x_1, \dots, x_m; \theta_k) = \prod_{i=1}^{m} f_k(x_i) \tag{6.24}$$

be the two likelihood functions L_{0m}, L_{1m} under the hypothesis H_0, H_1, respectively, after m observations (i.e. the realization $x_1, \dots, x_m$). Then, the general idea of Wald's sequential test is as follows : (i) we appropriately fix two positive numbers r, R $(r < 1 < R)$, (ii) we continue testing as long as the likelihood ratio $l_m \equiv L_{0m}/L_{1m}$ lies between the two limits r, R and (iii) terminate the process for the first index which violates one of the inequalities

$$r < \frac{L_{0m}}{L_{1m}} < R \tag{6.25}$$

If we call n this stopping index, n is the realization of a r.v. N and we have two possibilities (a) $l_n = L_{0n}/L_{1n} < r$ or (b) $l_n = L_{0n}/L_{ln} > R$; in case (a) we reject H_0 (accept H_1) while, on the contrary, we accept H_0 in case (b). It is evident at this point that the limit numbers r, R will be determined on the basis of the risk we are prepared to take in coming to one decision or the other and this, in other words, means that they will depend on $\alpha = P(H_1|H_0)$ and $\beta = P(H_0|H_1)$. We will come to this point shortly; for the moment we just say that if α and β are given, we speak of a test of strength (α, β). Among all tests of strength (α, β) we will tend to prefer the one with the smallest number of observations. In this light, a test which minimizes both $E_0(N)$ and $E_1(N)$ – that is, the average number of observations under H_0 or H_1, respectively – is called optimal and it can be shown (see Ref. [10]) that Wald's test is, in fact, optimal.

The considerations above imply in practice that r and R divide the sample space into three disjoint regions: the rejection region Ξ_1, the acceptance region Ξ_0 and an intermediate region (we can call it the 'doubtful' or 'indifference' region) D. As long as the values of the likelihood ratio fall in the doubtful region – that is, between r and R – the experiment continues. For a given strength (α, β), however, the problem remains of specifying the numbers r and R. It turns out that they cannot be determined exactly but we can nonetheless obtain a lower limit r' for r and an upper limit R' for R by considering the two probabilities of making a correct decision. In fact, from the relations

$$\begin{aligned} 1-\alpha &= \int_{\Xi_0} L_0(\mathbf{x})\,d\mathbf{x} \ge R \int_{\Xi_0} L_1(\mathbf{x})\,d\mathbf{x} = \beta R \\ 1-\beta &= \int_{\Xi_1} L_1(\mathbf{x})\,d\mathbf{x} \ge \frac{1}{r} \int_{\Xi_1} L_0(\mathbf{x})\,d\mathbf{x} = \frac{\alpha}{r} \end{aligned} \tag{6.26}$$

we get

$$\begin{aligned} r &\ge \frac{\alpha}{1-\beta} \equiv r' \\ R &\le \frac{1-\alpha}{\beta} \equiv R' \end{aligned} \tag{6.27}$$

which, as noted above, do not specify r and R uniquely. The usual choice is to take r' and R' as the two limiting boundaries and consider the resulting test as a valid approximation of the desired test of strength (α, β). The choice is satisfactory because, denoting by α' and β' the probabilities of error of the approximate test, we have

$$\alpha' + \beta' \le \alpha + \beta \tag{6.28}$$

In fact, writing the counterparts of eq. (6.26) for the primed quantities α', β', r', R' and taking eq. (6.27) into account we get $\alpha'(1-\beta) \le \alpha(1-\beta')$ and $\beta'(1-\alpha) \le \beta(1-\alpha')$. Adding these two inequalities leads to eq. (6.28).

Before closing this section, it is worth pointing out some specific features of the sequential method as compared to the 'standard' Neyman–Pearson's procedure. First of all, for a given statistical model, in Neyman–Pearson's approach we need to know the distribution of the test statistic – under H_0 and under H_1 – in order to define the critical region and calculate the power the test. No such information is needed in Wald's test because its strength is decided beforehand and the values of the likelihood-ratio can be calculated directly from the data without searching for any distribution.

Second, Wald's test is economical in terms of number of trials performed before taking a decision. In fact, for given values α and β it can be shown ([10] or [12]) that the ratios $E_0(N)/n$ and $E_1(N)/n$ – where n is here the sample size of a Neyman–Pearson's test – are generally less than unity, in some cases even reaching values close to 0.5. This observation, by implicitly implying that sooner or later we come to a decision, leads to a third noteworthy feature of Wald's test: the fact that it stops. This is established by a theorem(see Ref. [10]) stating that Wald's test will stop with probability one after a finite number of steps, thus preventing the possibility of endless sampling.

6.3.3 *Composite hypotheses: the likelihood ratio test*

Parametric testing problems with two simple hypotheses are rather rare in applications and, in general, at least one hypothesis is composite. Typical examples are the frequently encountered cases where the null hypothesis $H_0 : \theta = \theta_0$ must be tested again one of the possible alternatives (a) $H_1 : \theta > \theta_0$, (b) $H_1 : \theta < \theta_0$ or (c) $H_1 : \theta \neq \theta_0$ and one speaks of one-sided alternative in cases (a) and (b) – right- and left-sided, respectively – and of two-sided alternative in case (c).

With composite hypotheses , a uniformly most powerful (ump) test exists only for some special classes of problems but many of these, fortunately, occur quite often in practice. So, for instance, many statistical models for which there is a sufficient statistic T (for the parameter θ under test so that eq. (5.52) holds) have a monotone (in T) likelihood ratio; for these models it can be shown (see Ref. [10] or [15]) that a ump test to verify $H_0 : \theta = \theta_0$ against a one-sided alternative does exist. This 'optimal' test, moreover, coincides with the Neyman–Pearson's test for $H_0 : \theta = \theta_0$ against an arbitrarily fixed alternative H_1, where H_1 is in the form (a) or (b). Even more, the first test is also the ump test for the doubly composite case $H_0 : \theta \leq \theta_0; H_1 : \theta > \theta_0$ while the second is the ump test for $H_0 : \theta \geq \theta_0; H_1 : \theta < \theta_0$.

In spite of all these interesting and important results , it is not our intention to enter into such details and we refer the interested reader to more specialized literature. Here, after some examples of composite hypotheses cases , we will limit ourselves to the description of the general method called 'likelihood ratio test' which – although not leading to ump tests in most cases – has a number of other desirable properties.

Example 6.3(a) If we wish to test the hypotheses $H_0 : \theta = \theta_0; H_1 : \theta > \theta_0$ for the mean of a normal model with known variance σ^2, we can follow the same line of reasoning of Example 6.1(a) and obtain the rejection region (6.14a). Since this rejection region does not depend on the specific value θ_1 against which we compare H_0 (provided that $\theta_1 > \theta_0$), it turns out that this is the uniformly most powerful test for the case under investigation and, as noted above, for the pair of hypotheses $H_0 : \theta \le \theta_0; H_1 : \theta > \theta_0$ as well.

Similar considerations apply to the problem $H_0 : \theta = \theta_0; H_1 : \theta < \theta_0$ and we can conclude that the rejection region (6.14b) provides the ump test for this case and for $H_0 : \theta \ge \theta_0; H_1 : \theta < \theta_0$.

Example 6.3(b) Considering the normal model of Example 6.1(b) – that is, known mean and unknown variance $\sigma^2 = \theta^2$ – it is now evident that the rejection region (6.18a) provides the ump test for the problem $H_0 : \theta^2 \le \theta_0^2; H_1 : \theta^2 > \theta_0^2$ while (6.18b) applies to the case $H_0 : \theta^2 \ge \theta_0^2; H_1 : \theta^2 < \theta_0^2$.

In all the cases above, the probability of a type II error β (and therefore the power) will depend on the specific value of the alternative. Often, in fact, one can find graphs of β plotted against an appropriate variable with the sample size n as a parameter. These graphs are called operating characteristic curves (OC curves) and the variable on the abscissa axis depends on the type of test. So, for instance, the OC curve for the first test of Example 6.3(a) plots β versus $(\theta_1 - \theta_0)/\sigma$ for some values of n. Fig. 6.1 is one such graph for $\alpha = 0.05$ and the three values of sample size $n = 5$, $n = 10$ and $n = 15$. As it should be expected, β decreases as the difference $\theta_1 - \theta_0$ increases and, for a fixed value of this quantity, β is lower for larger sample sizes.

Similar curves can generally be drawn with little effort for the desired sample size by using widely available software packages such as, for instance,

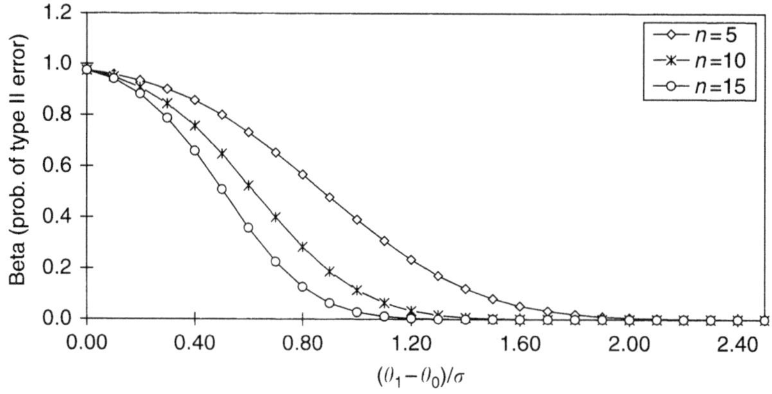

Figure 6.1 One-sided test (size $= 0.05 - H_1 : \theta > \theta_0$).

Excel®, Matlab® etc. The reader is invited to do so for the cases of Example 6.3(b).

Example 6.3(c) Referring back to Example 6.3(a) – normal model with known variance – let us make some considerations on the two-sided case $H_0 : \theta = \theta_0; H_1 : \theta \neq \theta_0$. At the significance level α, the rejection regions for the one-sided alternatives $H_1 : \theta > \theta_0$ and $H_1 : \theta < \theta_0$ are given by eqs (6.14a) and (6.14b), respectively. If we conveniently rewrite these equations as

$$\begin{aligned} \Xi_1^+ &= \{\mathbf{x} : \sqrt{n}(m - \theta_0)/\sigma \geq z_\alpha\} \\ \Xi_1^- &= \{\mathbf{x} : \sqrt{n}(m - \theta_0)/\sigma \leq -z_\alpha\} \end{aligned} \tag{6.29}$$

we may think of specifying the rejection region Ξ_1 of the two-sided test (at the level α) as $\Xi_{1,\alpha} = \Xi_{1,a}^- \cup \Xi_{1,b}^+$, where a, b are two numbers such that $a + b = \alpha$. Moreover, intuition suggests to take a 'symmetric' region by choosing $a = b = \alpha/2$ thus obtaining

$$\tilde{\Xi}_1 = \left\{\mathbf{x} : \frac{\sqrt{n}}{\sigma}|m - \theta_0| \geq z_{\alpha/2}\right\} \tag{6.30}$$

which, in other words , means that we reject the null hypothesis when θ_0 is sufficiently far – on one side or the other – from the sample mean m. As before, the term 'sufficiently far' depends on the risk involved in rejecting a true null hypothesis (or, equivalently, accepting a false alternative).

We do not do it here but these heuristic considerations leading to (6.30) can be justified on a more rigorous basis showing that, for the case at hand, eq. (6.30) is a good choice because it defines the ump test among the class of unbiased tests. In fact, it turns out that a ump test does not exist for this case because (at the level α) the two tests leading to (6.29) can be considered in their own right as tests for the alternative $H_1 : \theta \neq \theta_0$. In this light, we already know that (i) the Ξ_1^+-test is the most powerful in the region $\theta > \theta_0$ (ii) the Ξ_1^--test is the most powerful for $\theta < \theta_0$ and (iii) both their powers take on the value α at $\theta = \theta_0$. However, as tests against $H_1 : \theta \neq \theta_0$, they are biased. In fact, the power $W^+(\theta)$ of the first test is rather poor (i.e. low and such that $W^+(\theta) < \alpha$) for $\theta < \theta_0$ and the same holds true for $W^-(\theta)$ when $\theta > \theta_0$ so that, calling $\tilde{W}(\theta)$ the power of the test (6.30), we have the inequalities

$$\begin{aligned} &W^+(\theta) < \alpha < \tilde{W}(\theta) < W^-(\theta), \quad \theta < \theta_0 \\ &W^-(\theta) < \alpha < \tilde{W}(\theta) < W^+(\theta), \quad \theta > \theta_0 \end{aligned}$$

while, clearly, $W^-(\theta_0) = W^+(\theta_0) = \tilde{W}(\theta_0) = \alpha$. The fact that for two-sided alternative hypothesis there is no ump test but there sometimes exists a ump unbiased test is more general than the special case considered here and the interested reader can refer, for instance, to [15] for more details.

As pointed out above, the likelihood ratio method is a rather general procedure used to test composite hypotheses of the type (6.2). In general, it does not lead to ump tests but gives satisfactory results in many practical problems. As before, let $\mathbf{X} = (X_1, \ldots, X_n)$ be a random sample and let the (absolutely continuous) model be expressed in terms of the pdf $f(x)$. The likelihood ratio statistic is defined as

$$\lambda(\mathbf{X}) = \frac{\sup_{\theta \in \Theta_0} L(\mathbf{X}; \theta)}{\sup_{\theta \in \Theta} L(\mathbf{X}; \theta)} = \frac{L(\hat{\theta}_0)}{L(\hat{\theta})} \tag{6.31}$$

where $\hat{\theta}_0$ is the maximum likelihood (ML) estimate of the unknown parameter θ when $\theta \in \Theta_0$ and $\hat{\theta}$ is the ML estimate of θ over the entire parameter space Θ. Definition 6.31 shows that $0 \leq \lambda \leq 1$ because we expect λ to be close to zero when the null hypothesis is false and close to unity when H_0 is true. In this light, the rejection region is defined by

$$\Xi_1 = \{\mathbf{x} : \lambda(\mathbf{x}) \leq c\} \tag{6.32}$$

where the number c is determined by the significance level α and it is such that

$$\sup_{\theta \in \Theta_0} P_\theta\{\lambda(\mathbf{X}) \leq c\} = \alpha \tag{6.33}$$

which amounts to the condition $P(H_1|H_0) = P_\theta\{\lambda(\mathbf{X}) \leq c\} \leq \alpha$ for all $\theta \in \Theta_0$ (recall the definition of power (6.4) and eq. (6.5)). It is clear at this point that the likelihood ratio test generalizes Neyman–Pearson's procedure to the case of composite hypotheses and reduces to it when both hypotheses are simple. If the null hypothesis is simple – that is, of the form $H_0 : \theta = \theta_0$ – the numerator of the likelihood ratio is simply $L(\theta_0)$ and, since Θ_0 contains only the single element θ_0, no 'sup' appears both at the numerator of (6.31) and in eq. (6.33).

Example 6.4(a) In Example 6.3(a) we have already discussed the normal model with known variance when the test on the mean $\mu = \theta$ is of the form $H_0 : \theta \leq \theta_0; H_1 : \theta > \theta_0$. Let us now examine it in the light of the likelihood ratio method. For this case, clearly, $\Theta = \mathbb{R}$, $\Theta_0 = (-\infty, \theta_0]$ and $\Theta_1 = (\theta_0, \infty)$. The denominator of (6.31) is $L(M)$ because the sample mean M is the ML estimator of the mean (recall Section 5.5) over the entire parameter space. On the other hand, it is not difficult to see that the numerator of (6.31) is $L(\theta_0)$ so that

$$\lambda(\mathbf{X}) = \exp\left(-\frac{n}{2\sigma^2}(M - \theta_0)^2\right) \tag{6.34}$$

and the inequality in (6.32) is satisfied if

$$|m - \theta_0| \geq \sqrt{\frac{2\sigma^2}{n} \log(1/c)} = t(c)$$

where, as usual, m is the realization of M. Since $m > \theta_0$ in the rejection region, the condition (6.33) reads

$$\begin{aligned}\sup_{\theta \leq \theta_0} P\{M \geq \theta_0 + t(c)\} &= \sup_{\theta \leq \theta_0} P\left\{\frac{\sqrt{n}(M-\theta)}{\sigma} \geq \frac{\sqrt{n}(\theta_0 - \theta + t(c))}{\sigma}\right\} \\ &= P\left(Z \geq \frac{\sqrt{n}t(c)}{\sigma}\right) = \alpha\end{aligned}$$

because in Ξ_1 the r.v. $Z = \sqrt{n}(M-\theta)/\sigma$ is standard normal and the 'sup' of the probability above is attained when $\theta = \theta_0$. The conclusion is that, as expected, the rejection region is given by eq. (6.14), that is, the same as for the simple case $H_0 : \theta = \theta_0; H_1 : \theta = \theta_1$ with $\theta_1 > \theta_0$. Also, from the considerations above we know that eq. (6.14a) defines the ump for the problem at hand.

Example 6.4(b) The reader is invited to work out the details of the likelihood ratio method for the normal case in which the hypotheses on the mean are $H_0 : \theta = \theta_0; H_1 : \theta \neq \theta_0$ and the variance is known. We have in this case $\Theta = \mathbb{R}$, $\Theta_0 = \{\theta_0\}$ and $\Theta_1 = (-\infty, \theta_0) \cup (\theta_0, \infty)$ and, as above, $\lambda(\mathbf{X}) = L(\theta_0)/L(M)$ so that eq. (6.34) still applies. Now, however, no 'sup' needs to be taken in eq. (6.33) which, in turn, becomes $P\{|Z| \geq \sqrt{n}\, t(c)/\sigma\} = \alpha$ thus leading to the rejection region (6.30) where, we recall, $z_{\alpha/2}$ denotes the $\alpha/2$-upper quantile of the standard normal distribution. So, for instance, if $\alpha = 0.01$ we have $z_{\alpha/2} = z_{0.005} = 2.576$ while for a 10 times smaller probability of type I error – that is, $\alpha = 0.001$ – we find in tables the value $z_{\alpha/2} = z_{0.0005} = 3.291$.

The fact that

(a) testing hypotheses on the mean of a normal model with known variance leads to rejection regions where the quantiles of the standard normal distribution appear;
(b) testing hypotheses on the variance (Examples 6.1(b) and 6.3(b)) of normal model with known mean brings into play the quantiles of the χ^2 distribution with n degrees of freedom

may suggest a connection between parametric hypothesis testing and interval estimation problems (Section 5.6). The connection, in fact, does exist and is rather strong. Before considering the situation from a general point of view, we give two more examples that confirm this state of affairs.

Example 6.5(a) Consider a normal model with the same hypotheses on the mean as in Example 6.4(b) but with the important difference that the variance σ^2 is now unknown (note that in this situation the null hypothesis $H_0 : \theta = \theta_0$ is not simple because it does not uniquely determine the probability distribution). In this case, the parameter space $\Theta = \mathbb{R} \times \mathbb{R}^+$ is divided into the two sets $\Theta_0 = \{\theta_0\} \times \mathbb{R}^+$ and $\Theta_1 = [(-\infty, \theta_0) \cup (\theta_0, \infty)] \times \mathbb{R}^+$. Under the null hypothesis $\theta = \theta_0$ the maximum of the likelihood function is attained when $\sigma^2 = \sigma_0^2 = n^{-1} \sum_i (x_i - \theta_0)^2$ and the numerator of (6.31) is

$$L(\theta_0, \sigma_0^2) = (2\pi\sigma_0^2 \mathrm{e})^{-n/2}$$

Similarly, the denominator of (6.31) is given by $L(M, S^2) = (2\pi S^2 \mathrm{e})^{-n/2}$ because M and $S^2 = n^{-1} \sum_i (X_i - M)^2$, respectively, are the ML estimators of the mean and the variance. Consequently, if s^2 is the realization of the sample variance S^2, we get

$$\lambda = \left(\frac{\sigma_0^2}{s^2}\right)^{-n/2} = \left(1 + \frac{t^2}{n-1}\right)^{-n/2} \tag{6.35}$$

where we set $t = t(\mathbf{x}) = \sqrt{n-1}(m - \theta_0)/s$ and the second relation is easily obtained by noting that $\sigma_0^2 = s^2 + (m - \theta_0)^2$. Equation (6.35) shows that there is a one-to-one correspondence between λ and t^2 and therefore the inequality in (6.32) is equivalent to $|t| \geq c'$ (where c' is as appropriate). Since the r.v. $t(\mathbf{X})$ is distributed according to a Student distribution with $n-1$ degrees of freedom, the boundary c' of the rejection region (at the significance level α) must be the $\alpha/2$-upper quantile of this distribution. Denoting this upper quantile by $t_{\alpha/2;n-1}$ we have

$$\Xi_1 = \left\{ \mathbf{x} : \frac{\sqrt{n-1}}{s} |m - \theta_0| \geq t_{\alpha/2;n-1} \right\} \tag{6.36}$$

Alternatively, if one prefers to do so, Ξ_1 can be specified in terms of the $(1 - \alpha/2)$-lower quantile and/or using the unbiased estimator $\bar{S}^2$ instead of S^2 (and recalling that $\sqrt{n-1}/s = \sqrt{n}/\bar{s}$). As a numerical example, suppose we carried out $n = 10$ trials to test the hypotheses $H_0 : \theta = 35; H_1 : \theta \neq 35$ at the level $\alpha = 0.01$. The rejection region is then $\Xi_1 = \{\mathbf{x} : (3/s)|m - 35| \geq 3.25\}$ because the 0.005-upper quantile of the Student distribution with 9 degrees of freedom is $t_{0.005;9} = 3.250$. So, if the experiment gives, for instance, $m = 32.6$ and $s = 3.3$ we fall outside Ξ_1 and we accept the null hypothesis at the specified significance level.

In the lights of the comments above, the conclusion is as expected: testing the mean of a normal model with unknown variance leads to the appearance of the Student quantiles, in agreement with Example 5.9(b) where, for the same model, we determined a confidence interval for the mean (eq. (5.76)).

Example 6.5(b) At this point it should not be surprising if we say that testing the variance of a normal model with unknown mean involves the quantiles of the χ^2 distribution with $n-1$ degrees of freedom (recall Example 5.10(b)). It is left to the reader to use the likelihood ratio method to determine that the rejection region for the hypotheses $H_0 : \theta^2 = \theta_0^2; H_1 : \theta^2 \neq \theta_0^2$ on the variance is

$$\Xi_1 = \left\{ \mathbf{x} : \left(0 \leq \sqrt{n-1}\frac{\bar{s}^2}{\theta_0^2} \leq \chi^2_{1-\alpha/2;n-1} \right) \cup \left(\sqrt{n-1}\frac{\bar{s}^2}{\theta_0^2} \geq \chi^2_{\alpha/2;n-1} \right) \right\} \tag{6.37}$$

where $\chi^2_{1-\alpha/2;n-1}$ and $\chi^2_{\alpha/2;n-1}$, respectively, are the $(1-\alpha/2)$-upper quantile and the $\alpha/2$-upper quantile of the χ^2 distribution with $n-1$ degrees of freedom. So, for instance, if $n = 10$ and we are testing at the level $\alpha = 0.05$ we find $\chi^2_{1-\alpha/2;n-1} = \chi^2_{0.975;9} = 2.70$ and $\chi^2_{\alpha/2;n-1} = \chi^2_{0.025;9} = 19.023$. Therefore, we will accept the null hypothesis H_0 if $2.70 < 3\bar{s}^2/\theta_0^2 < 19.023$.

6.3.4 *Complements on parametric hypothesis testing*

In addition to the main ideas of parametric hypotheses testing discussed in the preceding sections, some further developments deserve consideration. Without claim of completeness, we do this here by starting from where we left off in Section 6.3.3: the relationship with confidence intervals estimation.

6.3.4.1 *Parametric tests and confidence intervals*

Let us fix a significance level α. If we are testing $H_0 : \theta = \theta_0$ against the composite alternative $H_1 : \theta \neq \theta_0$, the resulting rejection region Ξ_1 will depend, for obvious reasons , on the value θ_0 and so will the acceptance region $\Xi_0 = \Xi_1^C$. As θ_0 varies in the parameter space we can define the family of sets $\Xi_0(\theta) \subset \Xi$, where each $\Xi_0(\theta)$ corresponds to a value of θ. On the other hand, a given realization of the sample $\mathbf{x}$ will fall – or will not fall – in the acceptance region depending on the value of θ under test and those values of θ such that $\mathbf{x}$ does fall in the acceptance region define a subset $G \subset \Theta$ of the parameter space. By letting $\mathbf{x}$ free to vary in Ξ, therefore, we can define the family of subsets $G(\mathbf{x}) = \{\theta : \mathbf{x} \in \Xi_0(\theta)\} \subset \Theta$. The consequence of this (rather intricate at first sight) construction of subsets in the sample and parameter space is the fact that the events $\{\mathbf{X} \in H_0(\theta)\}$ and

$\{\theta \in G(\mathbf{X})\}$ are equivalent, and since the probability of the first event is $1-\alpha$ so is the probability of the second. Recalling Section 5.6, however, we know that this second event defines a confidence interval for the parameter θ. By construction, the confidence level associated to this interval is $1-\alpha=\gamma$.

The conclusion, as anticipated, is that parametric hypothesis testing and confidence interval estimation are two strictly related problems and the solution of one leads immediately to the solution of the other. Moreover, a ump test – if it exists – corresponds to the shortest confidence interval and *vice versa*.

Having established the nature of the connection between the two problems, we can now reconsider some of the preceding results from this point of view. Let us go back to Example 6.3(c) where the rejection region is given by eq. (6.30). This relation implies that the 'acceptance' subsets $\Xi_0(\theta)$ have the form

$$\Xi_0(\theta)=\left\{\mathbf{x}: m-\frac{\sigma}{\sqrt{n}}z_{\alpha/2}<\theta<m+\frac{\sigma}{\sqrt{n}}z_{\alpha/2}\right\} \tag{6.38}$$

Those $\mathbf{x}$ such that $\mathbf{x}\in\Xi_0(\theta)$ correspond to the values of θ which satisfy the inequality in (6.38) and these θ, in turn, define the corresponding set $G(\mathbf{x})$. From the discussion above it follows that $(M\pm z_{\alpha/2}\sigma/\sqrt{n})$ is a $1-\alpha$ confidence interval for the mean of a normal model with known variance. This is in agreement with eq. (5.75) of Example 5.9(a) because the $\alpha/2$-upper quantile $z_{\alpha/2}$ (of the standard normal distribution) is the $(1-\alpha/2)$-lower quantile and, since $1-\alpha=\gamma$, this is just the $(1+\gamma)/2$-lower quantile (which was denoted by the symbol $c_{(1+\gamma)/2}$ in Example 5.9(a)).

Similarly, for the same statistical model we have seen that the rejection region to test $H_0:\theta=\theta_0; H_1:\theta>\theta_0$ is given by (6.14a). This implies that the acceptance sets are $\Xi_0(\theta)=\{\mathbf{x}: m<\theta+z_\alpha\sigma/\sqrt{n}\}$ and the corresponding sets $G(\mathbf{x})$ are given by $G(\mathbf{x})=\{\theta:\theta>m-z_\alpha\sigma/\sqrt{n}\}$. The conclusion is that

$$\left(M-z_\alpha\frac{\sigma}{\sqrt{n}},+\infty\right) \tag{6.39a}$$

is a lower $(1-\alpha)$-confidence (one-sided) interval for the mean. The interval (6.39a) is called 'lower' because only the lower limit for θ (recall eq. (5.69a)) is specified; by the same token, it is evident that we can use the rejection region (6.14b) to construct the upper $(1-\alpha)$-confidence interval

$$\left(-\infty,M+z_\alpha\frac{\sigma}{\sqrt{n}}\right) \tag{6.39b}$$

The argument, of course, works in both directions and we can, for instance, start from the γ-CI (5.76) for the mean of a normal model with unknown variance to write the acceptance region for the test $H_0 : \theta = \theta_0; H_1 : \theta \neq \theta_0$ as

$$\Xi_0 = \left\{ \mathbf{x} : -t_{(1+\gamma)/2;n-1} < \frac{\sqrt{n}(m - \theta_0)}{\bar{s}} < t_{(1+\gamma)/2;n-1} \right\}$$

from which we get the rejection region $\Xi_1 = \Xi_0^C$ of eq. (6.36) by noting that (i) $\sqrt{n-1}/s = \sqrt{n}/\bar{s}$ and (ii) $(1+\gamma)/2 = 1 - \alpha/2$ so that the Student lower quantiles of (5.76) are the $1 - (1 - \alpha/2) = \alpha/2$ (Student) upper quantiles of eq. (6.36).

At this point it is left to the reader to work out the details of the parametric hypothesis counterparts of Examples 5.11 (a–c) by determining the relevant acceptance and rejection regions.

6.3.4.2 *Asymptotic behaviour of parametric tests*

The second aspect we consider is the asymptotic behaviour of parametric tests. As a first observation, we recall from Chapter 5 that a number of important sample characteristics are asymptotically normal with means and variances determined by certain population parameters. Consequently, when the test concerns one of these characteristics the first thing that comes to mind is, for large samples, to use the normal approximation by replacing any unknown population parameter by its (known) sample counterpart thus obtaining a rejection region determined by the appropriate quantile of the standard normal distribution. This is a legitimate procedure but it should be kept in mind that it involves two types of approximations (i) the normal approximation for the distribution of the characteristic under test and (ii) the use of sample values for the relevant unknown population parameters. So, in practice, it is often rather difficult to know whether our sample is large enough and our test has given a reliable result. As a rule of thumb, $n > 30$ is generally good enough when we are dealing with means while $n > 100$ is advisable for variances, medians , coefficients of skewness and kurtosis. For some other 'less tractable' characteristics, however, even samples as large as 300 or more do not always give a satisfactory approximation.

Let us now turn our attention to Neyman–Pearson's lemma on simple hypotheses (Proposition 6.1). The boundary value c in eq. (6.9) can only be calculated when we know the distribution of the statistic $l(\mathbf{X})$ under H_0 (and, similarly, the probability β of a type II error can be calculated when we know the distribution of $l(\mathbf{X})$ under H_1). Since this is not always possible, we can proceed as follows. If we define the r.v.s

$$Y_i = \log \frac{f_0(X_i)}{f_1(X_i)} \tag{6.40}$$

for $i = 1, 2, \ldots, n$ then $S_n = Y_1 + Y_2 + \cdots + Y_n$ is the sum of n iid variables. Depending on which hypothesis is true, its mean and variance are

$E_{\theta_0}(S_n) = na_0$ and $\mathrm{Var}_{\theta_0}(S_n) = n\sigma_0^2$ under H_0 or $E_{\theta_1}(S_n) = na_1$ and $Var_{\theta_1}(S_n) = n\sigma_1^2$ under H_1, where we called a_0, σ_0^2 and a_1, σ_1^2 the mean and variance (provided that they exist) of the variables Y_i under H_0 and H_1, respectively. At this point, the CLT (Proposition 4.22) tells us that, under H_0, the r.v. $Z = (S_n - na_0)/\sigma_0\sqrt{n}$ is asymptotically standard normal and, since definition (6.40) implies that the inequality in eq. (6.9) is equivalent to $S_n \le \log c$, for sufficiently large values of n we can write

$$P_{\theta_0}\left\{Z \le \frac{\log c - na_0}{\sigma_0\sqrt{n}}\right\} = \alpha \tag{6.41}$$

which, on the practical side, implies that we have an approximate rejection region: we reject the null hypothesis if the realization of the sample is such that

$$\frac{(y_1 + y_2 + \cdots + y_n) - na_0}{\sigma_0\sqrt{n}} \le c_\alpha \tag{6.42}$$

where c_α is the α-lower quantile of the standard normal distribution. Clearly, the goodness of the approximation depends on how fast the variable Z converges (in distribution) to the standard normal r.v.; if the rate of convergence is slow, a rather large sample is required to obtain a reliable test.

In regard to the more general likelihood ratio method, it is convenient to consider the monotone function $\Lambda(\mathbf{X}) = -2\log\lambda(\mathbf{X})$ of the likelihood ratio $\lambda(\mathbf{X})$. Provided that the regularity conditions for the existence, uniqueness and asymptotic normality of the ML estimate $\hat{\theta}$ of the parameter θ are met (see Sections 5.5 and 5.5.1), it can be shown that the asymptotic rejection region for testing the null hypothesis $H_0 : \theta = \theta_0$ is given by

$$\Xi_1 = \{\mathbf{x} : \Lambda(\mathbf{x}) \ge \chi^2_{1-\alpha;1}\} \tag{6.43}$$

where $\chi^2_{1-\alpha;1}$ is the $(1-\alpha)$-lower quantile of the χ^2 distribution with one degree of freedom. We do not prove this assertion here but we note that eq. (6.43) is essentially due to the fact that, under H_0, we have

(i) $\Lambda(\mathbf{X}) \to \chi^2(1)[D]$;
(ii) $P_{\theta_0}\{\Lambda(\mathbf{X}) \ge \chi^2_{1-\alpha;\,1}\} \to \alpha$

as $n \to \infty$. The result can also be extended directly to the case of a vector, say k-dimensional, parameter $\mathbf{q} = (\theta_1, \ldots, \theta_k)$ and it turns out that the rejection region is defined by means of the $(1-\alpha)$-lower quantile of the χ^2 distribution with k degrees of freedom. In addition, the procedure still applies if the null hypothesis specifies only a certain number, say r, of the k components of $\mathbf{q}$. In this case the numerator of (6.31) is obtained by maximizing L with respect to the remaining $k - r$ components and r is the number of degrees of freedom of the asymptotic distribution. Also, another application of the result stated by eq. (6.43) is the construction of confidence intervals for a general parametric model. In fact, eq. (6.43) implies that the (asymptotic) acceptance region is $\Xi_0 = \{\mathbf{x} : \Lambda(\mathbf{x}) < \chi^2_{\gamma;1}\}$, where $\gamma = 1 - \alpha$. In the light of the relation between hypothesis testing and CIs we have that $G(\mathbf{X}) = \{\theta : \Lambda(\mathbf{X}) < \chi^2_{\gamma;1}\}$ is an asymptotic γ-CI for the parameter θ because $P_\theta\{\theta \in G(\mathbf{X})\} \to \gamma$ as $n \to \infty$. In practice $G(\mathbf{X})$ – called a maximum likelihood confidence interval – can be used as an approximate CI when the sample size is large. All these further developments , however – together with the proof of the theorem above – are beyond our scope. For more details the interested reader may refer, for instance, to [1, 2, 10, 13, 19].

6.3.4.3 The p-value: significance testing

A third aspect worthy of mention concerns a slightly different implementation of the hypothesis testing procedure shown in the preceding sections. This modified procedure is often called 'significance testing' and is becoming more and more popular because it somehow overcomes the rigidity of hypothesis testing. The main idea of significance testing originates from the fact that the choice of the level α is, to a certain extent, arbitrary and the common values 0.05, 0.025 and 0.01 are often used out of habit rather than through careful analysis of the consequences of a type I error. So, instead of fixing α in advance we perform the experiment, calculate the value of the appropriate test statistic and report the so-called p-value (or 'observed significance level' and denoted by α_{obs}), defined as the smallest value of α for which we reject the null hypothesis. Let us consider an example.

Example 6.6 Suppose that we are testing $H_0 : \theta = 100$ against $H_1 : \theta < 100$ where the parameter θ is the mean of a normally distributed r.v. with known variance $\sigma^2 = 25$. Suppose further that an experiment on a sample of $n = 16$ products gives the sample mean $m = 97.5$. Since the rejection region for this case is given by eq. (6.14b), a test at level $\alpha = 0.05$ ($z_{0.05} = 1.645$) leads to reject H_0 in favour of H_1 and the same happens at the level $\alpha = 0.025$ ($z_{0.025} = 1.960$). If, however, we choose $\alpha = 0.01$ ($z_{0.01} = 2.326$) the conclusion is that we must accept H_0. This kind of situation is illustrative of the rigidity of the method; we are saying in practice that we tolerate 1 chance in 100 of making a type I error but at the same time we state that 2.5 chances in 100 is too risky.

One way around this problem is, as noted above, to calculate the p-value and move on from there. For the case at hand the relevant test statistic $\sqrt{n}(M - \theta_0)/\sigma$ is standard normal and attains the value $4(-2.5)/5 = -2.0$ which, we find on statistical tables, corresponds to the level $\alpha_{obs} = 0.0228$.

This is the p-value for our case, that is, the value of α that will just barely cause H_0 to be rejected. In other words, on the basis of the observed data we would reject the null hypothesis for any level $\alpha \geq \alpha_{obs} = 0.0228$ and accept it otherwise. Reporting the p-value, therefore, provides the necessary information to the reader to decide whether to accept or reject H_0 by comparing this value with his/her own choice of α: if one is satisfied with a level $\alpha = 0.05$ then he/she will not accept H_0 but if he/she thinks that $\alpha = 0.01$ is more appropriate for the case at hand, the conclusion is that H_0 cannot be rejected. (As an incidental remark, it should be noted that the calculation of the p-value for this example is rather easy but it may not be so if the test statistic is not standard normal and we must rely only on statistical tables. However, the use widely available software packages has made things much easier because p-values are generally given by the software together with all the other relevant results of the test.)

The example clarifies the general idea. Basically, this approach provides the desired flexibility; if an experiment results in a low p-value, say $\alpha_{obs} < 0.01$, then we can be rather confident in our decision of rejecting the null hypothesis because – had we tested it at the 'usual' levels 0.05, 0.025 or 0.01 – we would have rejected it anyway. Similarly, if $\alpha_{obs} > 0.1$ any one of the usual testing levels would have led to the acceptance of H_0 and we can feel quite comfortable with the decision of accepting H_0. A kind of 'grey area', so to speak, is when $0.01 < \alpha_{obs} < 0.1$ and, as a rule of thumb, we may reject H_0 for $0.01 < \alpha_{obs} < 0.05$ and accept it for $0.05 < \alpha_{obs} < 0.1$. It goes without saying, however, that exceptions to this rule are not rare and the specific case under study may suggest a different choice. In addition, we must not forget to always keep an eye on the probability β of a type II error.

6.3.4.4 *Closing remarks*

To close this section, a comment of general nature is not out of place: in some cases , taking a too large sample size may be as bad an error as taking a too small sample size. The reason lies in the fact as n increases we are able to detect smaller and smaller differences from the null hypothesis (this, in other words, means that that the 'discriminating power' of the test increases as n increases) and consequently we will almost always reject H_0 if n is large enough. In performing a test, therefore, we must keep in mind the difference between statistical significance and practical significance, where this latter term refers to both reasonableness and to the nominal specifications , if any, for the case under study. So, without loss of generality, consider the test of Example 6.6 and suppose that for all practical purposes it would not matter much if the mean of the population were within ± 0.5 units from the value $\theta_0 = 100$ under test. If we decided to take a sample of $n = 1600$ products obtaining, say, the sample mean $m = 99.7$, we would reject the null hypothesis at $\alpha = 0.05$, $\alpha = 0.025$ and $\alpha = 0.01$ (incidentally, the p-value in this case is $\alpha_{\text{obs}} = 0.0082$). This is clearly unreasonable because the statistically significant difference detected by the test (which leads to the rejection of H_0) does not correspond to a practical difference and we put ourselves in the same ironical situation of somebody who uses a microscope when a simple magnifying glass would do.

As stated at the beginning of this section, these complementary notes do not exhaust the broad subject of parametric hypothesis testing. In particular, the various methods classified under the name 'analysis of variance' (generally denoted by the acronym ANOVA) have not been considered. We just mention here that the simplest case of ANOVA consists in comparing the unknown means $\mu_1, \mu_2, \ldots, \mu_k$ of a number $k > 2$ of normal populations by testing the null hypothesis $H_0 : \mu_1 = \mu_2 = \cdots = \mu_k$ against the alternative that at least one of the equalities does not hold. As we can immediately see, this is an important case of parametric test where – it can easily be shown – pair wise comparison is not appropriate. For a detailed discussion of this interesting topic and of its ramifications the reader may refer to [6, 7, 9, 12, 18].

6.4 Testing the type of distribution (goodness-of-fit tests)

In the preceding sections we concerned ourselves with tests pertaining to one (or more) unknown parameter(s) of a known distribution law, thus tacitly implying that we already have enough evidence on the underlying probability distribution – that is, normal, Poisson, binomial or other – with which we are dealing. Often, however, the uncertainty is on the type of distribution itself and we would like to give more support to our belief that the sample $\mathbf{X} = (X_1, \ldots, X_n)$ is, in fact, a sample from a certain distribution law. In other words, this means that on the basis of n independent observations of a r.v. X with unknown distribution $F_X(x)$ we would like to test the null hypothesis $H_0 : F_X(x) = F(x)$ against $H_1 : F_X(x) \neq F(x)$, where $F(x)$ is a specified probability distribution. Two of the most popular tests for this purpose are Pearson's χ^2-test and Kolmogorov–Smirnov test.

6.4.1 *Pearson's χ^2-test and the modified χ^2-test*

Let us start with Pearson's test by assuming at first that $F(x)$ is completely specified – that is, it is not of the form $F(x; \theta)$, where θ is some unknown parameter of the distribution. Now, let $D_1, D_2, \ldots, D_r$ be a finite partition of the space D of possible values of X (i.e. $D = \cup_{j=1}^{r} D_j$ and $D_i \cap D_i = \emptyset$ for $i \neq j$) and let $p_j = P(X \in D_j | H_0)$ be the probabilities of the event $\{X \in D_j\}$ under H_0. These probabilities – which can be arranged in a r-dimensional vector $\mathbf{p} = (p_1, \ldots, p_r)$ – are known because they depend on $F(x)$ which, in turn, is assumed to be completely specified. In fact, for $j = 1, \dots, r$ we have

$$p_j = \sum_{k:x_k \in D_j} P(X = x_k) \tag{6.44a}$$

if X is discrete and

$$p_j = \int_{D_j} f(x)\,\mathrm{d}x \tag{6.44b}$$

if X is absolutely continuous with pdf $f(x) = F'(x)$. Turning now our attention to the sample $\mathbf{X} = (X_1, \dots, X_n)$, let N_j be the r.v. representing the number of its elements falling in D_j so that $\nu_j = N_j/n$ is a r.v. representing the relative frequency of occurrence pertaining to D_j. Clearly, $N_1 + \cdots + N_r = n$ or, equivalently

$$\sum_{j=1}^{r} \nu_j = 1 \tag{6.45}$$

With these definitions the hypotheses under test can be re-expressed as $H_0 : p_j = \nu_j$ and $H_1 : p_j \neq \nu_j$ for $j = 1, \dots, r$, thus implying that we should accept H_0 when the sample frequencies N_j are in reasonable agreement with the 'theoretical' (assumed) frequencies np_j. It was shown by Pearson that an appropriate test statistic for this purpose is

$$T \equiv \sum_{j=1}^{r} \frac{(N_j - np_j)^2}{np_j} = \sum_{j=1}^{r} \frac{N_j^2}{np_j} - n \tag{6.46}$$

for which, under H_0, we have

$$\begin{aligned} E(T) &= r - 1 \\ \mathrm{Var}(T) &= 2(r-1) + \frac{1}{n}\left(\sum_{j=1}^{r} \frac{1}{p_j} - r^2 - 2r + 2\right) \end{aligned} \tag{6.47}$$

and, most important, as $n \to \infty$

$$T \to \chi^2(r-1)\ [D] \tag{6.48}$$

Equation (6.48) leads directly to the formulation of Pearson's χ^2 goodness-of-fit test: at the significance level α, the approximate rejection region to test the null hypothesis $H_0 : p_j = \nu_j$ is given by

$$\Xi_1 = \{\mathbf{x} : t(\mathbf{x}) \geq \chi^2_{\alpha;r-1}\} \tag{6.49}$$

where $t(\mathbf{x})$ is the sample realization of the statistic T and $\chi^2_{\alpha;r-1}$ is the α-upper quantile of the χ^2 distribution with $r-1$ degrees of freedom.

Although we do not prove the results (6.47) and (6.48) – the interested reader can refer to [4] or [10] – a few comments are in order:

(i) Explicitly, $t(\mathbf{x}) = \sum_j (n_j - np_j)^2/np_j$ where $n_j (j = 1, \ldots, r)$ is the realization of the r.v. N_j; in other words, once we have carried out the experiment leading to the realization of the sample $\mathbf{x} = (x_1, \ldots, x_n)$, n_1 is the number of elements of $\mathbf{x}$ whose values fall in D_1, n_2 is the number of elements of $\mathbf{x}$ whose values fall in D_2, and so on.

(ii) The rejection region (6.49) is approximate because eq. (6.48) is an asymptotic relation; similarly, the equation $P(T \in \Xi_1 | H_0) = \alpha$ defining the probability of a type I error is strictly valid only in the limit of $n \to \infty$. However, the quality of the approximation is generally rather good for $n \geq 50$. For better results, many authors recommend to choose the D_j so that $np_j \geq 5$ for all j, or, at least, for more than 80–85% of all j (since we divide by np_j in calculating T, we do not want the terms with the smaller denominators to 'dominate' the sum(6.46)).

(iii) On the practical side, the choice of the D_j – which, for a one-dimensional random variable, are non-overlapping intervals of the real line – plays an important role. Broadly speaking, they should not be too few and they should not be too many, that is, r should be neither too small nor too large. A possible suggestion (although in no way a strict rule) is to use the formula

$$r \cong 2\left(\frac{2(n-1)^2}{z_\alpha^2}\right)^{0.2} \tag{6.50}$$

where z_α is the α-upper quantile of the standard normal distribution. So, for the most common levels of significance 0.05, 0.025 and 0.01 we have $r \cong 1.883(n-1)^{0.4}$, $r \cong 1.755(n-1)^{0.4}$ and $r \cong 1.639(n-1)^{0.4}$, respectively. Moreover, in order to comply with the indicative rule of point (ii) – that is, $np_j \geq 5$ for almost all j – adjacent end intervals (which cover the tails of the assumed distribution) are sometimes regrouped to ensure that the minimum absolute frequency is 5 or, at least, not much smaller than 5.

Regarding the width of the intervals, it is common practice to choose equal-width intervals, although this requirement is not necessary. Some authors, in fact, prefer to select the intervals so that the expected frequencies will be the same in each interval; excluding a uniform distribution hypothesis, it is clear that this choice will result in different interval widths.

Example 6.7(a) Suppose that in $n = 4000$ independent trials the events A_1, A_2, A_3 have been obtained $n_1 = 1905$, $n_2 = 1015$ and $n_3 = 1080$ times, respectively. At the level $\alpha = 0.05$ we want to test the null hypothesis $H_0 : p_1 = 0.5; p_2 = p_3 = 0.25$, where $p_j = P(A_j)$.

In this intentionally simple example the assumed distribution is discrete and $np_1 = 2000$, $np_2 = np_3 = 1000$. Since the sample realization of the statistic T is $t = 11.14$ and it is higher than the 0.05-upper quantile $\chi^2_{0.05;2} = 5.99$ we fall in the rejection region (6.49) and therefore, at the level $\alpha = 0.05$, we reject the null hypothesis.

Example 6.7(b) In $n = 12000$ tosses of a coin Pearson obtained $n_1 = 6019$ heads and $n_2 = 5981$ tails. Let us check at the levels 0.05 and 0.01 if this result is consistent with the hypothesis that he was using a fair coin (i.e. $H_0 : p_1 = p_2 = 0.5$). We get now $t = 0.120$ and this value is lower than both upper quantiles $\chi^2_{0.05;1} = 3.841$ and $\chi^2_{0.01;1} = 6.635$. Since we fall in the acceptance region in both cases, we accept H_0 and infer that the data agree with the hypothesis of unbiased coin.

The study of the power W of Pearson's χ^2 test when H_0 is not true is rather involved, also in the light of the fact that W cannot be computed unless a specified alternative H_1 is considered. Nonetheless, an important result is worthy of mention: for every fixed set of probabilities $\bar{\mathbf{p}} \neq \mathbf{p}$ the power function $W(\bar{\mathbf{p}})$ tends to unity as $n \to \infty$ [10]. This, in words, is expressed by saying that the test is 'consistent' (not to be confused with consistency for an estimator introduced in Section 5.4) and means that, under any fixed alternative H_1, the probability $1 - \beta$ of rejecting the null hypothesis when it is false tends to 1 as n increases. It is evident that consistency, for any statistical hypothesis test in general, is a highly desirable property.

A modified version of the χ^2 goodness-of-fit test can be used even when the assumed probability distribution $F(x)$ contains some unknown parameters, that is, it is of the form $F(x; \mathbf{q})$ where $\mathbf{q} = (\theta_1, \ldots, \theta_k)$ is a set of k unknown parameters. In this case the null hypothesis is clearly composite – in fact, it identifies a class of distributions and not one specific distribution in particular – and the statistic T itself will depend on $\mathbf{q}$ through the probabilities p_j, that is, $T = T(\mathbf{q})$ where

$$T(\mathbf{q}) = \sum_{j=1}^{r} \frac{[N_j - np_j(\mathbf{q})]^2}{np_j(\mathbf{q})} = \sum_{j=1}^{r} \frac{N_j^2}{np_j(\mathbf{q})} - n \tag{6.51}$$

So, if T is the appropriate test statistic for the case at hand – and it turns out that, in general, it is – we must first eliminate the indeterminacy brought about by $\mathbf{q}$. One possible solution is to estimate the parameter(s) $\mathbf{q}$ by some estimating method and use the estimate $\tilde{\mathbf{q}}$ in eq. (6.51). At this point, however, two objections come to mind. First, by so doing the probabilities p_j are no longer constants but depend on the sample (in fact, no matter which estimation method we choose, the sample must be used to calculate $\tilde{\mathbf{q}}$), thus implying that eq. (6.48) will probably no longer hold. Second, if eq. (6.48) does not hold but there exists nonetheless a limiting distribution for T, is this limiting distribution independent on the estimation method used to obtain $\tilde{\mathbf{q}}$?

The way out of this rather intricate situation was found in the 1920s by Fisher who showed that for an important class of estimation methods the χ^2 distribution is still the asymptotic distribution of T but eq. (6.48) must be modified to

$$T(\tilde{\mathbf{q}}) \to \chi^2(r-k-1)\ [D] \tag{6.52}$$

thus determining that the effect of the k unknown parameters is just a decrease – of precisely k units, one unit for each estimated parameter – of the number of degrees of freedom. Such a simple result does not correspond to a simple proof and the interested reader is referred to Chapter 30 of [4] for the details. There, the reader will also find the (rather mild) conditions on the continuity and differentiability of the functions $p_j(\mathbf{q})$ for eq. (6.52) to hold.

On the practical side, once we have obtained the estimate $\tilde{\mathbf{q}}$ by means of an appropriate method – we will come to this point shortly – eq. (6.52) implies that the 'large-sample' rejection region for the test is

$$\Xi_1 = \left\{\mathbf{x} : t(\mathbf{x}; \tilde{\mathbf{q}}) \ge \chi^2_{\alpha;r-k-1}\right\} \tag{6.53}$$

and, as above, the probability of rejecting a true null hypothesis (type I error) is approximately equal to α. In regard to the estimation method, we can argue that a 'good' estimate of $\mathbf{q}$ can be obtained by making $T(\mathbf{q})$ as small as possible (note that the smaller is T, the better it agrees with the null hypothesis) so that our estimate $\tilde{\mathbf{q}}$ can be obtained by solving for the unknowns $\theta_1, \ldots, \theta_k$ the system of k equations

$$\sum_{j=1}^{r}\left(\frac{n_j - np_j}{p_j} - \frac{(n_j - np_j)^2}{2np_j^2}\right)\frac{\partial p_j}{\partial \theta_i} = 0 \tag{6.54a}$$

where it is understood that $p_j = p_j(\mathbf{q})$ and $i = 1, 2, \ldots, k$. This is called the χ^2 minimum method of estimation and its advantage is that, under sufficiently general conditions, it leads to estimates that are consistent, asymptotically normal and asymptotically efficient. Its main drawback, however, is that finding the solution of (6.54a) is generally a difficult task. For large values of n, fortunately, it can be shown that the second term within parenthesis becomes negligible and therefore the estimate $\tilde{\mathbf{q}}$ can be obtained by solving the modified, and simpler, equations

$$\sum_{j=1}^{r}\left(\frac{n_j - np_j}{p_j}\right)\frac{\partial p_j}{\partial \theta_i} = \sum_{j=1}^{r}\left(\frac{n_j}{p_j}\right)\frac{\partial p_j}{\partial \theta_i} = 0 \tag{6.54b}$$

where the first equality is due to the condition $\sum_j p_j(\mathbf{q}) = 1$ for all $\mathbf{q} \in \Theta$. Equation (6.54b) express the so-called 'modified χ^2 minimum method' which, for large samples, gives estimates with the same asymptotic properties as the χ^2 method of eqs (6.54a) and (6.54b). This asymptotic behaviour of the 'modified χ^2 estimates' is not surprising if one notes that, for the observations grouped by means of the partition $D_1, D_2, \ldots, D_r$, $\tilde{\mathbf{q}}$ coincides with the ML estimate $\hat{\mathbf{q}}_g$ (the subscript g is for 'grouped'). In fact, once the partition $\{D_j\}_{j=1}^{r}$ has been chosen, the r.v.s N_j are distributed according to the multinomial probability law (eq. (3.46a)) because each observation x_i can fall in the interval D_j with probability p_j. It follows that the likelihood function of the grouped observations $L_g(n_1, \ldots, n_r; \mathbf{q})$ is given by

$$L_g(N_1 = n_1, \ldots, N_r = n_r; \mathbf{q}) = \frac{n}{n_1! \cdots n_r!}\prod_{j=1}^{r} p_j^{n_j}(\mathbf{q}) \tag{6.55}$$

and the ML estimate $\hat{\mathbf{q}}_g$ of $\mathbf{q}$ is obtained by solving the system of equations $\partial \log L_g / \partial \theta_i = 0$ $(i = 1, 2, \ldots, k)$ which, when written explicitly, coincides with (6.54b).

At this point an interesting remark can be made. If we calculate the ML estimate before grouping by maximizing the 'ungrouped' likelihood function $L(\mathbf{x}; \mathbf{q}) = f(x_1; \mathbf{q}) \cdots f(x_n; \mathbf{q})$ – which, we note, is different from (6.55) and therefore leads to an estimate $\hat{\mathbf{q}} \neq \hat{\mathbf{q}}_g$ – then $\hat{\mathbf{q}}$ is probably a better estimate than $\hat{\mathbf{q}}_g$ because it uses the entire information from the sample (grouping, in fact, leads to a partial loss of information). Furthermore, maximizing $L(\mathbf{x}; \mathbf{q})$ is often computationally easier than finding the solution of eq. (6.54b). Unfortunately, while it is true that (eq. (6.52)) $T(\hat{\mathbf{q}}_g) \to \chi^2(r-k-1)$ $[D]$, it has been shown that $T(\hat{\mathbf{q}}) \to \chi^2(r-k-1)$ $[D]$ does not hold in general and the asymptotic distribution of $T(\hat{\mathbf{q}})$ is more complicated than the χ^2 distribution with $r-k-1$ degrees of freedom. In conclusion, we need to group the data first and then estimate the unknown parameter(s); by so doing, Cramer [4] shows that the above results are valid for any set of asymptotically normal and asymptotically efficient estimates of the parameters (however obtained, that is, not necessarily by means of the modified χ^2 method).

Example 6.8 Suppose we want to test the null hypothesis that some observed data come from a normal distribution with unknown mean $\mu = \theta_1$ and variance $\sigma^2 = \theta_2^2$ so that $\mathbf{q} = (\theta_1, \theta_2)$ is the set of $k = 2$ unknown parameters. For $j = 1, \ldots, r$ let the grouping intervals be defined as $D_j = (x_{j-1}, x_j]$ where, in particular, $x_0 = -\infty, x_r = \infty$ and $x_j = x_1 + (j-1)\,d$ for $j = 1, \ldots, r-1, x_1$ and d being appropriate constants (in other words, the D_js are a partition of the real line in non-overlapping intervals which, except for the first and the last, all have a constant width d). Also, in order to simplify the notation, let

$$g(x;\mathbf{q}) = \frac{1}{\sqrt{2\pi}} \exp\left(-\frac{(x-\theta_1)^2}{2\theta_2^2} \right)$$

With these definitions we have the 'theoretical' probabilities

$$p_j(\mathbf{q}) = \frac{1}{\theta_2} \int_{D_j} g(x;\mathbf{q})\,\mathrm{d}x \tag{6.56}$$

so that, calculating the appropriate derivatives, the modified minimal χ^2 estimate' of $\mathbf{q}$ is obtained by solving the system of two equations

$$\begin{aligned} &\sum_{j=1}^{r} \frac{n_j}{p_j(\mathbf{q})} \int_{D_j} (x-\theta_1) g(x;\mathbf{q})\,\mathrm{d}x = 0 \\ &\sum_{j=1}^{r} \frac{n_j}{p_j(\mathbf{q})} \left(\int_{D_j} (x-\theta_1)^2 g(x;\mathbf{q})\,\mathrm{d}x - \theta_2^2 \int_{D_j} g(x;\mathbf{q})\,\mathrm{d}x \right) = 0 \end{aligned} \tag{6.57a}$$

which, after rearranging terms and taking $\sum n_j = n$ into account, become

$$\begin{aligned} \theta_1 &= \sum_j \left(\frac{n_j}{n}\right) \frac{\int xg\,\mathrm{d}x}{\int g\,\mathrm{d}x} \\ \theta_2^2 &= \sum_j \left(\frac{n_j}{n}\right) \frac{\int (x-\theta_1)^2 g\,\mathrm{d}x}{\int g\,\mathrm{d}x} \end{aligned} \tag{6.57b}$$

where all integrals are on D_j and, for brevity, we have omitted the functional dependence of $g(x;\mathbf{q})$. For small values of the interval width d and assuming $n_1 = n_r = 0$ (i.e. the extreme intervals contain no data) we can find an approximate solution of eq. (6.57b) by replacing each function under integral by its corresponding value at the midpoint ξ_j of the interval D_j. This leads to the 'grouped' estimate $\hat{\mathbf{q}}_g = (\hat{\theta}_1, \hat{\theta}_2^2)$, where

$$\hat{\theta}_1 \cong \frac{1}{n}\sum_{j=2}^{r-1} n_j \xi_j$$
$$\hat{\theta}_2^2 \cong \frac{1}{n}\sum_{j=2}^{r-1} n_j(\xi_j - \hat{\theta}_1)^2 \tag{6.58}$$

In general, the approximate estimates (6.58) are sufficiently good for practical purposes even if the extreme intervals are not empty but contain a small part of the data. So, if $\mathbf{y} = (y_1, \ldots, y_n)$ is the set of data obtained by the experiment (we call the data y_i, and not x_i as usual, to avoid confusion with the intervals extreme points defined above) we reject the null hypothesis of normality at the level α if

$$t(\hat{\mathbf{q}}_g) = \sum_j \frac{[n_j - np_j(\hat{\mathbf{q}}_g)]}{np_j(\hat{\mathbf{q}}_g)} \geq \chi^2_{\alpha;r-3} \tag{6.59}$$

where $\chi^2_{\alpha;r-3}$ is the α-upper quantile of the distribution $\chi^2(r-3)$.

As a numerical example, suppose we have $n = 1000$ observations of a r.v. which we suspect to be normal; also, let the minimum and maximum observed values be $y_{\min} = 36.4$ and $y_{\max} = 98.3$, respectively.

Let us choose a partition of the real line in $r = 9$ intervals with $d = 10$ and $D_1 = (-\infty, 35], D_2 = (35, 45]$ etc. up to $D_8 = (95, 105]$ and $D_9 = (105, \infty)$. With this partition, suppose further that our data give the absolute frequencies

$$\mathbf{n} = (n_1, \ldots, n_9) = (0, 5, 60, 233, 393, 254, 49, 6, 0) \tag{6.60}$$

from which, since the intervals midpoints are $\xi_2 = 40$, $\xi_3 = 50, \ldots, \xi_8 = 100$, we calculate (eq. (6.58)) the grouped estimates $\hat{m} = 70.02$ and $\hat{s}^2 = 102.20$ for the mean and the variance, respectively. A normal distribution with this mean and variance leads to the (approximate) set of theoretical frequencies

$$n\mathbf{p} = n(p_2, \ldots, p_8) \cong (4.8, 55.5, 241.5, 394.6, 242.4, 56.0, 4.9) \tag{6.61}$$

where the approximation lies in the fact that we calculated the integrals (6.56) by using, once again, the value of the function at the midpoints ξ_j.

Finally, using the experimental and theoretical values of eqs. (6.60) and (6.61) we get $t(\hat{\mathbf{q}}_g) = 2.36$ which is less than $\chi^2_{0.05;6} = 12.59$ therefore implying that we accept the null hypothesis of normality.

This example is just to illustrate the method and in a real case we should use a finer partition, say $r \cong 15$. In this case we would probably have to pool the extreme intervals to comply with the suggestion $np_j \geq 5$. So, for instance if we choose $d = 5$ and the first two non-empty intervals have the observed frequencies 1 and 4, we can pool these two intervals to obtain an interval of width $d' = 10$ with a frequency count of 5. In this case Cramer [4] suggests to calculate the estimates with the original grouping (i.e. before pooling), and then perform the test by using the quantiles of the distribution $\chi^2(r' - 3)$, where r' is the number of intervals after pooling.

6.4.2 *Kolmogorov–Smirnov test and some remarks on the empirical distribution function*

Similarly to the χ^2 goodness-of-fit test, the Kolmogorov–Smirnov test (KS test) concerns the agreement between an empirical distribution and an assumed theoretical one when this latter is continuous. The test is performed by using the PDFs rather than – as the χ^2 does – the pdfs. The relevant statistic is

$$D_n = \sup_{-\infty < x < \infty} |F_n(x) - F(x)| \tag{6.62}$$

where $F_n(x)$ is the empirical (or sample) PDF, $F(x)$ is the assumed PDF and the subscript n refers, as usual, to the sample size.

Before considering the test itself, however, a few remarks on $F_n(x)$ are in order. Given a realization $\mathbf{x} = (x_1, \ldots, x_n)$ of a sample $\mathbf{X} = (X_1, \ldots, X_n)$, let $X_{(1)}, X_{(2)}, \ldots, X_{(n)}$ be its order statistics (see Section 5.3.1) so that $x_{(1)} \leq x_{(2)} \leq \cdots \leq x_{(n)}$. Then, the empirical PDF $F_n(x)$ is constructed by setting

$$F_n(x) = \begin{cases} 0, & x < x_{(1)} \\ k/n, & x_{(k)} \leq x < x_{(k+1)} \\ 1, & x \geq x_{(n)} \end{cases} \tag{6.63a}$$

and can be also expressed as

$$F_n(x) = \frac{1}{n} \sum_{k=1}^{n} I\{X_{(k)} \leq x\} \tag{6.63b}$$

where $I\{X_{(k)} \leq x\}$ is the indicator function of the event $\{X_{(k)} \leq x\}$ (recall from Section 2.3.1, Example 2.3, that an indicator function is a r.v.). It is immediate to see that $F_n(x)$ has all the properties of a PDF (Section 2.3): that is, it ranges from 0 to 1, is non-decreasing and right-continuous. Moreover, it is a piecewise-constant step function which, if all the components of $\mathbf{x}$ are distinct, has a step of height $1/n$ at each point $x = x_{(k)}$.

Clearly, an empirical PDF can be constructed for any set of data but if – as we assume – $\mathbf{X}$ is a random sample from a parent r.v. X with PDF $F(x)$ we expect $F_n(x)$ to be a 'statistical image' of $F(x)$ and, more important, we expect $F_n(x)$ to get closer and closer to $F(x)$ as the sample size increases. A direct consequence of Bernoulli WLLN (Section 4.4), in fact, is that the relation

$$F_n(x) \to F(x)\ [P] \tag{6.64}$$

holds for all $x \in \mathbb{R}$ because $F_n(x)$ is the relative frequency of the event $\{X \leq x\}$ – defined as a 'success' – in n Bernoulli trials with probability of success $F(x)$. This last observation, in addition, tells us that the r.v. $nF_n(x)$ is binomially distributed (with parameter $F(x)$) and consequently

$$\begin{aligned} E\{F_n(x)\} &= F(x) \\ \operatorname{Var}\{F_n(x)\} &= \frac{F(x)[1-F(x)]}{n} \end{aligned} \tag{6.65}$$

thus implying that $F_n(x)$ is both a consistent and an unbiased estimator of $F(x)$ for all x.

Returning now to K–S test, the above results fully justify the reason why the statistic D_n is a very good candidate to test the null hypothesis $H_0 : F_X(x) = F(x)$. In fact, if H_0 is true we expect D_n to be close to zero. This is more so if one considers a further important result known as Glivenko–Cantelli theorem which we state without proof (the interested reader is referred, for instance, to [11]).

Proposition 6.2 (Glivenko–Cantelli theorem) *Let* $\mathbf{X} = (X_1, \ldots, X_n)$ *be a random sample from a parent r.v. X with PDF F(x). Then, for all* $x \in \mathbb{R}$

$$P\left\{\lim_{n\to\infty} \sup_x |F_n(x) - F(x)| = 0\right\} = 1 \tag{6.66a}$$

where $F_n(x)$ *is the empirical PDF of the sample. (Recalling Section 4.3, note that eq. (6.66a) can be equivalently stated by writing*

$$D_n \to 0\ [\text{a.s.}] \tag{6.66b}$$

uniformly in x as $n \to \infty$*).*

At this point it is clear that the rejection region for the test will be of the form $\Xi_1 = \{\mathbf{x} : D_n \geq t_\alpha\}$ where the number t_α – for different values of the significance level α – can be determined once we know the distribution of D_n. It was shown by Kolmogorov that for any fixed $t > 0$

$$\lim_{n\to\infty} P(\sqrt{n}D_n \le t) = K(t) \equiv 1 - 2\sum_{j=1}^{\infty}(-1)^{j+1}\exp(-2j^2t^2) \qquad (6.67)$$

which, in words, means that $\sqrt{n}D_n$ tends in distribution to a r.v. whose PDF is the function $K(t)$ defined on the r.h.s. of (6.67). On the practical side, therefore, one chooses the critical boundary t_α in the form $t_\alpha = u_\alpha/\sqrt{n}$ by means of the relation $1 - K(u_\alpha) = \alpha$; in fact, by so doing, we have

$$P(D_n \ge t_\alpha|H_0) = P(\sqrt{n}D_n \ge u_\alpha|H_0) \cong 1 - K(u_\alpha) = \alpha \qquad (6.68)$$

and the probability of a type I error is approximately equal to α. For the most frequently adopted levels of significance $\alpha = 0.05$ and $\alpha = 0.01$ we have $u_{0.05} = 1.358$ and $u_{0.01} = 1.628$ and these values can be used for samples larger than 35–40 units. For smaller samples the relation (6.67) does not provide a good approximation and values of u_α can easily be found in table form. So, for instance, we find the values $u_{0.05} = 0.409$ and $u_{0.01} = 0.489$ for $n = 10$ while $u_{0.05} = 0.294$ and $u_{0.01} = 0.352$ for $n = 20$.

The K–S test has the advantage of simplicity because the realization $D_n(\mathbf{x})$ of the statistic D_n is rather easy to calculate. Once this is done we reject the null hypothesis at the level α if $\sqrt{n}D_n(\mathbf{x}) \ge u_\alpha$.

A second useful application of the test is called 'K–S two-sample test' and it is used to check whether two independent samples of different size come from the same distribution . This is, in other words a so-called 'homogeneity test' because the null hypothesis is that the data are supposed to be homogeneous. Let n and m be the sample sizes and let us denote the two samples by $\mathbf{X} = (X_1, \ldots, X_n)$ and $\mathbf{Y} = (Y_1, \ldots, Y_m)$, respectively. If we define the statistic

$$D_{n,m} = \sup_{-\infty<x<\infty} |F_n(x) - F_m(x)| \qquad (6.69)$$

it is due to Smirnov to have shown that for any fixed $t > 0$

$$\lim_{n,m\to\infty} P\left(\sqrt{\frac{nm}{n+m}}D_{n,m} \le t\right) = K(t) \qquad (6.70)$$

where $K(t)$ is as in eq. (6.67). By the same line of reasoning as above, if H_0 is true we expect $D_{n,m}$ to be close to zero because the two empirical PDFs $F_n(x)$ and $F_m(x)$ are estimating the same (generally unknown) continuous distribution and therefore they should get closer and closer as the sample sizes increase. So, since the expected rejection region will be $\Xi_1 = \{\mathbf{x}, \mathbf{y} : D_{n,m} \ge t_\alpha\}$ for appropriate values of t_α, we can choose the critical boundary in the form $t_\alpha = u_\alpha \sqrt{(n+m)/(nm)}$ so that

$$P\left(\sqrt{\frac{nm}{n+m}} D_{n,m} \geq u_\alpha | H_0\right) \cong 1 - K(u_\alpha) = \alpha \tag{6.71}$$

and the test can be formulated as follows: for sufficiently large sample sizes we reject the null hypothesis of homogeneity if the realization $D_{n,m}(\mathbf{x}, \mathbf{y})$ of $D_{n,m}$ is such that $\sqrt{(nm)/(n+m)} D_{n,m}(\mathbf{x}, \mathbf{y}) \geq u_\alpha$. As a numerical example suppose that we construct the empirical PDFs for two samples of sizes $n = 80$ and $m = 60$ and we obtain a maximum absolute difference between the two of $D_{n,m}(\mathbf{x}, \mathbf{y}) = 0.12$. Then, since $\sqrt{(nm)/(n+m)} D_{n,m}(\mathbf{x}, \mathbf{y}) = 0.703$ and this is lower than $u_{0.05} = 1.358$, we do not reject the null hypothesis (at the level $\alpha = 0.05$).

For small sample sizes the approximation (6.70) is obviously not satisfactory and values of the critical boundary for different values of n and m can be found in statistical tables. Moreover, we point out that, as for K–S goodness-of-fit test, the K–S homogeneity test applies only if the assumed distribution is continuous.

In closing this section we make two final observation. The first regards the answer to the question if, as for the χ^2 test, the K–S goodness-of-fit test can be used when the assumed distribution depends on some unknown parameter, that is, when we wish to test the composite null hypothesis $H_0 : F_X(x) = F(x; \theta)$. The answer is yes and for regular problems (Section 5.4.1) it is possible to use the statistic

$$\hat{D}_n = \sup_{-\infty < x < \infty} |F_n(x) - F(x; \hat{\theta})| \tag{6.72}$$

where $\hat{\theta}$ is the ML estimate of θ. The limiting distribution of $\hat{D}_n$ is, as a matter of fact, known but unfortunately it differs from (6.67) and the situation turns out to be rather complicated. This is beyond our scopes and for more details we refer the interested reader, for instance, to [20].

Another difficult issue is the study of the power of K–S test. For this the reader can refer to Massey [16]. In this regard, however, an interesting observation is that in some cases where a comparison has been possible, the K–S test seems to be much more powerful than the χ^2 test. This in no way implies that we should discard the χ^2 test because this test is more advantageous in other respects (for instance, it can be used for discrete cases).

6.5 Miscellaneous complements

In the preceding section we have seen how the K–S test can be used to test homogeneity of two sets of data from a continuous distribution. If the underlying distribution is discrete or the observations are grouped there exists – among others – a χ^2 test for homogeneity. Furthermore, this test can be extended to the case of more than two samples. This is where we start. Then, in the order, we will consider the χ^2 test for independence (Section 6.5.2), two tests for randomness (Section 6.5.3) and make a few remarks on the identification of outliers (Section 6.5.4).

6.5.1 *The χ^2 test for homogeneity*

Let $\mathbf{Y}_1, \ldots, \mathbf{Y}_k$ be k ($k \geq 2$) independent samples of $n_1, \ldots, n_k$ observations respectively – that is, $\mathbf{Y}_1 = (Y_{11}, \ldots, Y_{1n_1})$, $\mathbf{Y}_2 = (Y_{21}, \ldots, Y_{2n_2})$, etc. The null hypothesis of homogeneity is that all the $n = n_1 + \cdots + n_k$ observations regard the same random variable whose possible values, for the purpose of the test, must have been preliminarily partitioned r classes.

Now, let N_{ij} be the r.v. representing the number of components of $\mathbf{Y}_j$ falling in the ith class so that for $j = 1, \ldots, k$ we have

$$\sum_{i=1}^{r} N_{ij} - n_j \tag{6.73}$$

If we denote by p_{ij} the (unknown) probability of ith outcome in the jth sample, our null hypothesis can be expressed as

$$H_0 : (p_{1j}, \ldots, p_{rj}) = (p_1, \ldots, p_r), \quad j = 1, \ldots, k \tag{6.74}$$

where $\mathbf{p} = (p_1, \cdots, p_r)$ is a vector of (unknown) probabilities such that $p_1 + \cdots + p_r = 1$. In this light, since $n_j p_i$ is the expected (i.e. if H_0 is true) number of data from the jth sample falling in the ith class, it is reasonable to take

$$T = T(\mathbf{p}) = \sum_{i=1}^{r} \sum_{j=1}^{k} \frac{(N_{ij} - n_j p_i)^2}{n_j p_i} \tag{6.75}$$

as the relevant statistic for the case at hand. The p_i, however, are not known and must be estimated from the data. Denoting by n_{ij} the realizations (obtained through the samples realizations $\mathbf{y}_1, \ldots, \mathbf{y}_k$) of the r.v.s N_{ij}, it is not difficult to determine that the 'grouped' ML estimate of p_i is its relative frequency $\hat{p}_i = n^{-1} \sum_{j=1}^{k} n_{ij} = \nu_i / n$, where we denote by $\nu_i \equiv \sum_{j=1}^{k} n_{ij}$ the total number of observations falling in the ith class. With these estimates, a direct generalization of the theorem leading to eq. (6.52) shows that under H_0

$$T(\hat{\mathbf{p}}) \to \chi^2\{(r-1)(k-1)\} \; [D] \tag{6.76}$$

so that the critical boundary is defined in terms of the α-upper quantiles $\chi^2_{\alpha;(r-1)(k-1)}$ of the χ^2 distribution with $(r-1)(k-1)$ degrees of freedom. In practice, with the observed data we calculate the realization $t(\hat{\mathbf{p}})$ of $T(\hat{\mathbf{p}})$ as

$$t(\hat{\mathbf{p}}) = n\sum_{i=1}^{r}\sum_{j=1}^{k}\frac{(n_{ij} - n_j\nu_i/n)^2}{n_j\nu_i} = n\left(\sum_{i=1}^{r}\sum_{j=1}^{k}\frac{n_{ij}^2}{n_j\nu_i} - 1\right) \tag{6.77}$$

(the expression on the r.h.s. is due to the relations $\sum_{ij} n_{ij} = n$ and $\sum_{ij} n_j n_i = n^2$) and, for sufficiently large samples, we reject the null hypothesis at the significance level α if $t(\hat{\mathbf{p}}) \geq \chi^2_{\alpha;(r-1)(k-1)}$. By so doing, the probability of a type I error will be approximately equal to α.

For the interested reader, we mention that the result (6.76) is due to a theorem stating that with k independent samples grouped into r classes the limiting distribution of T is the $\chi^2[(r-1)k - m]$ distribution, where m is the number of estimated parameters (by means of the modified χ^2 minimum method or the grouped ML method). Since in our case we had to estimate $m = r - 1$ parameters – the remaining rth parameter is determined by the 'constraint' condition $p_1 + \cdots + p_r = 1$ – we have $(r-1)k - m = (r-1)(k-1)$ degrees of freedom.

A slightly different situation arises if we have made some assumption on the underlying common distribution of the k samples – for example, we suspect it to be normal, or Poisson or other – and we must estimate some unknown parameters $\mathbf{q} = (\theta_1, \ldots, \theta_s)$ of this assumed distribution. The recommended procedure in this case is to first find the grouped (multinomial, see eq. (6.55)) ML estimate $\hat{\mathbf{q}}_g$ of $\mathbf{q}$ under H_0, calculate the r probabilities $p_i(\hat{\mathbf{q}}_g)$ and use them in eq. (6.75). By so doing we estimate s parameters and therefore, for sufficiently large samples, the test is carried out by using the appropriate quantile of the distribution $\chi^2[(r-1)k - s]$.

Returning to the original test – that is, no assumption on the type of distribution – the case $k = 2$ implies a two-sample test which can be compared to the K–S homogeneity test of Section 6.4.2 (if the underlying distribution is continuous). It is left to the reader to check that now the $t(\hat{\mathbf{p}})$ of eq. (6.77) can be more conveniently calculated as

$$t(\hat{\mathbf{p}}) = \frac{1}{R(1-R)}\left(\sum_{i=1}^{r} n_{i1}Q_i - n_1 R\right) \tag{6.78}$$

where $R = n_1/n = n_1/(n_1 + n_2)$ and $Q_i = n_{i1}/\nu_i = n_{i1}/(n_{i1} + n_{i2})$.

6.5.2 *The χ^2 test for independence*

Turning now our attention to a different type of test, it often happens that we are given a sample $((X_1, Y_1), \ldots, (X_n, Y_n))$ from a bivariate population with an unknown distribution $F(x, y)$ and we wish to check the null hypothesis of independence $H_0 : F(x, y) = F_X(x)F_Y(y)$. A rather common test for this case is yet another version of the χ^2 test. First, each component of the observations is grouped in, say, r classes for the X variable and s classes for the Y variable (if the variables are discrete one generally retains the 'natural' grouping due to the discreteness of the possible outcomes so that r and s are the numbers of possible values for X and Y, respectively). Then, letting $a_1, \ldots, a_r$ and $b_1, \ldots, b_s$ denote the possible outcomes for X and Y, we have rs (unknown) probabilities $p_{ij} = P(X = a_i, Y = b_j)$ which, in turn, will correspond to rs random variables N_{ij}, where N_{ij} is the number of times the event $(X = a_i, Y = b_j)$ occurs. If H_0 is true there will exist $r+s$ probabilities $\mathbf{p} = (p_1, \ldots, p_r, q_1, \ldots, q_s)$ such that $p_{ij} = p_i q_j$ and

$$\sum_{i=1}^{r} p_i = \sum_{j=1}^{s} q_j = 1 \tag{6.79}$$

At this point it is not difficult to see that the relevant test statistic is

$$T(\mathbf{p}) = \sum_{i=1}^{r}\sum_{j=1}^{s} \frac{(N_{ij} - np_i q_j)^2}{np_i q_j} \tag{6.80}$$

where, however, the p_i and q_j must be estimated from the data. Denoting by n_{ij} the realization of the r.v.s N_{ij} and letting $A_i = \sum_{j=1}^{s} n_{ij}$ and $B_j = \sum_{i=1}^{r} n_{ij}$ (the so-called marginal frequencies) be the total number of times in which the events a_i and b_j occur, respectively, the reader can check that under H_0 the grouped ML estimates are

$$\begin{aligned} \hat{p}_i &= A_i/n \\ \hat{q}_j &= B_j/n \end{aligned} \tag{6.81}$$

so that the realization $t(\hat{\mathbf{p}})$ of $T(\hat{\mathbf{p}})$ is obtained by calculating

$$t(\hat{\mathbf{p}}) = n\sum_{i=1}^{r}\sum_{j=1}^{s} \frac{(n_{ij} - A_i B_j/n)^2}{A_i B_j} = n\left(\sum_{i,j} \frac{n_{ij}^2}{A_i B_j} - 1\right) \tag{6.82}$$

which, in turn, must be compared to the α-upper quantiles of the χ^2 distribution with the appropriate number of degrees of freedom. Since we have a total of rs groups and we have estimated $r+s-2$ parameters (the subtracted 2 comes from the two constraint eq. (6.79)), the same theorem leading to eq. (6.52) shows that this number is $rs - (r+s-2) - 1 = (r-1)(s-1)$. The conclusion, therefore, is as follows: at (approximately) the significance level α and for sufficiently large samples, we reject the null hypothesis of independence if $t(\hat{\mathbf{p}}) \geq \chi^2_{\alpha;(r-1)(k-1)}$ and accept it otherwise.

A few things about this test should be pointed out:

(i) Despite the similarities of notation and the fact that we have the same numbers of degrees of freedom, it must not be confused with the χ^2 test for homogeneity discussed before. In fact the two tests refer to different situations and even if some symbols may be in common, they generally have different meanings in the two contexts.

(ii) The χ^2 test for independence is often carried out with the aid of 'contingency tables'. These are tables with rs entries arranged in r rows and s columns where the analyst enters the observed n_{ij} values. Then A_i $(i = 1, \ldots, r)$ is obtained by summing horizontally along the ith row and B_j $(j = 1, \ldots, s)$ is obtained by summing vertically along the jth column.

(iii) This type of test is frequently used to determine whether two characteristics X and Y can be considered independent. In this case n items – individuals, objects, products or else – are classified according to two attributes (the characteristics) and r, s are the 'levels' of the characteristic X and Y, respectively. Moreover, X and Y are not necessarily numerical characteristics but can be qualitative attributes. For example, a sample of people may be classified by sex (X) and by opinion on a certain political issue (Y) to test the null hypothesis that opinions on that issue are independent of sex. In this case we would obtain a 2×2 contingency table, the X and Y levels being $(a_1 = \text{female}; a_2 = \text{male})$ and $(b_1 = \text{in favour}; b_2 = \text{against})$, respectively.

(iv) A large value of $t(\hat{\mathbf{p}})$ tends to indicate that the hypothesis of independence is false. This fact, however, does not provide any direct information on the 'degree of dependence' (or association) between X and Y. For discrete or grouped variables the so-called mean-square contingency (introduced by Pearson) and defined as

$$\phi^2 = \sum_{i,j} \frac{(p_{ij} - p_i q_j)^2}{p_i q_j} = \sum_{i,j} \frac{p_{ij}^2}{p_i q_j} - 1 \tag{6.83}$$

(the meaning of the symbols is the same as in eq. (6.80)) is a measure of this quantity because – letting $u = \min(r, s)$ or $u = r = s$ if $r = s$ – we have $0 \le \phi^2 \le u - 1$ and $\phi^2 = 0$ if and only if X and Y are independent.

Consequently, $\phi^2/(u-1)$ is one such measure on a scale from 0 to 1. Using the estimates of eq. (6.81), the sample realization f^2 of ϕ^2 is $f^2 = t(\hat{\mathbf{p}})/n$ so that

$$0 \le \frac{f^2}{u-1} = \frac{t(\hat{\mathbf{p}})}{n(u-1)} \le 1 \tag{6.84}$$

can be regarded as a measure of the degree of dependence indicated by the observed data. In this regard, in Chapter 30 of Ref. [4] the reader can find an interesting example taken from data of the Swedish census: although $t(\hat{\mathbf{p}})$ is about 20 times higher than the appropriate upper quantile of $\chi^2[(r-1)(k-1)]$ – and consequently H_0 is rejected – the degree of dependence between the relevant variables is rather low. In fact, the calculations give $(u-1)^{-1}f^2 = 0.0075$, thus showing that a value of $t(\hat{\mathbf{p}})$ much higher than $\chi^2_{\alpha;(r-1)(k-1)}$ is not necessarily an indication of strong dependence between X and Y; it simply means that we should reject the null hypothesis.

6.5.3 *Testing for randomness*

In most cases considered so far we assumed that the sample $\mathbf{X} = (X_1, \ldots, X_n)$ is a random sample from a parent random variable X, that is, that the components X_i of $\mathbf{X}$ are iid variables. Denoting by $F(x)$ the distribution of X, this assumption can be expressed in mathematical terms by writing $F_{\mathbf{X}}(\mathbf{x}) = F(x_1)\cdots F(x_n)$ and, in general, its validity is suggested by the nature of the problem under study. There are cases, however, in which $F_{\mathbf{X}}(\mathbf{x}) = F(x_1)\cdots F(x_n)$ is a (null) hypothesis that should be tested before proceeding with further analysis. With this idea in mind, we can intuitively argue that randomness is, to a certain extent, an index of disorder and that, under H_0, each component of the sample X_i should somehow 'enjoy equal rights'. Consequently, we would tend to reject H_0 if our data show 'too much order' of some kind. A test for randomness along this line of reasoning is as follows.

Let us arrange the components of the sample in increasing order; in the ordered series we say that X_i and X_j form an 'inversion' if, for $i < j$, X_i comes to the right of X_j. Then, if we let $T(\mathbf{X})$ be the statistic representing the number of inversions of the sample where N_1 is the number of inversions formed by X_1, N_2 the number of inversions formed by X_2, etc., we can write $T(\mathbf{X}) = N_1 + \cdots + N_{n-1}$. In the two extreme cases of 'maximum order' $X_1 < X_2 < \cdots < X_n$ and $X_n < X_{n-1} < \cdots < X_1$ we have $T(\mathbf{X}) = 0$ and $T(\mathbf{X}) = n(n-1)/2$, respectively, and therefore we will reject H_0 if $T(\mathbf{X})$ is either too high or too low. Since it can be shown that under H_0 we have

$$E(T) = n(n-1)/4$$
$$\mathrm{Var}(T) = (2n^3 + 3n^2 - 5n)/72$$

$E(T)$ is the midpoint of the interval $D = [0, n(n-1)/2]$ and we can assume as our rejection region the set of all integers in D such that $|t - n(n-1)/4| \geq t_\alpha$, where $t = T(\mathbf{x})$ is the observed value of $T(\mathbf{X})$ and the critical boundary t_α must be determined, as usual, from the condition $P(T \in \Xi_1|H_0) \leq \alpha$. For small samples these values can generally be found on statistical tables. For large samples, however, there is an asymptotic version of the test based on the fact that, under H_0, the normalized statistic

$$\tilde{T}(\mathbf{X}) = \left(T(\mathbf{X}) - \frac{n(n-1)}{4} \right) \frac{6}{\sqrt{n^3}} \tag{6.85}$$

converges in distribution to a standard normal r.v. Since this version provides an acceptable approximation already for $n > 10$ we can use the quantiles of the standard normal distribution to obtain an approximate test for randomness. At the level α we reject H_0 if $|\tilde{T}(\mathbf{x})| \ge z_{\alpha/2}$ where $z_{\alpha/2}$ is the $\alpha/2$-upper quantile of the standard normal distribution.

A second test for randomness is based on the number R of 'runs' of the sample. If, according to some criterion, we divide our sample into two classes, say A and B, a 'run' is a sequence of data of one type preceded and followed by a sequence of data of the second type (clearly, the first run is preceded by no data and the last run is followed by no data). Denoting by N_A and N_B $(N_A + N_B = n)$ the r.v.s representing the number of sample components in A and B, respectively, the test is based on the fact that, as $n \to \infty$, the statistic $R(\mathbf{X})$ converges in distribution to a normal r.v. with mean and variance

$$\begin{aligned} \mu_R &= E(R) = 1 + \frac{2N_A N_B}{n} \\ \sigma_R^2 &= \mathrm{Var}(R) = \frac{2N_A N_B(2N_A N_B - n)}{n^2(n-1)} \end{aligned} \tag{6.86}$$

So, if n_A, n_B are the sample realizations of N_A, N_B, we use them to calculate the estimates m_R and s_R^2 of μ_R and σ_R^2 and reject the null hypothesis if

$$\left| \frac{r - m_R}{s_R} \right| \ge z_{\alpha/2} \tag{6.87}$$

where $r = R(\mathbf{x})$ and, as above, $z_{\alpha/2}$ is the $\alpha/2$-upper quantile of the standard normal distribution.

A few remarks worthy of mention:

(i) The test is clearly approximate but can be used already for $n_A, n_B \ge 10$.
(ii) If the data are numerical values the two classes A and B can be, for instance, the number of observations below and above the sample median. However the test can be used also with non-numerical values when a twofold classification is possible; for example, when evaluating the quality of a batch of products we can classify them as 'acceptable' and 'unacceptable'.
(iii) The basic idea leading to eq. (6.87) is quite similar to the rationale of the previous test: in a truly random sample the number of runs should not be too low or too high. Too few runs may suggest a clustering

or grouping of some kind while too many runs may indicate a high-frequency oscillatory behaviour and in both cases the null hypothesis of randomness is certainly questionable.

(iv) A simple example: in 25 tosses of a coin we obtain 10 tails and 15 heads in the order T-HHHHH-T-H-T-H-TT-H-T-H-TTT-HHH-T-HHH and we wish to test for randomness at the level $\alpha = 0.05$. The sequence shows $r = 14$ runs (runs are separated by a hyphen) and we calculate from eq. (6.86) $m_R = 13$ and $s_R^2 = 5.5$. Since $(r - m_R)/s_R = 0.42 < z_{0.025} = 1.96$, we cannot, at the stated significance level, reject the null hypothesis.

6.5.4 *Identification of outliers*

At the end of Section 5.4 we briefly mentioned the fact that sometimes our data may be contaminated by outliers. These, we recall, are unexpectedly high or low values which, at first sight, do not seem to belong to the sample. In some cases they may be true data of exceptional events but the most common situation by far is that they are 'wrong' data due to recording, transmission or copying errors. In order to identify them, one possible solution that comes to mind is to use the sample mean m and standard deviation s to calculate the n standardized quantities

$$z_i = \frac{x_i - m}{s} \tag{6.88}$$

and consider as outliers those x_i such that, say, $|z_i| \geq 2.5$ or $|z_i| \geq 3$ (the choice of the cut-off value is, to a certain extent, arbitrary; the values 2.5 or 3 come from the assumption of normally distributed data because, if there are no outliers, the probability that $|z_i| \geq 2.5$ or $|z_i| \geq 3$ is very small. These indicative values, however, can be used even in cases of moderate departures from normality). The problem is that the above argument is flawed if our data do contain outliers. In fact, since both m and s are very sensitive to outliers – or, in other words, are not robust estimators of 'location' and 'spread' – the method will generally find no outliers even when they are there.

The problem can be fixed if, instead of m and s, we use the robust estimators $\bar{m}$ and $\bar{s}$ in eq. (6.88), where $\bar{m}$ is the sample median and $\bar{s}$ is the so-called 'median of absolute deviations from the median' (MAD) and defined as

$$\bar{s} = 1.483 \underset{j=1,\ldots,n}{\text{median}} |x_j - \bar{m}| \tag{6.89}$$

(1.483 is a correction factor for consistency with the usual 'spread' parameter of a normal distribution and need not concern us here). In other words, calculating the quantities

$$\bar{z}_i = \frac{x_i - \bar{m}}{\bar{s}} \tag{6.90}$$

and identify as outliers those x_i such that $|\bar{z}_i| \geq 2.5$ or $|\bar{z}_i| \geq 3$ is a much better method than the 'wrong' solution of eq. (6.88).

As a simple illustrative example, suppose that the original data are the set of five measurements $(2.30, 2.36, 2.37, 2.42, 2.44)$ but suppose further that a misplaced decimal number in the second entry has transformed the data into the 'contaminated' set $(2.30, 23.6, 2.37, 2.42, 2.44)$. For the original data we have $m = 2.378$; $s = 0.055$ and $\bar{m} = 2.37$; $\bar{s} = 0.074$. Correctly, both criteria $|z_i| \geq 2.5$ and $|\bar{z}_i| \geq 2.5$ do not identify any outliers.

For the contaminated data, however, we get $m = 6.63$; $s = 9.489$ and $\bar{m} = 2.42$; $\bar{s} = 0.074$. Then, the z_i set (eq. (6.88)) becomes $(-0.46, 1.79, -0.45, -0.44, -0.44)$ and does not identify any outliers; on the other hand, the $\bar{z}_i$ set (eq. (6.90)) becomes $(-1.62, 285.64, -0.67, 0.00, 0.27)$ and correctly identifies the second entry as an outlier.

At this point it should now be clear why the method of eq. (6.88) does not work. Referring to the example, we can see that the presence of the outlier has two effects: (a) moves m towards the outlier and (b) makes the standard deviation 'explode' (i.e. it becomes unreasonably high), thereby preventing the $|z_i|$ values from becoming too large. This is not the case for the robust estimators $\bar{m}$ and $\bar{s}$ which remain relatively – if not totally – unaffected by the wrong reading 23.6. A useful concept in this respect is the 'breakdown point' of an estimator, defined as the smallest fraction of observations that have to be replaced – generally in the least favourable way – to carry the estimator over all bounds. The breakdown point of m and s is $1/n$ because, in order to obtain this effect, it is enough to replace one observation by a large value. Not so for $\bar{m}$ and $\bar{s}$ whose breakdown point is 50% because we have to replace at least $n/2$ observations by outliers to be certain that the median is among them. The simple and yet far-reaching idea of breakdown has many important consequences and applications in many statistical procedures. For more details the reader can refer to Chapter 16 of Ref. [21] on which we based this short section on the identification of outliers.

We close this chapter with a final comment of general nature. The various tests – parametric and non-parametric – discussed in the present and preceding sections are just a few of the many tests that have been devised to cope with the wide variety of problems encountered in practice. With the main intention to explain and illustrate the fundamental ideas of hypothesis testing, it is evident that a number of widely adopted tests have not been considered. So, in regard to goodness-of-fit, for instance, we did not mention graphical methods or any of the many specific tests for normality. In fact, besides being outside our scopes, a detailed list of the available choices is, in any case, a rather difficult task (just to give a general idea, in a list of non-parametric tests classified according to (a) type of hypothesis to be tested and (b) type of data, the author personally counted 52 tests). The reader interested to this aspect will be able to find a sufficient number of other possibilities in the references and suggested reading at the end of the chapter.

6.6 Summary and comments

In practical cases the analyst is often faced with the problem of assessing the validity of some assumption regarding the problem under investigation. The usually adopted strategy in these cases is, in principle, quite easy: he/she formulates a working hypothesis, collects the necessary data and then, by any appropriate means, checks if the data are consistent with his/her hypothesis. If so, the hypothesis is accepted and it is rejected otherwise.

A closer look at this procedure, however, shows that things are not so simple because a number of subtle questions must be answered in order to put it into effect. Starting from the immediate consideration that, unless the entire population is examined, he/she will never know if the hypothesis is true or false, the consequences of a wrong decision must be taken into account. Also, the working hypothesis must be checked against some alternative hypothesis and the two hypotheses must be mutually exclusive in order to avoid ambiguities and, whenever possible, obtain a 'yes or no' answer. Last but not least, it should be specified what exactly is the 'appropriate means' by which we decide that the observed data agree, or disagree, with the hypothesis. The search for the answers to these and other questions brings us into the realm of an important branch of statistics known as 'hypothesis testing' whose main definitions and ideas – null and alternative hypothesis, type I and type II errors, rejection and acceptance regions, and so on – are given in Section 6.2.

A first classification distinguishes between parametric and non-parametric hypotheses and parametric hypotheses are the subject of all the Sections (from 6.3.1 to 6.3.4) of Section 6.3, including – in Section 6.3.2 – some remarks on a particular technique known as Wald's sequential analysis. Parametric hypotheses, moreover, can be simple or composite and the main result for the former type is given by Neyman–Pearson's lemma, which is stated and proved in Section 6.3.1 together with some worked-out examples on absolutely continuous and discrete distributions. For the latter type it is generally not possible to find a ump test – the definition of power of a parametric test is given in Section 6.3 – but the widely adopted technique known as 'likelihood ratio test' leads to acceptable results in most practical cases. Giving also a number of practical examples, the likelihood ratio test is explained in Section 6.3.3.

Finally, Section 6.3.4 deals with some important complements which should not be ignored by the reader willing to consider the subject in more detail. They are, in the order: (a) the strict relation between parametric tests and confidence intervals, (b) the asymptotic behaviour of parametric tests and (c) the notion of p-value which, somehow, provides more flexibility with respect to the 'standard' procedure.

Turning to non-parametric tests, one of the most frequently encountered situation occurs when we are not sure about the type of distribution underlying a given phenomenon and we wish to check whether a given distribution – normal, Poisson, or else – agrees with the observed data or, in other words, we wish to test how well an assumed distribution fits the data. In this case one speaks of goodness-of-fit tests and two important methods in this respect are Pearson's χ^2 test and K–S test, considered in Sections 6.4.1 and 6.4.2, respectively. The former is more general and, by comparing the appropriately partitioned data with their theoretical (assumed) frequencies, can be applied to both continuous and discrete distributions. The latter, on the other hand, applies only to continuous distributions and uses a statistic based on the difference between the theoretical and empirical PDFs. In regard to empirical PDFs, moreover, Section 6.4.2 includes a result known as Glivenko–Cantelli theorem which, besides being theoretically important in its own right, is at the basis of the idea behind the K–S test itself.

Section 6.5 closes the chapter by giving other types of tests which may often be useful in practical situations. In this light, Sections 6.5.1 and 6.5.2 deal with two more χ^2 tests: one for homogeneity and one for independence and it is pointed out that – in spite of the similarities – the two tests should not be confused because they refer to different contexts. Subsequently, Section 6.5.3 provides two tests for randomness to be used whenever we have doubts on the fact that our data are a random sample from a parent r.v., while Section 6.5.4 warns against the possibility of having a set of data 'contaminated' by outliers. When in doubt, it is always wise to use robust statistics in order to detect their presence because non-robust statistics as, for instance, the mean and variance, are generally of limited use in these cases. Nonetheless, the decision as to whether an outlier is due to a mistake – for instance, a recording or transcription error – or to the observation of a truly exceptional event (a rather rare occurrence indeed, but not impossible) is the analyst's responsibility.

References and further reading

[1] Azzalini, A., '*Inferenza Statistica: una Presentazione Basata sul Concetto di Verosimiglianza*', Springer-Verlag Italia, Milano (2001).

[2] Borovkov, A.A., '*Mathematical Statistics. Estimation of Parameters, Testing of Hypotheses*', Nauka, Moscow (1984).

[3] Bradley, J.V., '*Distribution-free Statistical Tests*', Prentice-Hall, Englewood Cliffs, NJ (1968).

[4] Cramér, H., '*Mathematical Methods of Statistics*', 19th edn., Princeton University Press, Princeton (1999).

[5] Di Crescenzo, A., Ricciardi, L.M., '*Elementi di Statistica*', Liguori Editore, Napoli (2000).

[6] Duncan, A., *'Quality Control and Industrial Statistics'*, 5th edn., Irwin, Homewood, Illinois (1986).
[7] Dunn, O., Clark, V., *'Applied Statistics: Analysis of Variance and Regression'*, Wiley, New York (1974).
[8] Green, J.R., Margerison, D., *'Statistical Treatment of Experimental Data'*, Elsevier, Amsterdam(1977).
[9] Guenther, W.C., *'Analysis of Variance'*, Prentice-Hall, Englewood Cliffs, NJ (1964).
[10] Ivchenko, G., Medvedev, Yu., *'Mathematical Statistics'*, Mir Publishers, Moscow (1990).
[11] Karr, A.F., *'Probability'*, Spinger-Verlag, New York (1993).
[12] Keeping, E.S., *'Introduction to Statistical Inference'*, Dover, New York (1995).
[13] Klimov, G., *'Probability Theory and Mathematical Statistics'*, Mir Publisher, Moscow (1986).
[14] Kottegoda, N.T., Rosso, R., *'Statistics, Probability and Reliability for Civil and Environmental Engineers'*, McGraw-Hill, New York (1997).
[15] Lehmann, E.L., *'Testing Statistical Hypotheses'*, Wiley, New York (1986).
[16] Massey, F.J. Jr., 'The Kolmogorov–Smirnov test for goodness of fit', *J. Am. Stat. Ass.*, 46, 68–78 (1951).
[17] Mendenhall, W., Wackerly, D.D., Scheaffer, R.L., *'Mathematical Statistics with Applications'*, 4th edn., PWS-KENT Publishing Company, Boston (1990).
[18] Milton, J.S., Arnold, J.C., *'Introduction to Probability and Statistics: Principles and Applications for Engineering and the Computing Sciences'*, McGraw-Hill, New York (1990).
[19] Serfling, R.J., *'Approximation Theorems in Mathematical Statistics'*, J. Wiley & Sons, New York (1980).
[20] Tyurin, Yu.N., *'On the Limiting Distribution of Kolmogorov–Smirnov Statistic for a Composite Hypothesis'*, Izv. AN SSSR, Ser. Math., 6, Vol. 48, pp. 1314–1343 (1984).
[21] Wadsworth, H.M. (ed.), *'Handbook of Statistical Methods for Engineers and Scientists'*, McGraw-Hill, New York (1990).
[22] Wald, A., *'Sequential Analysis'*, John Wiley & Sons, New York (1947).
[23] Wald, A., Wolfowitz, J., 'On a Test whether Two Samples are from the Same Population', *Ann. Math. Stat.*, 11, 147–162 (1940).

7 Regression, correlation and the method of least squares

7.1 Introduction

The study of relations between variables is fundamental in every branch of science and, in this respect, Statistics is no exception. An important distinction, however, must be made from the outset: while in disciplines like, for instance, Physics and Engineering these relations generally have an intrinsic cause–effect meaning, it may not necessarily be so in Statistics. Let us consider an example. When an engineer writes the equation $F = kx$ for an elastic spring of constant k, he/she is expressing a well-defined cause-effect linear relation between the force F applied to the spring and its elongation x (with respect to its rest position). If, however, we obtain a statistically satisfactory linear relation between, say, the number of new-born babies in Italy and the number of New Yorkers who quit smoking over the last 10 years, it would be rather hard to believe that this linear relationship has some intrinsic meaning. If, nonetheless, somebody is willing to assume that a meaning does exist, he/she must look for it outside the realm of Statistics.

This state of affair is due to the fact that, as we pointed out before, statistical significance is different from real-world significance and only in some cases the two concepts may coincide. So, before undertaking a study on the functional relation between, say, two variables X and Y we must answer the basic question: what are we trying to accomplish by fitting the sets of X and Y data by means of a mathematical function?

The answer, in fact, varies and depends on the objectives of the investigation. In some cases the type of functional relation is known from the theory and the fitting is used to obtain numerical values for the parameters. For example, by applying different forces to a spring and measuring its elongation at each level of force, we may simply want to determine a reliable value of the spring constant k in order to use it for further work. If, on the other hand, even the parameters of an equation are known, the purpose of the fit may be to confirm the theory from which the equation – and the parameter(s) value(s) – have been derived. In other cases we may have no *a priori* idea on the type of functional relation and a curve-fit – when, by ingenuity and/or trial and error, we find a satisfactory one – may provide clues for a new theory or suggest the presence of a formerly unexpected factor.

A third possibility is to fit the data merely to represent them by means of an empirical relation, thus summarizing a large body of information in a compact formula which is easily accessible at all times and readily communicable. Therefore, if we have an answer to the question above we are ready to accept, in this last case, that even a high statistical significance – that is a good fit – may not correspond to any relationship of cause and effect, exactly as in the (rather extreme) example on new-born Italian babies and no-longer-smoking New Yorkers.

A final word on terminology. The term 'regression', generally intended in the sense of 'average relationship', dates back to the work of Sir Francis Galton who, in the 1880s, plotted the average heights of children against the average heights of their parents. He discovered that, on average, the offspring of tall parents were not as tall as their parents while the offspring of short parents were not as short as their parents, thereby concluding that human height tends to 'regress' towards the average height of the population. The term remained and, although outdated and mostly of historical significance only, is still widely used when referring to statistical curve-fitting procedures.

7.2 The general linear regression problem

In practical situations we often have to study the behavior of a *response* (random) variable Y which is assumed to depend on a number, say k, of non-random *predictor* variables $x_1, \dots, x_k$ whose values, in turn, change from trial to trial. For instance, the x_j $(j = 1, \dots, k)$ may be the controlled input characteristics of a device whose measured response Y (the output) varies as the x_j vary. The form of the functional relation between Y and the x_j is sometimes suggested by the problem under investigation but for a number of reasons – convenience being often one of them – the assumption

$$Y = \sum_{j=1}^{k} \beta_j x_j \tag{7.1a}$$

(where $\beta_1, \dots, \beta_k$ is a set of k unknown parameters) is rather frequent in practice. As it is written, eq. (7.1a) is a deterministic relation with nothing random about it. The statistical nature of the problem, however, arises through the presence of an 'error term' associated – in the case we are considering – with the response. In fact, even in the typical case of a laboratory situation where we can control the x variables without appreciable error, the response is measured with an experimental error ε which fluctuates unpredictably from trial to trial. The fact that ε is a random variable makes the response Y a random variable and we should rewrite (7.1a) in the form of the statistical linear model

$$Y = \sum_{j=1}^{k} \beta_j x_j + \varepsilon \tag{7.1b}$$

On the basis of this model, an experiment consisting of n trials ($n \geq k$) in which we measure Y as a function of the x_j leads to the n equations

$$Y_i = \sum_{j=1}^{k} \beta_j x_{ij} + \varepsilon_i \tag{7.2a}$$

which, in turn, can be compactly arranged in matrix form by writing

$$\mathbf{Y} = \mathbf{Xb} + \mathbf{e} \tag{7.2b}$$

where $\mathbf{Y}$ and $\mathbf{e}$ are the $n \times 1$ response and error column vectors $(Y_1 \ldots Y_n)^T$ and $(\varepsilon_1 \ldots \varepsilon_n)^T$ respectively, $\mathbf{b}$ is the $k \times 1$ vector of parameters $(\beta_1 \ldots \beta_k)^T$ and $\mathbf{X}$ is the $n \times k$ matrix whose ith row represents the x values in the ith trial. The common assumptions on the model are as follows

(1) the errors are uncorrelated with $E(\varepsilon_i) = 0$ and $\mathrm{Var}(\varepsilon_i) = \sigma^2$ for all $i = 1, \ldots, n$, where σ^2 – often called the *residual variance* – is generally unknown. These conditions are written in matrix form as $E(\mathbf{e}) = \mathbf{0}$ and $\mathbf{K}(\varepsilon) = E(\mathbf{ee}^T) = \sigma^2 \mathbf{I}_n$ where $\mathbf{I}_n = \mathrm{diag}(1, \ldots, 1)$ is the $n \times n$ identity matrix and we denote by the symbol $\mathbf{K}(\varepsilon)$ the covariance matrix of the ε_i
(2) the matrix $\mathbf{X}$ is such that $\mathrm{rank}(\mathbf{X}) = k$.

The problem then consists in estimating the unknown $k + 1$ parameters $\beta_1, \ldots, \beta_k, \sigma^2$. Now, although the assumptions above give immediately $E(\mathbf{Y}) = \mathbf{Xb}$ and $\mathbf{K}_\mathbf{Y} = \sigma^2 \mathbf{I}_n$, we cannot use the maximum likelihood method because nothing has been said about the distribution of Y and therefore we cannot write its likelihood function. In order to avoid, for the moment, additional assumptions on the distribution of $\mathbf{e}$ or Y, a solution to the problem is given by the *method of least squares* (LS method) which suggests to obtain an estimate of $\mathbf{b}$ by minimizing the quadratic form

$$Q(\mathbf{b}) = \mathbf{e}^T\mathbf{e} = [\mathbf{Y} - E(\mathbf{Y})]^T[\mathbf{Y} - E(\mathbf{Y})] = (\mathbf{Y} - \mathbf{Xb})^T(\mathbf{Y} - \mathbf{Xb}) \tag{7.3}$$

thus meaning that we have to solve the system of the k so-called 'normal' equations $\partial Q(\mathbf{b})/\partial \beta_j = 0$. It is not difficult to show that these equations can be expressed in matrix form as

$$\mathbf{Ab} = \mathbf{X}^{\mathrm{T}}\mathbf{Y} \tag{7.4}$$

where, for brevity, we defined $\mathbf{A} = \mathbf{X}^{\mathrm{T}}\mathbf{X}$. A direct consequence of our assumptions is that $\mathbf{A}$ is a (symmetric) positive-definite $k \times k$ matrix of rank k and can be inverted to obtain the LS estimate $\hat{\mathbf{b}}$ of $\mathbf{b}$ as

$$\hat{\mathbf{b}} = \mathbf{A}^{-1}\mathbf{X}^{\mathrm{T}}\mathbf{Y} \tag{7.5}$$

This solution is unique and the fact that it represents a minimum of $Q(\mathbf{b})$ can be seen by noting that the matrix of second derivatives is $2\mathbf{A}$ and therefore, by the arguments above, is positive-definite.

The estimate (7.5) has an interesting geometric interpretation. If we denote by $\mathbf{x}_1, \ldots, \mathbf{x}_k$ the columns of $\mathbf{X}$, the linear combination $\mathbf{Xb} - \beta_1\mathbf{x}_1 + \cdots + \beta_k\mathbf{x}_k$ spans a k-dimensional linear subspace of $\mathbb{R}^n$ as $\mathbf{b}$ varies in $\mathbb{R}^k$. Let us call this subspace S. Equation (7.2b) and assumption (1) imply $E(\mathbf{Y}) = \mathbf{Xb}$ and therefore $E(\mathbf{Y}) \in S$ which, in turn, means that the estimating procedure of least squares selects that particular vector $\mathbf{X}\hat{\mathbf{b}} \in S$ which minimizes the Euclidean distance between $\mathbf{Y}$ and the subspace S. Since it is known from the geometry of finite dimensional vector spaces that this state of affairs implies the orthogonality condition $(\mathbf{Y} - \mathbf{X}\hat{\mathbf{b}})\perp S$, it follows that $(\mathbf{Y} - \mathbf{X}\hat{\mathbf{b}})^{\mathrm{T}}\mathbf{X} = 0$ (the columns of $\mathbf{X}$ form a basis of S) and this, in turn, shows that $\hat{\mathbf{b}}$ is in fact a solution of eq. (7.4). In different words, this fact can be expressed by saying that

$$\hat{\mathbf{Y}} \equiv \mathbf{X}\hat{\mathbf{b}} = (\mathbf{X}\mathbf{A}^{-1}\mathbf{X}^{\mathrm{T}})\mathbf{Y} \equiv \mathbf{P}_S\mathbf{Y} \tag{7.6}$$

is the projection of $\mathbf{Y}$ on the subspace S. In fact, it is easy to show that the $n \times n$ matrix $\mathbf{P}_S = \mathbf{X}\mathbf{A}^{-1}\mathbf{X}^{\mathrm{T}}$ is both symmetric (i.e. $\mathbf{P}_S^{\mathrm{T}} = \mathbf{P}_S$) and idempotent (i.e. $\mathbf{P}_S^2 = \mathbf{P}_S$) and therefore is a projection matrix. Moreover, from these considerations it follows that $\mathbf{P}_{S^\perp} = \mathbf{I}_n - \mathbf{P}_S$ is the projection matrix on the $(n - k)$-dimensional subspace (of $\mathbb{R}^n$) $S^\perp$ orthogonal to S. The fact that the squared lengths ('norms' in mathematical terminology) of the vectors $\mathbf{P}_S\mathbf{Y}$ and $\mathbf{P}_{S^\perp}\mathbf{Y}$ sum up to give the length (norm) of $\mathbf{Y}$ is just a multi-dimensional version of Pythagoras' theorem.

Let us consider now the properties of the estimate $\hat{\mathbf{b}}$. We have:

Proposition 7.1 *The LS estimate $\hat{\mathbf{b}}$ is unbiased and its covariance matrix is $\mathbf{K}(\hat{\mathbf{b}}) = \sigma^2\mathbf{A}^{-1}$. Moreover, $\hat{\mathbf{b}}$ is the estimate with the minimum variance in the class of all unbiased estimates of $\mathbf{b}$ which depend linearly on $\mathbf{Y}$ (this second part of the proposition is known as Gauss–Markov theorem).*

Unbiasedness is immediate, in fact from eq. (7.5) we get

$$E(\hat{\mathbf{b}}) = \mathbf{A}^{-1}\mathbf{X}^{\mathrm{T}}E(\mathbf{Y}) = \mathbf{A}^{-1}\mathbf{X}^{\mathrm{T}}\mathbf{X}\mathbf{b} = \mathbf{A}^{-1}\mathbf{A}\mathbf{b} = \mathbf{b} \tag{7.7}$$

On the other hand, using eq. (3.35) we get

$$\mathbf{K}(\hat{\mathbf{b}}) = \mathbf{A}^{-1}\mathbf{X}^{\mathrm{T}}\mathbf{K}_{\mathbf{Y}}(\mathbf{A}^{-1}\mathbf{X}^{\mathrm{T}})^{\mathrm{T}} = \sigma^2\mathbf{A}^{-1}\mathbf{X}^{\mathrm{T}}\mathbf{X}\mathbf{A}^{-1} = \sigma^2\mathbf{A}^{-1} \tag{7.8}$$

In order to prove the minimum variance property, let $\mathbf{a}$ be a general linear (in $\mathbf{Y}$) unbiased estimate of $\mathbf{b}$. Then $\mathbf{a}$ is of the form $\mathbf{a} = \mathbf{LY}$ where $\mathbf{L}$ is a general $k \times n$ matrix of constants. Since the unbiasedness relation $E(\mathbf{a}) = \mathbf{L}E(\mathbf{Y}) = \mathbf{LXb} = \mathbf{b}$ must hold for all $\mathbf{b}$, it follows $\mathbf{LX} = \mathbf{I}_k$. Moreover, eq. (3.35) gives $\mathbf{K}(\mathbf{a}) = \mathbf{L}\mathbf{K}_{\mathbf{Y}}\mathbf{L}^{\mathrm{T}} = \sigma^2\mathbf{L}\mathbf{L}^{\mathrm{T}}$ and we can use the identity

$$\mathbf{L}\mathbf{L}^{\mathrm{T}} = (\mathbf{A}^{-1}\mathbf{X}^{\mathrm{T}})(\mathbf{A}^{-1}\mathbf{X}^{\mathrm{T}})^{\mathrm{T}} + (\mathbf{L} - \mathbf{A}^{-1}\mathbf{X}^{\mathrm{T}})(\mathbf{L} - \mathbf{A}^{-1}\mathbf{X}^{\mathrm{T}})^{\mathrm{T}}$$

to determine that all the diagonal elements of $\mathbf{L}\mathbf{L}^{\mathrm{T}}$ attain (simultaneously) their minimum when $\mathbf{L} = \mathbf{A}^{-1}\mathbf{X}^{\mathrm{T}}$, that is, when the estimate $\mathbf{a}$ coincides with $\hat{\mathbf{b}}$. This is due to the fact that all the matrices on the r.h.s. of the identity have the form $\mathbf{H}\mathbf{H}^{\mathrm{T}}$ and therefore (a) all their diagonal terms are non-negative and (b) the minimum is for $\mathbf{L} = \mathbf{A}^{-1}\mathbf{X}^{\mathrm{T}}$ because only the second term depends on $\mathbf{L}$.

In the light of eq. (7.8), we can make an observation on the consistency of $\hat{\mathbf{b}}$. Recalling Proposition 5.6, consistency holds if $\mathrm{Var}(\hat{\beta}_j) \to 0$ as $n \to \infty$ for all $j = 1, \ldots, k$ or, equivalently, if the diagonal elements of $\mathbf{A}^{-1}$ tend to zero as $n \to \infty$. This circumstance depends clearly on the nature of the matrix $\mathbf{X}$ and, *a priori*, not much can be said without making any assumption on its asymptotic behavior. For our purposes, however, it suffices to say that in most practical cases it turns out that $\hat{\mathbf{b}}$ is also a consistent estimate of $\mathbf{b}$.

At this point, the only problem left unanswered by the LS solution is the estimate of the residual variance σ^2. Starting from the obvious relation $\sigma^2 = \mathrm{Var}(\varepsilon_i) = E(\varepsilon_i^2)$, uncorrelation (of the ε_i) gives

$$\mathrm{Var}\left(\sum \varepsilon_i\right) = \sum \mathrm{Var}(\varepsilon_i) = \sum E\left(\varepsilon_i^2\right) = E(\mathbf{e}^{\mathrm{T}}\mathbf{e}) = n\sigma^2 \tag{7.9}$$

Since $\mathbf{e} = \mathbf{Y} - \mathbf{Xb}$, the last equality on the r.h.s. suggests to use the residual vector $\hat{\mathbf{e}} \equiv \mathbf{Y} - \mathbf{X}\hat{\mathbf{b}}$ and consider the estimate $\hat{\sigma}^2$ (of σ^2) given by

$$n\hat{\sigma}^2 = \hat{\mathbf{e}}^{\mathrm{T}}\hat{\mathbf{e}} = Q(\hat{\mathbf{b}}) = (\mathbf{Y} - \mathbf{X}\hat{\mathbf{b}})^{\mathrm{T}}(\mathbf{Y} - \mathbf{X}\hat{\mathbf{b}}) = \mathbf{Y}^{\mathrm{T}}(\mathbf{I}_n - \mathbf{P}_S)\mathbf{Y} \tag{7.10}$$

where we took eq. (7.6) into account in the last equality. The result

$$E\{\mathbf{Y}^{\mathrm{T}}(\mathbf{I}_n - \mathbf{P}_S)\mathbf{Y}\} = \sigma^2(n - k) \tag{7.11}$$

which we will prove shortly, shows that $E(\hat{\sigma}^2) = n^{-1}(n-k)\sigma^2$ thus implying that $\hat{\sigma}^2$ is a biased estimate (although asymptotically unbiased). The bias, however, can be easily removed by taking

$$\hat{s}^2 \equiv \frac{n\hat{\sigma}^2}{n-k} = \frac{Q(\hat{\mathbf{b}})}{n-k} = \frac{\hat{\mathbf{e}}^T\hat{\mathbf{e}}}{n-k} \tag{7.12a}$$

as the estimate of σ^2. This, in the end, shows that our unbiased estimate $\hat{s}^2$ is given by the sum of squared residuals $\sum \hat{\varepsilon}_i^2$ divided by $n-k$. Also, for computational purposes, note that eq. (7.12a) can be conveniently written as

$$\hat{s}^2 = \frac{1}{n-k}(\mathbf{Y}^T\mathbf{Y} - \hat{\mathbf{b}}^T\mathbf{X}^T\mathbf{Y}) \tag{7.12b}$$

In fact, from $\hat{\mathbf{e}}^T\hat{\mathbf{e}} = \mathbf{Y}^T\mathbf{Y} - \mathbf{Y}^T\mathbf{X}\hat{\mathbf{b}} - \hat{\mathbf{b}}^T\mathbf{X}^T\mathbf{Y} + \hat{\mathbf{b}}^T\mathbf{A}\hat{\mathbf{b}}$ eq. (7.12b) follows because $\mathbf{Y}^T\mathbf{X}\hat{\mathbf{b}} = \hat{\mathbf{b}}^T\mathbf{X}^T\mathbf{Y}$ and $\mathbf{A}\hat{\mathbf{b}} = \mathbf{X}^T\mathbf{Y}$ (eq. (7.4)).

Let us now turn to the proof of eq. (7.11). Starting from the general relation $E(\mathbf{Y}^T\mathbf{B}\mathbf{Y}) = \mathbf{m}_\mathbf{Y}^T\mathbf{B}\mathbf{m}_\mathbf{Y} + \mathrm{tr}(\mathbf{B}\mathbf{K}_\mathbf{Y})$ (whose proof is left to the reader), where we denoted by $\mathbf{B}$ any $k \times k$ non-random matrix and by $\mathbf{m}_\mathbf{Y}$ the vector $E(\mathbf{Y}) = \mathbf{X}\mathbf{b}$, we get in our case

$$E\{\mathbf{Y}^T(\mathbf{I}_n - \mathbf{P}_S)\mathbf{Y}\} = \mathbf{m}_\mathbf{Y}^T\mathbf{P}_{S^\perp}\mathbf{m}_\mathbf{Y} + \mathrm{tr}[(\mathbf{I}_n - \mathbf{P}_S)\mathbf{K}_\mathbf{Y}]$$

The first term on the r.h.s. is zero because $\mathbf{m}_\mathbf{Y} \in S$ and $\mathbf{P}_{S^\perp}$ projects on the space $S^\perp$; the second term, in turn, becomes

$$\mathrm{tr}(\mathbf{I}_n\mathbf{K}_\mathbf{Y}) - \mathrm{tr}(\mathbf{P}_S\mathbf{K}_\mathbf{Y}) = n\sigma^2 - \sigma^2\mathrm{tr}(\mathbf{X}\mathbf{A}^{-1}\mathbf{X}^T)$$

and eq. (7.11) follows by virtue of a well-known property on the trace of products of matrices which gives $\mathrm{tr}(\mathbf{X}\mathbf{A}^{-1}\mathbf{X}^T) = \mathrm{tr}(\mathbf{A}^{-1}\mathbf{X}^T\mathbf{X}) = \mathrm{tr}(\mathbf{A}^{-1}\mathbf{A}) = k$.

To close this section, four important remarks are in order:

(i) In the above scheme the variables x_j may be functionally dependent. A rather common case, for instance, is $x_1 = x, x_2 = x^2, \ldots, x_k = x^k$ and the expression $\mathbf{X}\mathbf{b} = \beta_1 x + \beta_2 x^2 + \cdots + \beta_k x^k$ is a polynomial of degree k in x. In the general linear model of eq. (7.2b), in fact, linearity is meant in the parameters $\beta_1, \ldots, \beta_k$, not in the variables.

(ii) Denoting by $\overline{Y}$ the mean of the Y_i, that is, $\overline{Y} = n^{-1}\sum_i Y_i$, it is easy to show that the total variability S_{YY} of the Y variable can be written as

$$S_{YY} \equiv \sum_i (Y_i - \overline{Y})^2 = \mathbf{Y}^T\mathbf{Y} - n\overline{Y}^2 \tag{7.13}$$

Then, since $\hat{\mathbf{e}}^T\hat{\mathbf{e}} = \mathbf{Y}^T\mathbf{Y} - \hat{\mathbf{b}}^T\mathbf{X}^T\mathbf{Y}$ is the part of variability due to error (and not 'explained', so to speak, by the regression), we can substitute this equation into eq. (7.13) to get

$$S_{YY} = \hat{\mathbf{e}}^T\hat{\mathbf{e}} + \hat{\mathbf{b}}^T\mathbf{X}^T\mathbf{Y} - n\overline{Y}^2 \tag{7.14a}$$

where the quantity $SS_R = \hat{\mathbf{b}}^T\mathbf{X}^T\mathbf{Y} - n\overline{Y}^2$ is the part of variability due to the regression. In many statistical books eq. (7.14a) is often written

$$S_{YY} = SS_E + SS_R \tag{7.14b}$$

thereby meaning that the total variability of Y is the sum of two terms: (a) the sum of squared residuals $SS_E = \hat{\mathbf{e}}^T\hat{\mathbf{e}}$ plus (b) the regression sum of squares SS_R. If the linear model provides a good fit of the data we expect SS_E to be small in comparison to SS_R – and therefore in comparison to S_{YY}. In this regard, in fact, the quantity called the coefficient of multiple regression (or of multiple determination) and defined as

$$R^2 \equiv \frac{SS_R}{S_{YY}} \tag{7.15}$$

is a measure of the proportion of the Y variability explained by the LS fitting procedure.

(iii) In certain problems we may be more interested in some linear combinations $\mathbf{t} = (t_1 \dots t_m)^T$, $m \le k$, of the parameters rather than in the parameters $\beta_1, \dots, \beta_k$ themselves. Then $\mathbf{t}$ is of the form $\mathbf{t} = \mathbf{Tb}$, where $\mathbf{T}$ is a given $m \times k$ matrix of constants. In this case the LS estimate $\hat{\mathbf{t}}$ of $\mathbf{t}$ is given by $\hat{\mathbf{t}} = \mathbf{T}\hat{\mathbf{b}} = \mathbf{TA}^{-1}\mathbf{X}^T\mathbf{Y}$ and has the same properties of unbiasedness and efficiency (in the class of linear unbiased estimates of $\mathbf{t}$) mentioned in Proposition 7.1. Moreover, its covariance matrix is given by $\mathbf{K}(\hat{\mathbf{t}}) = \sigma^2\mathbf{TA}^{-1}\mathbf{T}^T$.

(iv) *Constrained estimates.* In the case considered so far we found min $Q(\mathbf{b})$ by letting $\mathbf{b}$ free to vary in the entire space $\mathbb{R}^k$. Then our solution $\hat{\mathbf{b}}$ is a so-called unconstrained LS estimate of $\mathbf{b}$. In some situations, however, the possible values of $\mathbf{b}$ are restricted by linear constraints of the form

$$\mathbf{Cb} = \mathbf{c} \tag{7.16}$$

where $\mathbf{C}$ is a given $m \times k$ ($m \le k$) matrix of constants of rank k and $\mathbf{c}$ is a given m-dimensional vector. The LS solution to this problem is found by searching for that vector $\hat{\mathbf{b}}_C$ (C is for constrained) which satisfies the minimum condition

$$Q(\hat{\mathbf{b}}_C) = \min_{\mathbf{b}:\mathbf{Cb}=\mathbf{c}} Q(\mathbf{b}) \tag{7.17}$$

Without getting into the details of the calculations (the method of Lagrange multipliers is used in this case) the final result is

$$\hat{\mathbf{b}}_C = \hat{\mathbf{b}} - \mathbf{A}^{-1}\mathbf{C}^T\mathbf{D}^{-1}(\mathbf{C}\hat{\mathbf{b}} - \mathbf{c}) \tag{7.18}$$

where $\mathbf{D} = \mathbf{C}\mathbf{A}^{-1}\mathbf{C}^T$ (of size $m \times m$) is positive definite and $\hat{\mathbf{b}}$ is the unconstrained LS estimate of eq. (7.5).

7.2.1 *Simple linear regression*

An important special case of the general model (7.1b) is the simple linear model $Y = \beta_1 + \beta_2 x + \varepsilon$ in which the underlying assumption is that the response Y depends only on one predictor variable and the functional relation between $E(Y)$ and x is a straight line with intercept β_1 and slope β_2. By carrying out n trials ($n > 2$; $n = 2$ is a trivial case) in which we measure the values $Y_1, \ldots, Y_n$ in correspondence to $x_1, \ldots, x_n$, respectively, the explicit form of eq. (7.2b) is

$$\begin{pmatrix} Y_1 \\ Y_2 \\ \vdots \\ Y_n \end{pmatrix} = \begin{pmatrix} 1 & x_1 \\ 1 & x_2 \\ \vdots & \vdots \\ 1 & x_n \end{pmatrix} \begin{pmatrix} \beta_1 \\ \beta_2 \end{pmatrix} + \begin{pmatrix} \varepsilon_1 \\ \varepsilon_2 \\ \vdots \\ \varepsilon_n \end{pmatrix} \tag{7.19}$$

while the (2×2) matrix $\mathbf{A} = \mathbf{X}^T\mathbf{X}$ is

$$\mathbf{A} = \begin{pmatrix} n & \sum_i x_i \\ \sum_i x_i & \sum_i x_i^2 \end{pmatrix} = \begin{pmatrix} n & n\bar{x} \\ n\bar{x} & \sum_i x_i^2 \end{pmatrix} \tag{7.20}$$

where we denoted by x the arithmetic mean of the x_i, that is, $x = n^{-1}\sum_i x_i$. Also,

$$\mathbf{A}^{-1} = \frac{1}{\det(\mathbf{A})} \begin{pmatrix} \sum_i x_i^2 & -n\bar{x} \\ -n\bar{x} & n \end{pmatrix} \tag{7.21a}$$

where

$$\det(\mathbf{A}) = n\sum_i x_i^2 - \left(\sum_i x_i\right)^2 = n\sum_i (x_i - \bar{x})^2 \tag{7.21b}$$

If, as in the preceding section, $\overline{Y} = n^{-1}\sum_i Y_i$ is the mean of the observed values Y_i, eq. (7.5) gives the LS estimate of $\mathbf{b}$ as (we omit the summation index i for brevity)

$$\hat{\mathbf{b}} = \begin{pmatrix} \hat{\beta}_1 \\ \hat{\beta}_2 \end{pmatrix} = \frac{n}{\det(\mathbf{A})} \begin{pmatrix} \overline{Y} \sum x_i^2 - \bar{x} \sum x_i Y_i \\ \sum x_i Y_i - n\bar{x}\overline{Y} \end{pmatrix} \tag{7.22a}$$

For computational purposes, the two estimates can be more conveniently expressed as

$$\hat{\beta}_2 = \frac{\sum (x_i - \bar{x})(Y_i - \overline{Y})}{\sum (x_i - \bar{x})^2}, \qquad \hat{\beta}_1 = \overline{Y} - \hat{\beta}_2 \bar{x} \tag{7.22b}$$

where it should be noted that the slope $\hat{\beta}_2$ comes first because we need it to calculate the intercept $\hat{\beta}_1$.

Equation (7.8), in turn, gives the covariance matrix of $\hat{\mathbf{b}}$; its diagonal elements are

$$\operatorname{Var}(\hat{\beta}_1) = \frac{\sigma^2 \sum x_i^2}{n \sum (x_i - \bar{x})^2}, \qquad \operatorname{Var}(\hat{\beta}_2) = \frac{\sigma^2}{\sum (x_i - \bar{x})^2} \tag{7.23a}$$

while we get for the off-diagonal terms

$$\operatorname{Cov}(\hat{\beta}_1, \hat{\beta}_2) = \operatorname{Cov}(\hat{\beta}_2, \hat{\beta}_1) = -\frac{\sigma^2 \bar{x}}{\sum (x_i - \bar{x})^2} \tag{7.23b}$$

The last step consists in estimating the residual variance σ^2. Equation (7.12b) with $k = 2$ gives

$$\hat{s}^2 = \frac{\sum Y_i^2 - \hat{\beta}_1 \overline{Y} - \hat{\beta}_2 \sum x_i Y_i}{n-2} = \frac{\sum (Y_i - \overline{Y})^2 - \hat{\beta}_2^2 \sum (x_i - \bar{x})^2}{n-2} \tag{7.24}$$

where the last expression on the r.h.s. has been obtained by taking into account the second of eqs (7.22b) and the second of (7.22a). This form is more convenient for computation. Clearly, when σ^2 is not known and we estimate it by means of $\hat{s}^2$, the numerical values of $\operatorname{Var}(\hat{\beta}_1), \operatorname{Var}(\hat{\beta}_2)$ and $\operatorname{Cov}(\hat{\beta}_1, \hat{\beta}_2)$ are obtained by substituting $\hat{s}^2$ in eqs (7.23a) and (7.23b), respectively; the square roots $\sqrt{\operatorname{Var}(\hat{\beta}_1)}, \sqrt{\operatorname{Var}(\hat{\beta}_2)}$ are often referred to as 'standard errors' of the estimates.

A few observations at this point can be useful. First, denoting by the special symbols S_{xx} and S_{xY} the two sums

$$S_{xx} \equiv \sum_i (x_i - \bar{x})^2$$
$$S_{xY} \equiv \sum_i (x_i - \bar{x})(Y_i - \overline{Y}) \tag{7.25}$$

then $\hat{\beta}_2 = S_{xY}/S_{xx}$ and the regression line can be compactly written as

$$\hat{Y} = \overline{Y} + \frac{S_{xY}}{S_{xx}}(x - \bar{x}) \tag{7.26}$$

where $\hat{Y}$ is the calculated value corresponding to x. If x were itself a r.v. – and we should write X in this case – then S_{XY} would be the sample counterpart of $n\mathrm{Cov}(X, Y)$ and eq. (7.26) would be the counterpart of the first of eqs (3.98). In eq. (7.26) as it is, however, the analogy is only formal because we assumed that x is a non-random variable. Some comments on the case when x is a random variable in its own right are delayed to Section 7.3.2. For the moment we limit ourselves to a remark taken from Chapter 13 of [27]: '...there is general agreement that the classical regression approach (note of the author: i.e. the one considered so far) can be used safely when the variation in the X values is small relative to the variation in the Y values'.

A second observation of more practical nature is as follows: for a fixed residual variance σ^2, eqs (7.23a) show that both $\mathrm{Var}(\hat{\beta}_1)$ and $\mathrm{Var}(\hat{\beta}_2)$ decrease as S_{xx} increases. Consequently, in a controlled experiment – that is, an experiment in which the analyst has control on the x variable – choosing the x_i values far apart reduces the variances of the estimated regression parameters. When the fitting has been obtained, moreover, other two practical recommendations are worthy of mention:

(i) It is always advisable not to extrapolate, that is, use the regression beyond the range of values for which it was established. This is true in general – not just for a straight line – because any inference we can make on predicted values apply only within the region covered by the experimental data. Unless there is strong evidence in its favor, any extrapolation result is generally not justified and must be handled very cautiously.

(ii) Always check the sequence of signs of the observed residuals $\hat{\varepsilon}_i = Y_i - \hat{Y}_i$ (and, obviously, their values). If they follow a definite pattern there may be evidence of the fact that a curve – rather than a straight line – could provide a better model for the data. In addition, the type of curve is often suggested by the pattern of residuals itself.

The third and final point concerns the coefficient of determination of eq. (7.15). It can be shown (the proof is rather cumbersome and we leave it to the reader as an exercise) that

$$S_{YY} \equiv \sum(Y_i - \overline{Y})^2 = \sum(Y_i - \hat{Y}_i)^2 + \sum(\hat{Y}_i - \overline{Y})^2 \tag{7.27}$$

where $\hat{Y}_i = \hat{\beta}_1 + \hat{\beta}_2 x_i$ for $i = 1, \ldots, n$. Equation (7.27) is nothing but eq. (7.14b) written in a different form. The fact that the second term is SS_R follows from the chain of equalities

$$\begin{aligned}\sum(\hat{Y}_i - \overline{Y})^2 &= \sum \hat{Y}_i^2 + n\overline{Y}^2 - 2\overline{Y}\sum \hat{Y}_i = \sum \hat{Y}_i^2 - n\overline{Y}^2 \\ &= \hat{\mathbf{Y}}^{\mathrm{T}}\hat{\mathbf{Y}} - n\overline{Y}^2 = \hat{\mathbf{b}}^{\mathrm{T}}\mathbf{A}\hat{\mathbf{b}} - n\overline{Y}^2 = \hat{\mathbf{b}}^{\mathrm{T}}\mathbf{X}^{\mathrm{T}}\mathbf{Y} - n\overline{Y}^2 = SS_R\end{aligned}$$

where we took into account: (a) $\sum \hat{Y}_i = n\overline{Y}$ in the second equality (this relation follows immediately by substituting the second of (7.22b) in $\sum \hat{Y}_i = n\hat{\beta}_1 + n\hat{\beta}_2\bar{x}$), (b) $\hat{\mathbf{Y}} = \mathbf{X}\hat{\mathbf{b}}$ in the fourth, (c) $\mathbf{A}\hat{\mathbf{b}} = \mathbf{X}^{\mathrm{T}}\mathbf{Y}$ (eq. (7.4)) in the fifth equality and, clearly, $\hat{\mathbf{Y}} = (\hat{Y}_1, \ldots, \hat{Y}_n)^{\mathrm{T}}$. Owing to eq. (7.27) the determination coefficient can be written as

$$R^2 = \frac{SS_R}{S_{YY}} = \frac{\sum(\hat{Y}_i - \overline{Y})^2}{\sum(Y_i - \overline{Y})^2} = \hat{\beta}_2 \frac{S_{xY}}{S_{YY}} = \frac{S_{xY}^2}{S_{xx}S_{YY}} \tag{7.28a}$$

where the last two expressions on the r.h.s. (the easy proof of the equalities is left to the reader) are also often used for computation. It is evident that $0 \le R^2 \le 1$. The case $R^2 = 0$ corresponds to $\hat{Y}_i = \overline{Y}$ for all i thus implying that the role of the x_i is irrelevant because all of them lead to the same predicted value $\overline{Y}$. Moreover, since in this case $\hat{Y}_i = \hat{Y}_j$ for all $i, j = 1, \ldots, n$ the equality $\hat{\beta}_1 + \hat{\beta}_2 x_i = \hat{\beta}_1 + \hat{\beta}_2 x_j$ for $i \neq j$ (assuming $x_i \neq x_j$) gives necessarily $\hat{\beta}_2 = 0$. In other words, in this case we do not need x in order to 'explain' Y. On the other extreme, $R^2 = 1$ if and only if $SS_R = S_{YY}$, i.e. if $SS_E = 0$ and therefore $\hat{Y}_i = Y_i$. This is an index of a perfect linear relation between Y and x because all the Y_i are perfectly predicted – through the x_i – by the regression line. Finally, note that the square root of the determination coefficient, that is,

$$R = \frac{S_{xY}}{\sqrt{S_{xx}S_{YY}}} \tag{7.28b}$$

is the sample counterpart of the correlation coefficient (3.22).

7.3 Normal regression

In the general linear model considered in Section 7.2 – and consequently in the special simple regression case of Section 7.2.1 – the only assumptions on the errors ε_i were $E(\varepsilon_i) = 0$, $\mathrm{Var}(\varepsilon_i) = \sigma^2$ for all $i = 1, \ldots, n$ and $\mathrm{Cov}(\varepsilon_i, \varepsilon_j) = 0$ for $i \neq j$. Stronger inferences on the estimated parameters, however, can be made if we add a further assumption. The most common assumption in this regard – and one speaks of normal regression in this case – is that the errors are normally distributed, that is, $\varepsilon_i \approx N(0, \sigma^2)$ for all i or, in matrix form, $\mathbf{e} \approx N(\mathbf{0}, \sigma^2\mathbf{I}_n)$. This condition, in turn, clearly implies that the Y_i themselves are normally distributed as $\mathbf{Y} \approx N(\mathbf{Xb}, \sigma^2\mathbf{I}_n)$.

In this light we can now consider the (joint) likelihood function of the Y_i; recalling the matrix expression of eq. (3.68a) we have

$$L(\mathbf{Y}; \mathbf{q}) = \frac{1}{(2\pi\sigma^2)^{n/2}} \exp\left(-\frac{1}{2\sigma^2} Q(\mathbf{Y}, \mathbf{b})\right) \tag{7.29}$$

where $Q(\mathbf{Y}, \mathbf{b}) = (\mathbf{Y} - \mathbf{Xb})^{\mathrm{T}}(\mathbf{Y} - \mathbf{Xb})$ is the quadratic form of eq. (7.3) and $\mathbf{q}$ denotes the $(k+1)$-dimensional vector of unknown parameters $\mathbf{q} = (\beta_1, \ldots, \beta_k, \sigma^2)$. For any $\sigma^2 > 0$ it is evident that maximizing the likelihood function (7.29) with respect to $\mathbf{b}$ is equivalent to minimizing $Q(\mathbf{Y}, \mathbf{b})$ and therefore we can conclude that in case of normal regression the LS estimate $\hat{\mathbf{b}}$ of $\mathbf{b}$ is the same as its ML (maximum likelihood) estimate.

This fact, in turn, leads to an immediate observation. Since (first part of Proposition 7.1) we know that $\hat{\mathbf{b}}$ is an unbiased estimator, Gauss–Markov theorem (i.e. the second part of Proposition 7.1) can be strengthened by taking into account Proposition 5.2 on ML estimators: in the case of normal regression not only $\hat{\mathbf{b}}$ is the minimum variance estimator of $\mathbf{b}$ in the class of linear (in $\mathbf{Y}$) unbiased estimators, but it is the minimum variance estimator in the class of all (not only linear) unbiased estimators of $\mathbf{b}$.

Turning now to the ML estimate of the residual variance σ^2 we can substitute $\hat{\mathbf{b}}$ in eq. (7.29) and maximize $\ln L$ with respect to σ^2. The result is the biased estimator

$$\hat{\sigma}^2 = \frac{Q(\hat{\mathbf{b}})}{n} \tag{7.30}$$

In fact, since from eqs (7.10) and (7.11) we get $E\{Q(\hat{\mathbf{b}})\} = \sigma^2(n-k)$, the bias of $\hat{\sigma}^2$ is

$$E(\hat{\sigma}^2) - \sigma^2 = \frac{1}{n} E\{Q(\hat{\mathbf{b}})\} - \sigma^2 = -\frac{k}{n}\sigma^2 \tag{7.31}$$

a result that shows – as it often happens with ML estimators – that $\hat{\sigma}^2$ is, however, asymptotically unbiased.

At this point, we are in a position to fully exploit the additional assumption $\mathbf{e} \approx N(\mathbf{0}, \sigma^2\mathbf{I}_n)$ – or, better, its direct consequence $\mathbf{Y} \approx N(\mathbf{Xb}, \sigma^2\mathbf{I}_n)$ – well beyond the fact of just determining that the LS estimate of $\mathbf{b}$ coincides with its ML estimate. In fact, we can use a number of important results concerning the distribution of functions of normal r.v.s (see Appendix C) to make more stringent inferences on the point estimates $\hat{\mathbf{b}}$ and $\hat{\sigma}^2$.

Let us start with $\hat{\mathbf{b}}$: since $\hat{\mathbf{b}}$ given by eq. (7.5) is a linear function of a normal vector, it is a normal vector itself (Section 3.3.2). This fact, together with the results of Proposition 7.1, leads to $\hat{\mathbf{b}} \approx N(\mathbf{b}, \sigma^2\mathbf{A}^{-1})$ so that, in particular, each individual $\hat{\beta}_j$ is normally distributed with mean β_j and variance $\sigma^2 a_{jj}$ – where we denoted by a_{jj} the diagonal elements of the matrix $\mathbf{A}^{-1}$. Consequently, for all $j = 1, \ldots, k$ we have

$$\frac{\hat{\beta}_j - \beta_j}{\sigma\sqrt{a_{jj}}} \approx N(0, 1) \tag{7.32}$$

If, as it may happen in some cases, σ^2 is known, then (recall Example 5.9(a)) a γ-confidence interval for β_j is $(\hat{\beta}_j \pm c_{(1+\gamma)/2}\sigma\sqrt{a_{jj}})$ where $c_{(1+\gamma)/2}$ is the lower $(1+\gamma)/2$-quantile of the standard normal distribution. More often, however, σ^2 is not known and, as we did above, must be estimated from the data. Using the unbiased estimate $\hat{s}^2$ of σ^2 given by eq. (7.12) then $\hat{s}^2 a_{jj}$ is the (unbiased) estimate of $\mathrm{Var}(\hat{\beta}_j)$ and

$$T_j \equiv \sqrt{\frac{n-k}{a_{jj}Q(\hat{\mathbf{b}})}}(\hat{\beta}_j - \beta_j) \approx St(n-k) \tag{7.33}$$

(recall Section 5.6 and note that T_j is a pivot quantity for the case at hand). Equation (7.33), in turn, implies that the confidence interval will be expressed in terms of Student quantiles. In fact, the γ-CI for β_j is in this case

$$\left(\hat{\beta}_j \pm t_{(1+\gamma)/2;n-k}\sqrt{\frac{a_{jj}Q(\hat{\mathbf{b}})}{n-k}}\right) = \left(\hat{\beta}_j \pm t_{(1+\gamma)/2;n-k}\hat{s}\sqrt{a_{jj}}\right) \tag{7.34}$$

where $t_{(1+\gamma)/2;n-k}$ is the lower $(1+\gamma)/2$-quantile of the Student distribution with $n-k$ degrees of freedom.

Also, in regard to interval estimates of individual parameters, we can obtain a γ-CI for the residual variance σ^2 (when it is unknown). The result is

$$\left(\frac{Q(\hat{\mathbf{b}})}{\chi^2_{(1+\gamma)/2;n-k}}, \frac{Q(\hat{\mathbf{b}})}{\chi^2_{(1-\gamma)/2;n-k}}\right) = \left(\frac{(n-k)\hat{s}^2}{\chi^2_{(1+\gamma)/2;n-k}}, \frac{(n-k)\hat{s}^2}{\chi^2_{(1-\gamma)/2;n-k}}\right) \tag{7.35}$$

where $\chi^2_{(1\pm\gamma)/2;n-k}$ are, respectively, the lower $(1 \pm \gamma)/2$ quantiles of the distribution $\chi^2(n-k)$. Although we do not prove eq. (7.35), it may be worth noting that it is a consequence of the result

$$\frac{Q(\hat{\mathbf{b}})}{\sigma^2} \approx \chi^2(n-k) \tag{7.36}$$

which, in turn, is proven by using a theorem on quadratic forms of normal variables (see Ref. [12, Chapter 1]. The quadratic form in question here is $\mathbf{r}^{\mathrm{T}}\mathbf{P}_{S^\perp}\mathbf{r}$ where $\mathbf{r} \equiv \mathbf{e}/\sigma$ is the vector of normalized errors ($\mathbf{r} \approx N(\mathbf{0}, \mathbf{I}_n)$ and $r_i = \varepsilon_i/\sigma \approx N(0,1)$ for all $i = 1, \ldots, n$) and $\mathbf{P}_{S^\perp}$ is the projection matrix on $S^\perp$ introduced in Section 7.2. The equality $Q(\hat{\mathbf{b}})/\sigma^2 = \mathbf{r}^{\mathrm{T}}\mathbf{P}_{S^\perp}\mathbf{r}$ follows easily if we substitute $\mathbf{Y} = \mathbf{Xb} + \sigma\mathbf{r}$ in $Q(\hat{\mathbf{b}}) = \mathbf{Y}^{\mathrm{T}}\mathbf{P}_{S^\perp}\mathbf{Y}$ (third and fifth terms in eq. (7.10)) and then note that (a) $\mathbf{b}^{\mathrm{T}}\mathbf{X}^{\mathrm{T}}\mathbf{P}_{S^\perp}\mathbf{r} = \mathbf{r}^{\mathrm{T}}\mathbf{P}_{S^\perp}\mathbf{Xb}$ and (b) $\mathbf{P}_{S^\perp}\mathbf{Xb} = \mathbf{0}$ because $\mathbf{Xb} \in S$ and $\mathbf{P}_{S^\perp}$ projects on the $(n-k)$-dimensional space $S^\perp$.

In the light of the above results, it may not be out of place at this point to recall the interpretation of confidence intervals (Section 5.6). Equation (7.34), for instance, does not mean that the true value of β_j belongs to this interval – which has nothing random in it – with probability γ. Assuming that our model is correct, eq. (7.34) means that by repeating the data collection experiment and the fitting many times – thus obtaining many estimates of β_j and an equal number of CIs of the form (7.34) – these intervals will contain the true value of β_j in $100\gamma\%$ of the cases. The same 'long-run interpretation', clearly, applies to the true value of σ^2 and the CI of eq. (7.35).

Now, for all $j = 1, \ldots, k$ eq. (7.34) provides a γ-confidence interval for each individual β_j. A more general result, however, can be obtained if we are interested in a γ-confidence region $C_\gamma \subset \mathbb{R}^k$ which concerns the whole vector $\mathbf{b}$, that is, all the k parameters β_j simultaneously (in this regard note that the pivot quantities T_j are not, in general, independent). The desired result is

$$C_\gamma = \left\{ \mathbf{b} : (\hat{\mathbf{b}} - \mathbf{b})^{\mathrm{T}}\mathbf{A}(\hat{\mathbf{b}} - \mathbf{b}) < \frac{kQ(\hat{\mathbf{b}})}{n-k} F_{\gamma;\,k,\,n-k} \right\} \tag{7.37a}$$

where $F_{\gamma;k,n-k}$ is the γ-lower quantile of the Fisher distribution $\mathrm{Fsh}(k, n-k)$. The confidence region C_γ, therefore, is the inner side of an ellipsoid centered in $\hat{\mathbf{b}}$ with boundary defined by the equation

$$(\hat{\mathbf{b}} - \mathbf{b})^{\mathrm{T}}\mathbf{A}(\hat{\mathbf{b}} - \mathbf{b}) = \frac{kQ(\hat{\mathbf{b}})}{n-k} F_{\gamma;\,k,\,n-k} \tag{7.37b}$$

As we did for the CI (7.35), we do not prove rigorously eq. (7.37a) here; nonetheless, for the interested reader a few comments on why we get this result are worthy of mention. The main reason is that

$$\frac{n-k}{k}\left(\frac{U}{Q(\hat{\mathbf{b}})}\right) \approx \mathrm{Fsh}(k, n-k) \tag{7.38}$$

where we defined $U \equiv Q(\mathbf{b}) - Q(\hat{\mathbf{b}})$ and the proof of the relation $U = (\mathbf{b}-\hat{\mathbf{b}})^{\mathrm{T}}\mathbf{A}(\mathbf{b}-\hat{\mathbf{b}}) = (\hat{\mathbf{b}}-\mathbf{b})^{\mathrm{T}}\mathbf{A}(\hat{\mathbf{b}}-\mathbf{b})$ is left to the reader. Since, in addition to the result of eq. (7.36), it can be shown [12] that (a) the r.v.s $Q(\hat{\mathbf{b}})$ and U are independent and (b) $U/\sigma^2 \approx \chi^2(k)$, eq. (7.38) – and consequently (7.37) – descends from eq. (7.36) and the fact that the ratio of two independent χ^2-distributed r.v.s follows a Fisher distribution with the appropriate number of degrees of freedom (which depend on the degrees of freedom of the two variables in the ratio; see Appendix C).

On the basis of this last result, let us go back for a moment to remark (iii) at the end of Section 7.2. There, the interest was on m ($m \le k$) linear combinations $\mathbf{t} = \mathbf{Tb}$ of the k parameters β_j. We have already pointed out that $\hat{\mathbf{t}} = \mathbf{T}\hat{\mathbf{b}}$ is the desired estimate of $\mathbf{t}$ but now the assumption of normality gives $\hat{\mathbf{t}} \approx N(\mathbf{t}, \sigma^2\mathbf{D})$, where $\mathbf{D} = \mathbf{TA}^{-1}\mathbf{T}^{\mathrm{T}}$. Then, if we consider the quadratic form $U_{\mathrm{T}} = (\mathbf{t}-\hat{\mathbf{t}})^{\mathrm{T}}\mathbf{D}^{-1}(\mathbf{t}-\hat{\mathbf{t}})$, the counterparts of points (a) and (b) above are (a′) the r.v.s $Q(\hat{\mathbf{b}})$ and U_{T} are independent and (b′) $U_{\mathrm{T}}/\sigma^2 \approx \chi^2(m)$. Consequently, we have

$$\frac{n-k}{m}\left(\frac{U_{\mathrm{T}}}{Q(\hat{\mathbf{b}})}\right) \approx \mathrm{Fsh}(m, n-k) \tag{7.39}$$

which, in turn, is the counterpart of eq. (7.38) for the case at hand. The conclusion is that the γ-confidence region $C_{\mathrm{T};\gamma} \subset \mathbb{R}^m$ for $\mathbf{t}$ is

$$C_{\mathrm{T};\gamma} = \left\{\mathbf{t} : (\mathbf{T}\hat{\mathbf{b}}-\mathbf{t})^{\mathrm{T}}\mathbf{D}^{-1}(\mathbf{T}\hat{\mathbf{b}}-\mathbf{t}) < \frac{mQ(\hat{\mathbf{b}})}{n-k}F_{\gamma;m,n-k}\right\} \tag{7.40}$$

If $m = k$ and $\mathbf{T} = \mathbf{I}_k$ then $\mathbf{Tb} = \mathbf{b}$, $\mathbf{D}^{-1} = \mathbf{A}$ and we obtain the γ-confidence ellipsoid of eq. (7.37). If, on the other hand, $m = 1$ then $\mathbf{T}$ is a $1 \times k$ matrix of constants $t_1, \ldots, t_k$ which can be arranged in a k-dimensional vector $\mathbf{c} = (t_1 \ldots t_k)^{\mathrm{T}}$. Then $\mathbf{t} = \mathbf{Tb} = \sum_j t_j\beta_j = \mathbf{c}^{\mathrm{T}}\mathbf{b}$ is a single linear combination of the parameters and the matrix $\mathbf{D} = \mathbf{c}^{\mathrm{T}}\mathbf{A}^{-1}\mathbf{c}$ reduces to a scalar, which we will call d. With these definitions the region (7.40) becomes the γ-confidence interval

$$\left(\mathbf{c}^{\mathrm{T}}\hat{\mathbf{b}} \pm \sqrt{\frac{dQ(\hat{\mathbf{b}})}{n-k}F_{\gamma;1,n-k}}\right) \tag{7.41a}$$

Moreover, since Fisher's distribution $\mathrm{Fsh}(1, n-k)$ is related to the square of the Student distribution $\mathrm{St}(n-k)$ and the equality $F_{\gamma;1,n-k} = t^2_{(1+\gamma)/2;n-k}$ between their lower quantiles holds, eq. (7.41a) becomes

$$\left(\mathbf{c}^{\mathrm{T}}\hat{\mathbf{b}} \pm t_{(1+\gamma)/2;n-k}\sqrt{\frac{dQ(\hat{\mathbf{b}})}{n-k}}\right) = \left(\mathbf{c}^{\mathrm{T}}\hat{\mathbf{b}} \pm t_{(1+\gamma)/2;n-k}\,\hat{s}\sqrt{d}\right) \tag{7.41b}$$

which, in turn, reduces to (7.34) when $\mathbf{c}^{\mathrm{T}} = (0, \ldots, 0, 1, 0, \ldots, 0)$, with the only non-zero element in the jth position. In this case, in fact, it is easy to determine that $\mathbf{c}^{\mathrm{T}}\hat{\mathbf{b}} = \hat{\beta}_j$ and $d = a_{jj}$. Another special case with $m = 1$ occurs when $\mathbf{c} = \mathbf{x}_0$, where $\mathbf{x}_0 = (x_{01}, \ldots, x_{0k})^{\mathrm{T}}$ is a given set of values for the predictor variables. Then $\hat{Y}_0 = \mathbf{x}_0^{\mathrm{T}}\hat{\mathbf{b}}$ is an estimate of $E(Y_0)$ – that is, the mean value of Y corresponding to $\mathbf{X}_0$ – and the γ-CI for $E(Y_0)$ is obtained from eq. (7.41b) as

$$\left(\mathbf{x}_0^{\mathrm{T}}\hat{\mathbf{b}} \pm t_{(1+\gamma)/2;n-k}\,\hat{s}\sqrt{\mathbf{x}_0^{\mathrm{T}}\mathbf{A}^{-1}\mathbf{x}_0}\right) \tag{7.42}$$

If, on the other hand, we are interested in a prediction interval on Y_0 itself and not, as above, on $E(Y_0)$, it only takes a small effort to determine

$$\left(\mathbf{x}_0^{\mathrm{T}}\hat{\mathbf{b}} \pm t_{(1+\gamma)/2;n-k}\,\hat{s}\sqrt{\mathbf{x}_0^{\mathrm{T}}\mathbf{A}^{-1}\mathbf{x}_0 + 1}\right) \tag{7.43}$$

where it should be noted that this interval differs from (7.42) because it regards a single future trial at the given value $\mathbf{x}_0$. The fact that the interval (7.43) is wider than (7.42) reflects the circumstance that there is less precision in predicting a particular value of Y than in estimating its mean.

At this point, if we wish to test a given null hypothesis on the estimated regression parameters, we can recall the considerations at the beginning of Section 6.3.4. Let us consider, for instance, the γ-CI (7.35) for σ^2. If we wish to test $H_0 : \sigma^2 = \sigma_0^2$ against $H_1 : \sigma^2 \neq \sigma_0^2$ (where σ_0^2 is a specified value), then the acceptance region at the significance level $\alpha = 1 - \gamma$ is given by

$$\Xi_0 = \left\{ \frac{\sigma_0^2}{n-k}\chi^2_{\alpha/2;n-k} < \hat{s}^2 < \frac{\sigma_0^2}{n-k}\chi^2_{1-\alpha/2;n-k} \right\} \tag{7.44}$$

therefore implying that the rejection region is $\Xi_1 = \Xi_0^C$.

Similarly, when σ^2 is unknown we can use the CI (7.34) to test the pair of hypotheses $H_0 : \beta_j = \beta_{0j}; H_1 : \beta_j \neq \beta_{0j}$ (a frequently performed test is with $\beta_{0j} = 0$). The acceptance region for the test is now

$$\Xi_0 = \left\{ \beta_{0j} - t_{1-\alpha/2;n-k}\hat{s}\sqrt{a_{jj}} < \hat{\beta}_j < \beta_{0j} + t_{1-\alpha/2;n-k}\hat{s}\sqrt{a_{jj}} \right\} \tag{7.45a}$$

and consequently the rejection region is

$$\Xi_1 = \left\{ \left| \frac{\hat{\beta}_j - \beta_{0j}}{\hat{s}\sqrt{a_{jj}}} \right| \geq t_{1-\alpha/2;n-k} \right\} \tag{7.45b}$$

where $t_{1-\alpha/2;n-k}$ is the $(1-\alpha/2)$-lower quantile of $\mathrm{St}(n-k)$ (which, we note, is the same as the $\alpha/2$-upper quantile of the same distribution).

More generally, the null hypothesis H_0 will be in the form of a linear restriction which limits the values of the parameters $\beta_1, \ldots, \beta_k$ in a specified subset $B_0 \subset \mathbb{R}^k$. Explicitly this means that we will have $H_0 : \mathbf{b} \in B_0$ where $B_0 = \{\mathbf{b} : \mathbf{Tb} = \mathbf{t}_0\}$, $\mathbf{t}_0$ is a given vector and $\mathbf{T}$ is a non-random $m \times k$ $(m \leq k)$ restriction matrix of rank m. In general H_0 is a composite hypothesis because σ^2 is not known.

Since the γ-confidence region for $\mathbf{t} = \mathbf{Tb}$ is given by eq. (7.40) which, for our present purposes can be rewritten as

$$C_{\mathrm{T};\gamma} = \left\{ \mathbf{t} : \frac{U_{\mathrm{T}}(\mathbf{Y};\mathbf{t}_0)}{Q(\hat{\mathbf{b}})} < \frac{m}{n-k}F_{1-\alpha;m,n-k} \right\} \tag{7.46}$$

it follows that all realizations $\mathbf{y}$ of $\mathbf{Y}$ satisfying the inequality within brackets lead to the acceptance of the null hypothesis. The conclusion, therefore is as follows: at the significance level a the rejection region for the test $H_0 : \mathbf{b} \in B_0$ is given by

$$\Xi_1 = \left\{ \mathbf{y} : \left(\frac{n-k}{m}\right) \frac{U_{\mathrm{T}}(\mathbf{y};\mathbf{t}_0)}{Q(\hat{\mathbf{b}})} \geq F_{1-\alpha;m,n-k} \right\} \tag{7.47}$$

where $F_{1-\alpha;m,n-k}$ is the $(1-\alpha)$-lower quantile of the distribution $\mathrm{Fsh}(m, n-k)$.

By appropriately constructing the **Y** and **X** matrices, many problems can be cast in the form of a normal regression scheme, thereby taking advantage of the above considerations on confidence intervals and tests on linear combinations of the estimated parameters. The following Examples 7.1(a) and (b) illustrate this kind of situation.

Example 7.1(a) Suppose that we have $r \geq 2$ groups of independent and normally distributed Y-type observations. If each group is of size $n_i (i = 1, \ldots, r)$ and the underlying model is such that, for $j = 1, \ldots, n_i$

$$E\left(Y_j^{(i)}\right) = \beta_1^{(i)} + \beta_2^{(i)} x_j^{(i)}, \qquad i = 1, \ldots, r$$

we may be interested, for instance, in testing the null hypothesis that the r slopes $\beta_2^{(i)}$ are all equal, i.e. $H_0 : \beta_2^{(1)} = \beta_2^{(2)} = \cdots = \beta_2^{(r)}$.

Let $n = n_1 + \cdots + n_r$; if we form (i) a collective n-dimensional vector $\mathbf{Y}^{\mathrm{T}} = \left(Y_1^{(1)}, \ldots, Y_{n_1}^{(1)}, Y_1^{(2)}, \ldots, Y_{n_r}^{(r)}\right) \equiv (Y_1, \ldots, Y_n)$ whose first n_1 elements are the Y-observations of the first group followed by the n_2 Y-observations of the second group, etc., and (ii) a collective $2r$-dimensional vector of parameters $\mathbf{b}^{\mathrm{T}} = \left(\beta_1^{(1)}, \beta_2^{(1)}, \beta_1^{(2)}, \ldots, \beta_2^{(r)}\right) \equiv (\beta_1, \ldots, \beta_{2r})$ the model can be written in the matrix form (7.2b) by forming the $n \times 2r$ matrix **X** whose first n_1 rows are $\left(1, x_1^{(1)}, 0, \ldots, 0\right), \ldots, \left(1, x_{n_1}^{(1)}, 0, \ldots, 0\right)$ followed by the n_2 rows $\left(0, 0, 1, x_1^{(2)}, 0, \ldots, 0\right), \ldots, \left(0, 0, 1, x_{n_2}^{(2)}, 0, \ldots, 0\right)$, etc., and the last n_r rows are $\left(0, \ldots, 0, 1, x_1^{(r)}\right), \ldots, \left(0, \ldots, 0, 1, x_{n_r}^{(r)}\right)$.

At this point the estimate of the parameters is given by eq. (7.5). Then, noting that the null hypothesis can be rewritten as $H_0 : \beta_4 - \beta_2 - 0, \beta_6 - \beta_2 - 0, \ldots, \beta_{2r} - \beta_2 = 0$ and consequently cast in matrix form as $\mathbf{Tb} = \mathbf{0}$ where $\mathbf{T}$ is the $(r-1) \times r$ matrix whose first column is $(-1, -1, \ldots, -1)$ and the only non-zero element of the sth $(2 \leq s \leq r)$ column is in the $(s-1)$th place from the top, the rejection region for the test is given by eq. (7.47) where, for this case, $m = r - 1$ and $k = 2r$.

Example 7.1(b) Suppose that we have two samples $(z_1, \ldots, z_n)$ and $(t_1, \ldots, t_m)$ from the distributions $N(\mu_1, \sigma^2)$ and $N(\mu_2, \sigma^2)$, respectively, and we wish to test the null hypothesis $H_0 : \mu_1 = \mu_2$ against $H_1 : \mu_1 \neq \mu_2$. If we form the collective $(n + m)$-dimensional vector $\mathbf{Y}^{\mathrm{T}} = (z_1, \ldots, z_n, t_1, \ldots, t_m)$ and denote the two means by the symbols β_1, β_2 instead of μ_1, μ_2 we can (i) obtain the estimates of the unknown parameters $\mathbf{b} = (\beta_1, \beta_2)^{\mathrm{T}}$ and σ^2 and (ii) find a rejection region for the test. The problem, in fact, is in the form (7.2b) if we define the $(n+m) \times 2$ **X** matrix with (1, 0) in the first n rows and (0, 1) in the remaining m rows. Then, since

$$\mathbf{A} = \mathbf{X}^{\mathrm{T}}\mathbf{X} = \begin{pmatrix} n & 0 \\ 0 & m \end{pmatrix} \qquad \mathbf{X}^{\mathrm{T}}\mathbf{Y} = \begin{pmatrix} \sum_i z_i \\ \sum_i t_i \end{pmatrix} = \begin{pmatrix} n\bar{z} \\ m\bar{t} \end{pmatrix}$$

we use eqs (7.5) and (7.12b) to obtain the estimates $\hat{\mathbf{b}} = (\bar{z}, \bar{t})^{\mathrm{T}}$ and

$$\hat{s}^2 = \frac{\sum_i (z_i - \bar{z})^2 + \sum_i (t_i - \bar{t})^2}{n + m - 2}$$

At this point we rewrite the null hypothesis as $H_0 : \beta_1 - \beta_2 = 0$ and cast it in the matrix form $\mathbf{Tb} = 0$ where $\mathbf{T} = (1, -1)$; then since $\mathbf{D} = \mathbf{TA}^{-1}\mathbf{T}^{\mathrm{T}} = n^{-1} + m^{-1}$, we recall eqs (7.41a) and (7.41b) and arrive at the γ-CI for $\mathbf{T}\hat{\mathbf{b}} = \bar{z} - \bar{t}$

$$\left((\bar{z} - \bar{t}) \pm t_{(1+\gamma)/2;\, n+m-2}\, \hat{s}\sqrt{n^{-1} + m^{-1}} \right)$$

which, noting that $\bar{z} = M_1, \bar{t} = M_2$, is exactly the CI (5.87) of Example 5.11(a) when the variances of the two normal populations are equal (i.e. $\sigma_1^2 = \sigma_2^2$). On the basis of the γ-CI above, it is left to the reader to write explicitly the acceptance and rejection region for $H_0 : \beta_1 - \beta_2 = 0$ at the significance level $\alpha = 1 - \gamma$.

By a direct extension of this line of reasoning, it can be shown that the techniques known as 'analysis of variance' (ANOVA, briefly mentioned at the end of Section 6.3.4) can also be formulated in the form of linear regression problems. For more details the reader can refer, for instance, to [12] or [2].

7.3.1 *Back to simple linear regression*

It is clear from the preceding section that the assumption of normality has a number of interesting and far-reaching consequences. Under this assumption, we can now go back to the simple one-predictor model $Y = \beta_1 + \beta_2 x + \varepsilon$ of Section 7.2.1 and consider the new developments in this specific setting. The simple linear model, in fact, is so frequently used in practice that deserves special attention.

The first general observation is that the LS estimates (7.22b) coincide with the ML estimates of the slope and intercept, respectively and, as a consequence, they have all the desirable properties of ML estimators considered in Chapter 5. A second observation is that now we can use eq. (7.34), to define individual γ-confidence intervals for the estimated parameters β_1 and β_2. Assuming that σ^2 is unknown and this parameter too has been estimated by the data, we get

$$\left(\hat{\beta}_1 \pm t_{(1+\gamma)/2;n-2}\frac{\hat{s}}{\sqrt{n}}\sqrt{\frac{\sum_i x_i^2}{\sum_i (x_i - \bar{x})^2}}\right) \tag{7.48}$$

for the intercept and

$$\left(\hat{\beta}_2 \pm t_{(1+\gamma)/2;n-2}\,\hat{s}\sqrt{\frac{1}{\sum_i (x_i - \bar{x})^2}}\right) \tag{7.49}$$

for the slope. In both (7.48) and (7.49) $\hat{s}$ is the square root of the estimate (7.24) and $t_{(1+\gamma)/2;n-2}$ is the lower $(1+\gamma)/2$-quantile of the distribution $\mathrm{St}(n-2)$. The above intervals, as we know, are strictly related to the acceptance and rejection regions (7.45a) and (7.45b) defined by statistical tests on β_1 and β_2. In particular, when we obtain an estimate $\hat{\beta}_2$ close to zero, it is always advisable to test whether the slope is significantly different from zero because, if this is not the case, there is no linear relation between the variables (which does not imply, however, that there is no relation at all) and the assumed model should be changed. Owing to eq. (7.45b) the rejection region for the test $H_0 : \beta_2 = 0; H_1 : \beta_2 \neq 0$ is

$$\Xi_1 = \left\{\mathbf{y} : \frac{\left|\hat{\beta}_2\right|\sqrt{\sum_i (x_i - \bar{x})^2}}{\hat{s}} = \frac{\left|\hat{\beta}_2\right|\sqrt{S_{xx}}}{\hat{s}} \geq t_{1-\alpha/2;n-2}\right\} \tag{7.50}$$

In other cases we may obtain an estimate $\hat{\beta}_1$ of the intercept close to zero and then we should test the pair of hypotheses $H_0 : \beta_1 = 0; H_1 : \beta_1 \neq 0$. The rejection region for this test is

$$\Xi_1 = \left\{\mathbf{y} : \frac{\left|\hat{\beta}_1\right|}{\hat{s}}\sqrt{\frac{nS_{xx}}{\sum_i x_i^2}} \geq t_{1-\alpha/2;n-2}\right\} \tag{7.51}$$

and if the result of the test is the acceptance of H_0 it means that the original model $Y = \beta_1 + \beta_2 x + \varepsilon$ must be substituted by the new model $Y = \beta_2 x + \varepsilon$ (or, equivalently, $E(Y) = \beta_2 x$). In this case we must recalculate the estimate $\hat{\beta}_2$ of the slope by using the formula

$$\hat{\beta}_2 = \frac{\sum_i x_i Y_i}{\sum_i x_i^2} \tag{7.52}$$

which is just the result of eq. (7.5) when $\mathbf{X}$ is a $n \times 1$ matrix.

In regard to the residual variance, the appropriate confidence interval is obtained by simply setting $k = 2$ in eq. (7.35). Starting from this interval, the rejection region to test $H_0 : \sigma^2 = \sigma_0^2; H_1 : \sigma^2 \neq \sigma_0^2$ is given by eq. (7.44). If, in addition, we are interested in estimating the mean $E(Y_0)$ by means of the quantity $\hat{Y}_0 = \hat{\beta}_1 + \hat{\beta}_2 x_0$ it suffices to note that in this case $\mathbf{x}_0 = (1, x_0)^T$; then, substituting in $\mathbf{x}_0^T \mathbf{A}^{-1} \mathbf{x}_0$ the γ-CI of eq. (7.42) becomes

$$\left(\hat{Y}_0 \pm t_{(1+\gamma)/2; n-2}\, \hat{s} \sqrt{\frac{1}{n} + \frac{(x_0 - \bar{x})^2}{S_{xx}}} \right) \tag{7.53}$$

because $\mathbf{x}_0^T \mathbf{A}^{-1} \mathbf{x}_0 = (\det A)^{-1} \left(\sum x_i^2 - 2n\bar{x}x_0 + nx_0^2 \right)$ and it is left to the reader to show that this expression equals the quantity under square root in (7.53). Using this relation it is immediate to explicitly write also the γ-prediction interval (7.43) for Y_0 itself. The point worthy of notice is that the width of the interval (7.53) – and, similarly, of the interval corresponding to (7.43) – increases as $(x_0 - \bar{x})$ increases, that is, as the distance of x_0 from the sample mean $\bar{x}$ increases. This circumstance implies, as noted before, that it is generally unwise to extrapolate and use the regression line outside the x-values from which it has been obtained. By extrapolating, in fact, the interval (7.53) may become so large that our inference (on $E(Y_0)$ or on Y_0) could turn out to be useless.

As noted in the preceding section, the intervals (7.48) and (7.49) apply individually, that is, each interval concerns one parameter. However, the two parameters β_1, β_2 are not necessarily independent (recall eq. (7.23b)) and we may be interested in the joint confidence region for β_1, β_2. Carrying out the appropriate calculations on eq. (7.37) for the case at hand we obtain the desired region as the interior of the ellipse

$$n(\hat{\beta}_1 - \beta_1)^2 + 2n\bar{x}(\hat{\beta}_1 - \beta_1)(\hat{\beta}_2 - \beta_2) + (\hat{\beta}_2 - \beta_2)^2 \sum_i x_i^2 = 2\hat{s}^2 F_{\gamma; 2, n-2} \tag{7.54}$$

which, solving for a sufficient number of β_1, β_2 can easily be drawn on a graph with axes β_1 and β_2. In the general case the ellipse is slanted in such a way that its major axis runs in the north-west to south-east direction (this is reasonable because if another sample of size n leads to an estimate of the slope $\hat{\beta}_2' > \hat{\beta}_2$ we should expect that $\hat{\beta}_1' < \hat{\beta}_1$, a circumstance which reflects – eq. (7.23b) – the negative correlation between the estimates for the slope and the intercept). If, before the regression procedure is carried out, we structure our data so that $\bar{x} = 0$, it follows from eq. (7.23b) that $\text{Cov}(\hat{\beta}_1, \hat{\beta}_2) = 0$ and (7.54) is an ellipse with its major axes parallel to the β_1 and β_2 axes.

Another point to be made concerns the quantity R of eq. (7.28b) which, being the sample counterpart of ρ of eq. (3.22), is – we recall from

Section 3.3 – a measure of the strength of the linear relation between the variables Y and x (note that the term 'linear model' means linear in the parameters but in simple linear regression the type of model itself implies that the response variable Y depends linearly on the predictor variable x). In this regard, therefore, one may be interested in testing the null hypothesis $H_0 : \rho = 0$ against $H_1 : \rho \neq 0$. The distribution of R is rather complicated but we give without proof the following result: for large n the statistic

$$G = \frac{1}{2} \ln \left(\frac{1+R}{1-R} \right) \tag{7.55}$$

is approximately normal with mean and variance

$$E(G) = \frac{1}{2} \ln \left(\frac{1+\rho}{1-\rho} \right) \qquad \mathrm{Var}(G) = \frac{1}{n-3} \tag{7.56a}$$

thus implying that the r.v.

$$Z = \frac{\sqrt{n-3}}{2} \ln \frac{(1+R)(1-\rho)}{(1-R)(1+\rho)} \tag{7.56b}$$

is approximately standard normal. The consequence is that the rejection region to test the pair of hypotheses $H_0 : \rho = \rho_0; H_1 : \rho \neq \rho_0$ at (approximately) the significance level α is

$$\Xi_1 = \left\{ \mathrm{y} : \left| \frac{\sqrt{n-3}}{2} \ln \frac{(1+R)(1-\rho_0)}{(1-R)(1+\rho_0)} \right| \geq z_{\alpha/2} \right\} \tag{7.57a}$$

where $z_{\alpha/2}$ is the upper $\alpha/2$-quantile of the standard normal distribution. In the special (but of frequent practical interest) case $\rho_0 = 0$ mentioned above, the rejection region (7.57a) becomes

$$\Xi_1 = \left\{ \mathrm{y} : \left| \frac{\sqrt{n-3}}{2} \ln \frac{1+R}{1-R} \right| \geq z_{\alpha/2} \right\} \tag{7.57b}$$

and the approximation is generally acceptable even for moderate sample sizes (say, $n \geq 30$).

7.3.2 Simple linear regression with two random variables

As a preliminary step, let us summarize the main ideas on simple linear regression. In Section 7.2.1, by only making two basic assumptions on the first and second moments of the 'error' ε, we considered the model $Y = \beta_1 + \beta_2 x + \varepsilon$ as a special case of the general model (7.1b). Then, in

Section 7.3.1 we examined the same simple linear model with the additional assumption of normality and pointed out that this circumstance allows the analyst to determine confidence intervals and acceptance/rejection regions for testing the regression parameters. In the course of the discussion, moreover, we noted some formal similarities with the results of Section 3.4.2 (for instance eq. (7.26) as compared to the first of eqs (3.98) and the parallel between R and ρ). There, however, the context was different because we were considering two random variables X and Y with a joint pdf $f_{XY}(x, y)$. The point we want to make here is that the discussion of Sections 7.2.1 applies even when our data are n realizations of a two-dimensional random vector (X, Y) and the conditional expectation $E(Y|X = x)$ is a linear function of x. If, in addition, the vector (X, Y) is jointly normal or approximately so, the inferences of Section 7.3.1 are also valid.

Example 7.2 Suppose that X and Y are jointly Gaussian with pdf given by eqs (3.61a) and (3.61b). Then we have shown in Section 3.4.2 that $E(Y|X = x)$ is a linear function of x and eqs (3.98) hold.

As a particular example consider the joint pdf of eq. (3.13b). Since the marginal pdf $f_X(x)$ is given by eq. (3.14a), it is rather easy to determine the conditional pdf $f_{Y|X}(y|x) = f_{XY}(x, y)/f_X(x)$ and show that $E(Y|x) = -x/2$, that is, $E(Y|X = x)$ is a linear function of x. This equation is nothing but the explicit form of the first of (3.98) because we have $E(X) = E(Y) = 0$, $\rho = -1/2$ and $\sigma_X = \sigma_Y = \sqrt{2}$. It is left to the reader to determine that in this case we also get

$$\sigma_{Y|X}^2 = \int \{y - E(Y|X)\}^2 f(y|x)dy = 3/2$$

and therefore, since the conditional variance $\sigma_{Y|X}^2$ does not depend on x, the second of (3.98) holds as well. In fact, using the values given above, we have $\sigma_Y^2(1 - \rho^2) = 2(1 - 1/4) = 3/2$.

Clearly, not all two-dimensional joint distributions lead to a linear (in x) conditional expectation, but for those that do – that is, when eq. (3.100) applies – eqs (3.103a) hold and consequently

$$\begin{aligned} E(Y|x) &= E(Y) + \frac{\operatorname{Cov}(X, Y)}{\operatorname{Var}(X)}(x - E(X)) \\ &= E(Y) + \rho\frac{\sigma_Y}{\sigma_X}(x - E(X)) \end{aligned} \tag{7.58}$$

In regard to the second of eq. (3.98) we can recall the discussion near the end of Section 3.4.2. Among all those bivariate distributions for which $E(Y|X = x)$ is linear in x, not all of them satisfy this equation because, in general, the conditional variance $\sigma^2_{Y|X}$, defined as

$$\sigma^2_{Y|x} \equiv \int \{y - E(Y|X)\}^2 f(y|x)\,dy \tag{7.59}$$

is a function of x. In the general case, therefore, the quantity $\sigma^2_Y(1-\rho^2)$ does not equal $\sigma^2_{Y|X}$ but it represents the weighted average

$$\sigma^2_{Y(\text{avg})} \equiv \int \sigma^2_{Y|X} f_X(x)\,dx \tag{7.60}$$

For those distributions for which $\sigma^2_{Y|X}$ does not depend on x, however, we have $\sigma^2_{Y|X} = \sigma^2_Y(1-\rho^2)$ (i.e. the second of (3.98)). These particular distributions are called 'homoscedastic' (from Greek words meaning 'equal scattering'). As noted in Section 3.4.2 and in Example 7.2, the bivariate Gaussian is one of them.

With the above considerations of probabilistic nature in mind, let us now turn our attention to their statistical counterparts. The first immediate observation is that eq. (7.26) is the sample counterpart of (7.58) and that the point estimates (7.22b) correspond to eq. (3.103a). In other words, whenever the assumption that our data are a sample from a bivariate distribution such that $E(Y|x) = a + bx$, we estimate this theoretical regression line of Y on X (i.e. eq. (7.58)) by means of eq. (7.26). Moreover, it is also clear at this point why we said that $R = S_{XY}/\sqrt{S_{XX}S_{YY}}$ is the sample quantity used to estimate of the correlation coefficient $\rho = \text{Cov}(X,Y)/\sigma_X\sigma_Y$.

In regard to the conditional variance, we have the relations

$$\sigma^2_{Y(\text{avg})} = \sigma^2_Y(1-\rho^2) \qquad \sigma^2_{Y|X} = \sigma^2_Y(1-\rho^2) \tag{7.61}$$

where the first is more general than the second and reduces to it for homoscedastic distributions. In both cases, however, the quantity $\sigma^2_Y(1-\rho^2)$ is a measure of the variability of Y about the regression line: a weighted average of this variability in the first case and the true (conditional) variability in the second. Equations (7.61) show that, because of correlation, this variability is $\le \sigma^2_Y$, that is, less than or equal to the total variability of Y. In this light, it is hardly surprising that the sample counterpart of (7.61) is

$$SS_R = S_{YY}(1-R^2) \tag{7.62}$$

which, in turn, tells us that the part of variability (of the observed Y values) explained by the estimated regression line is less than or equal to the total (observed) variability S_{YY}.

Equation (7.62) is easily obtained. In fact, from eq. (7.14b) we get $SS_R = S_{YY}(1 - SS_E/S_{YY})$; then, noting that the same equation (7.14b) gives

$$\frac{SS_R}{S_{YY}} = 1 - \frac{SS_E}{S_{YY}}$$

and that the term on the l.h.s. equals R^2 (see eq. (7.28)), eq. (7.62) follows.

Clearly, the most common assumption on the underlying probabilistic model is that of bivariate Gaussian population but the considerations above show that the simple regression procedure applies even in more general contexts. Nonetheless, the assumption of normality – or, when not fully justified, of moderate departures from normality – is necessary in order to ensure the reliability of the inferences given in Section 7.3.1.

In case of two random variables, moreover, it makes sense to speak of regression of X on Y and the above considerations – if $f_{XY}(x,y)$ is such that $E(X|Y = y)$ is linear in y – apply without changes by simply inverting the roles of X and Y. Two points, however, are worthy of mention:

(i) The fact that $E(Y|X = x)$ is a linear function of x does not necessarily imply that $E(X|Y = y)$ is a linear function of y. In the particular case of a bivariate Gaussian distribution we already know, however, that both $E(X|Y = y)$ and $E(Y|X = x)$ are straight lines.

(ii) Even when the two regression curves (of Y on X and X on Y) are straight lines they are, in general, different, although they both pass through the point $(E(X), E(Y))$. In this regard, recall eqs (3.103a) and (3.103b).

7.4 Final remarks on regression

The considerations and results of the preceding sections by no means exhaust such a vast subject as linear regression. This was not our intention from the start because – as it can immediately be seen from the references – entire books are dedicated to it. For the interested reader, nonetheless, we think it may be useful to mention some further aspects that deserve due attention in their own right.

Let us start with simple regression first. Referring specifically to the developments of Sections 7.2.1 and 7.3.1, it is not rare to find cases in which the assumption of homoscedasticity – that is, that the variance of Y is a constant – is not justified even when the simple linear model is correct. Often, the nature of the problem itself may suggest this type of situation but even when it is not so we can have an idea of this state of affairs by first carrying out a simple regression on the data and then plotting the $\hat{\varepsilon}_i = Y_i - \hat{Y}_i = Y_i - \hat{\beta}_1 - \hat{\beta}_2 x_i$ versus x. If, by visual inspection, there is evidence of definite pattern we can deduce that either

(i) the simple linear model is not appropriate (recall point (ii) at the end of Section 7.2.1) or
(ii) the linear model is the right one but the assumption of homoscedasticity is not valid.

Case (ii), in general, can be distinguished form case (i) because the emerging pattern looks random in nature but shows that the residuals $\hat{\varepsilon}_i$ increase (or decrease) steadily with increasing x. The remedy to this kind of situation is to perform a so-called 'weighted' simple linear regression in which the underlying model is still $E(Y_i) = \beta_1 - \beta_2 x_i$ but, instead of $\mathrm{Var}(Y_i) = \sigma^2$, we can write $\mathrm{Var}(Y_i) = \sigma^2 / w_i (i = 1, \ldots, n)$, where the weights $w_1, \ldots, w_n$ are assumed to be known positive constants. It can be shown that the LS estimates are in this case

$$\hat{\beta}_2 = \frac{\sum w_i \sum w_i x_i Y_i - \sum w_i x_i \sum w_i Y_i}{\sum w_i \sum w_i x_i^2 - \left(\sum w_i x_i\right)^2}$$
$$\hat{\beta}_1 = \frac{\sum w_i Y_i - \hat{\beta}_2 \sum w_i x_i}{\sum w_i} \tag{7.63}$$

and they reduce to their unweighted counterparts (eq. (7.22b)) whenever $w_1 = w_2 = \cdots = w_n$. Two things should be noted in eq. (7.63). The first is that they do not change if all the w_i's are multiplied by a constant a, thus implying that is not necessary to know their absolute values but it is sufficient to know their relative magnitudes. The second is that, as noted above, the case in which $\mathrm{Var}(Y_i)$ depends linearly (or approximately so) on x_i is rather frequent. Then, depending on whether $\mathrm{Var}(Y_i)$ increases or decreases with increasing x, we can choose the weights as $w_i = 1/x_i$ or $w_i = x_i$ and, accordingly, we have $\mathrm{Var}(Y_i) = \sigma^2 x_i$ or $\mathrm{Var}(Y_i) = \sigma^2 / x_i$. In this regard, however, see for instance [8] for more details.

Another remark on the weighted method is that it is often needed when the original variables are transformed in order to obtain a linear relation. In Engineering and Physics practice, in fact, it is quite common to transform a known physical law (relating the variables Y and x) into a linear equation.

Consider, for instance, the relation $I(x) = I_0 \exp(-\alpha x)$ which gives the light intensity $I(x)$ at a depth x into an absorbing medium with absorption coefficient α (I_0 is the intensity of light incident on the medium, i.e., at $x = 0$). By taking logarithms on both sides we get the linear equation $\ln I = \ln I_0 - \alpha x$ so that, by measuring I at different values of x, we can use the simple regression procedure to obtain an estimate of α. If, however, the assumption of homoscedasticity may be justified for the original Y-variable (I in our case) it is not necessarily so for the transformed variable ($\ln I$) and therefore a weighted simple regression would be more appropriate. A simple strategy consists in performing an unweighted analysis first and then check the plot of residuals $\hat{\varepsilon}_i$. If there are signs of a steadily increasing (or decreasing) pattern, then a weighted analysis is required.

The second comment about simple regression is a word of caution on the use of R – that is, the sample estimate of the correlation coefficient ρ – as a measure of the strength of the linear relation between Y and x. For small samples, in fact, it is not unusual to find high values of values of $|R|$, say $|R| \geq 0.8$, even when the variables are uncorrelated. It can be calculated (see, for instance, [5]), that when the variables are uncorrelated, the probability of obtaining $|R| = 0.8$ with $n = 5$ is $P_{0.8}^{(n=5)} = 0.104$. This, in other words means that even with uncorrelated data we get $|R| = 0.8$ in approximately one trial out of ten. Clearly, the above probability rapidly decreases as n increases and $P_{0.8}$ is already less than 0.05 for $n = 7$ (in fact $P_{0.8}^{(n=7)} = 0.031$, but note that for $|R| = 0.7$ we must have at least $n = 9$ in order to have a probability $P_{0.7} < 0.05$). Nonetheless, the conclusion is that, for small samples, the coefficient R is not totally reliable and we should be very cautious in assigning too much significance to it when (approximately) $n \leq 10$.

A rather debated point is what to do when both variables are subject to error and the error on the regressor variable is not negligible in comparison to the error on the predicted variable. We do not address this problem here but suggest to the interested reader to consult the Refs [10, 14, 17–21, 24, 25].

A connection between simple regression models and general models – or between different general regression models – is easily established when we have little or no *a priori* information on what model could be 'the best' for the data at our disposal. While, in general, it may not be difficult to determine whether the two-predictor model $E(Y) = \beta_1 + \beta_2 x_1 - \beta_3 x_2$ or the second degree polynomial model $E(Y) = \beta_1 + \beta_2 x - \beta_3 x^2$ could be better than the simple model $E(Y) = \beta_1 + \beta_2 x$, it is evident that the choice becomes rapidly very difficult as the number of β-parameters increases. So, in complex cases where a number of variables can potentially have an influence on Y the questions arise: which and how many predictors do we choose? are all of them really influential or some can be neglected without an appreciable loss in the quality of the fitted model? As it often happens, part of the answer to these and similar questions comes from a close examination of the nature of the problem and therefore it is outside the realm of statistics. Non-statistical considerations, moreover, may also play a role in establishing what the term 'best model' means for the specific problem at hand. If, for instance, simplicity is paramount in a given situation, we may choose, say, a two-regressor model leading to acceptable results even if we know that a three or four-regressor model could be significantly better according to some criterion of strict statistical nature.

A rather immediate idea that comes to mind when comparing two or more regression models is to choose the one with the highest value of the determination coefficient R^2 which, we recall, represents the fraction of the

Y variability explained by the fitted model. Regrettably, R^2 always tends to favour the equation with the greater number of parameters because it never decreases as more variables are added to the model. In order to take the number of parameters into account, the 'adjusted' coefficient R^2_{adj} is often used. Its definition is

$$R^2_{\text{adj}} = \frac{(n-1)R^2 - (k-1)}{n-k} \tag{7.64}$$

where k is the number of β-parameters and R^2 is the ordinary determination coefficient of eq. (7.15). As k increases, the terms $k - 1$ at the numerator and $n - k$ at the denominator compensate for the natural increase of R^2 due to a higher number of parameters. A similar approach consists in using the Akaike information criterion (AIC, see Ref. [1]) where one calculates the AIC parameter

$$\text{AIC} = SS_E + 2k \tag{7.65}$$

for the competing models and chooses the model with the smallest AIC. In fact, the term SS_E – the sum of squares of the $\hat{\varepsilon}_i = Y_i - \hat{Y}_i$ – tends to decrease as k increases and consequently $2k$ is a 'penalty' term which increases by two units for each additional β-parameter.

The R^2_{adj} and AIC methods have the advantage of simplicity and immediateness but it is clear that more sophisticated techniques have been devised to address the problem of model selection and adequacy. In case of normal regression, one of them is called 'backward elimination': we start with a 'full' model by including all the relevant variables arranged in decreasing order of importance and then – starting form the least important – we delete any variable whose marginal contribution is not significant according to a partial test of the type $H_0 : \beta_j = 0$ (see Section 7.3). Incidentally, we notc that this is equivalent to calculating the CI (7.34) at the desired value of γ and delete the variable corresponding to that specific β_j whenever the resulting CI includes zero. It is evident, however, that in most cases both the determination of the 'full' model and the order of importance in which the variables enter the model are, to a certain extent, subjective and involve an educated guess on the analyst's part. The opposite of backward elimination is called 'forward selection' – the term is self-explanatory – and we sequentially add one variable at a time until there is some evidence of no or little improvement from one model to the next. Whatever method is used, however, simultaneous tests on more than one β_j as well as various forms of plots (in particular, plots of the residuals $\hat{\varepsilon}_i$; see, for instance, Refs [7] and [8]) are often of great help. In general, there is probably no 'best' method and, as it virtually happens in most cases, it is always wise to analyze the data in more than one way.

Two final issues that deserve special attention are (1) the presence of outliers (in this regard, recall also Section 6.5.4) and (2) a problem called 'multicollinearity', where the former may occur in both simple and general linear regression while the latter is typical of general regression, that is, when the number of predictor variables is at least two. Let us consider them briefly by noting, however, that our scope here is merely to make the reader aware of their existence and draw his/her attention to the fact that both problems (1) and (2) are potentially dangerous. For more details and/or remedial measures, the interested reader can consult the references at the end of the chapter.

An immediate idea of the effect of outliers can be had by comparing the estimates $\hat{\beta}_1, \hat{\beta}_2$ from a 'clean' set of data with the same estimates from a 'contaminated' sample, where the contaminated sample is obtained by intentionally changing one of the original Y_i to a much higher or much lower value. Since both estimates may change dramatically, the seriousness of the problem becomes evident. In simple regression, however, we have the advantage of visual inspection but – owing to the higher dimensionality – it is generally not so in more complex situations. Moreover, the common belief that regression outliers can be identified by looking at the least-squares residuals must be taken with a grain of salt because it is not difficult to construct examples where a spurious datum has a smaller residual than some of the 'good' data.

Even from these short notes, it is clear that the problem is rather delicate and generally does not lend itself to an easy solution. This is more so in the case of multiple outliers because there exists a so-called 'masking effect' where one outlier may mask another. Therefore, besides the use of specific tests for outlier detection, if we suspect our data to be contaminated the general suggestion is to resort to techniques of robust regression (see, for instance, Refs [23] or [26]) which are often based on the minimization of quantities other than the sum of squared errors $\sum \varepsilon_i^2$.

Multicollinearity, on the other hand, is a different problem and is limited to cases where we have two or more regressors. Its source lies in the existence of near-linear dependencies among the regressor variables and its main effect, in general, is that the results of the least-squares fit may turn out to be unacceptable because the matrix $\mathbf{A} = \mathbf{X}^{\mathrm{T}}\mathbf{X}$ is ill-conditioned. Ill-conditioning, in fact, implies that the LS estimates are not stable and this, in turn, may cause some of the variances $\mathrm{Var}(\hat{\beta}_j)$ – or all of them – to be very large, with ample variations between the estimates of different experiments.

In this light, it is clear that the analyst first task is to detect the presence of multicollinearity. Then, if this is the case, he/she is faced with the problems of (i) assessing its extent and its possible effects on the problem at hand and (ii) applying appropriate remedial measures whenever point (i) indicates that multicollinearity can be harmful in his/her specific case. Fortunately, a number of methods and techniques have been devised to address all these aspects and a careful use of these techniques can be of great help in limiting – or even eliminating altogether in some cases – the potential dangers of multicollinearity. Specific references in this regard are [4, 9, 16].

7.5 Summary and comments

The study of relations between variables is fundamental in every branch of science and Statistics, in this respect, is no exception. While in disciplines like Physics and Engineering, however, these relations – in the form of mathematical equations – generally express a cause–effect relationship and reflect laws of Nature, it is not necessarily so in Statistics. This circumstance, in essence, is due to the fact that statistical significance and real-world significance are two distinct concepts which may coincide in some instances but may not coincide in others. In any case, once the main scope of the investigation is clear and we have reasons to believe that, on average, a 'response' r.v. Y depends on a number – say k – of 'predictor' variables $x_1, \dots, x_k$, it often turns out that a general linear model of the form (7.1b) is the appropriate relation that establishes a connection between Y and the x_j. In this context, however, it should be noted that the attribute 'linear' refers to the parameters $\beta_1, \dots, \beta_k$ and not to the predictors x_j.

Then, the practical part of the analysis consists in (i) carrying out $n \geq k$ trials in which we measure the response corresponding to specified values of the predictor variables and (ii) use these observed data to estimate the $k+1$ parameters $\beta_1, \dots, \beta_k, \sigma^2$, where the residual variance σ^2 is assumed to be the same for all errors ε_i $(i = 1, \dots, n)$ and these, in turn, are assumed to be uncorrelated and with zero mean. Under the additional assumption of maximum rank for the $n \times k$ predictor matrix $\mathbf{X}$, the so-called method of least squares (LS method) leads to the LS estimates (7.5) for the β parameters while the variance σ^2 must be estimated separately by means of eq. (7.12). The whole procedure is explained in Section 7.2 where it is also shown (Proposition 7.1) that the LS estimates have some desirable properties. In addition, some further remarks introduce the concept of multiple-determination coefficient and generalize the method to the case in which the β parameters must satisfy a certain number $m \leq k$ of constraint equations.

Within the framework of the LS method, Section 7.2.1 considers the special – and frequently encountered in applications – case of one predictor variable, known as simple linear regression problem. Again, the term linear refers to the (two) β parameters but it should be noted that now we have linearity in the predictor variable as well.

If, in addition to zero mean and uncorrelation, we assume the errors to be normally distributed, it is possible to make stronger inferences on the estimated parameters. In this case one speaks of normal regression and the first result worthy of notice is that the LS estimators coincide with the ML estimators. Normal regression is the subject of Sections 7.3 and 7.3.1 where – somehow paralleling the developments of Sections 7.2 and

7.2.1 – Section 7.3 considers the general case while Section 7.3.1 deals with the simple one-predictor model. By virtue of a number of important results on the distribution of functions of normally-distributed r.v.s, both sections show how it is possible to determine confidence intervals for the individual parameters and/or confidence regions regarding simultaneous inferences on two or more parameters (or on linear combinations of them). These interval estimates, in turn, are directly related to acceptance and rejection regions for testing hypotheses on the parameters and in this light it is also drawn attention to the fact that the general normal regression scheme – by appropriately setting up the relevant matrices involved in the calculations – can be used to tackle a variety of problems such as, for instance, some problems generally classified under the acronym ANOVA (analysis of variance).

In Section 7.3.2 it is shown that simple regression still applies if our data are a sample from a bivariate distribution such that the conditional expectation of Y given X is a linear function of x. In particular, the bivariate Gaussian distribution – which is often assumed as the underlying probabilistic model – satisfies this requirement and, in addition, is homoscedastic. Along this line of reasoning, it shown that the statistical relations of simple regression are the sample counterparts of the probabilistic relations given in Section 3.4.2.

Finally, Section 7.4 draws attention to a number of topics which have not been considered in the main discussion but are worthy of mention in their own right. Besides briefly introducing the method of weighted simple linear regression and making some specific remarks on various aspects of both simple and general regression, the reader is particularly warned against the harmful effects of (1) the presence of outliers in the data and (2) the problem – in general regression – known as 'multicollinearity'. These two problems, in fact, are potentially very dangerous and may lead to highly unreliable estimates of the regression parameters. Fortunately, many authors have carefully studied them and the reader can easily find in current literature a number of methods and techniques to detect their presence and, in case, to adopt appropriate remedial measures.

References and further reading

[1] Akaike, H.A., 'A New Look at the Statistical Model Identification', *IEEE Transactions on Automatic Control*, 19, 716–723 (1974).

[2] Azzalini, A., '*Inferenza Statistica: una Presentazione Basata sul Concetto di Verosimiglianza*', Springer-Verlag Italia (2001).

[3] Barnes, J.W., '*Statistical Analysis for Engineers and Scientists: a Computer-based Approach*', McGraw-Hill, New York (1994).

[4] Belsley, D.A., Kuh, E., Welsch, R.E., '*Regression Diagnostics: Identifying Influential Data and Sources of Multicollinearity*', Wiley, New York (1980).

[5] Bevington, P.R., Robinson, D.K., '*Data Reduction and Error Analysis for the Physical Sciences*', McGraw-Hill, New York (1992).

[6] Crow, E.L., Davis, F.A, Maxfield, M.W., *'Statistics Manual'*, Dover Publications, New York (1960).

[7] Daniel, C., Wood, F.S., *'Fitting Equations to Data'*, 2nd edn., Wiley, New York (1980)

[8] Draper, N.R., Smith, H., *'Applied Regression Analysis'*, 2nd edn., Wiley, New York (1981).

[9] Farrar, D.E., Glauber, R.R., *'Multicollinearity in Regression Analysis: the Problem Revisited'*, *Rev. Econ. Stat.*, 49, 92–107 (1967).

[10] Fuller, W.A., *'Measurement Error Models'*, Wiley & Sons, New York (1987).

[11] Green, J.R., Margerison, D., *'Statistical Treatment of Experimental Data'*, Elsevier, Amsterdam (1977).

[12] Ivchenko, G., Medvedev, Yu., *'Mathematical Statistics'*, Mir Publishers, Moscow (1990).

[13] Ivchenko, G., Medvedev, Yu., Chistyakov, A., *'Problems in Mathematical Statistics'*, Mir Publishers, Moscow (1991).

[14] Keeping, E.S., *'Introduction to Statistical Inference'*, Dover Publications, New York (1995).

[15] Kottegoda, N.T., Rosso, R., *'Statistics, Probability and Reliability for Civil and Environmental Engineers'*, McGraw-Hill, New York (1997).

[16] Lawless, J.F., *'Ridge and Related Estimation Procedures: Theory and Practice'*, *Commun. Stat.*, A7, 139–164 (1978).

[17] Macdonald, J.R., Thompson, W.J., *'Least Squares Fitting when Both Variables are Subject to Error: Pitfalls and Possibilities'*, *Am. J. Physics*, 60, 66–73 (1992).

[18] Madansky, A., 'The Fitting of Straight Lines when Both Variables are Subject to Error', *J. Am. Stat. Assoc.*, 54, 173–205 (1959).

[19] Mandel, J., 'Fitting Straight Lines when Both Variables are Subject to Error', *J. Qual. Technol.*, 16, 1–14 (1984).

[20] Mandel, J., *'The Statistical Analysis of Experimental Data'*, Dover Publications, New York (1984).

[21] Neter, J., Wasserman, W., Kutner, M. *'Applied Linear Statistical Models'*, Richard D. Irwin, Homewood, IL (1990).

[22] Rao, C.R., *'Linear Statistical Inference and its Applications'*, 2nd edn., John Wiley & Sons, New York (1973).

[23] Rousseeuw, P.J., Leroy, A.M., *'Robust Regression and Outlier Detection'*, Wiley-Interscience, New York (1987).

[24] Sampson, A.R., 'A Tale of Two Regressions', *J. Am. Stat. Assoc.*, 69, 682–689 (1974).

[25] Seber, G.A.F., *'Linear Regression Analysis'*, Wiley, New York (1977).

[26] Siegel, A.F., 'Robust Regression Using Repeated Medians', *Biometrika*, 69, 242–244 (1982).

[27] Wadsworth, H.M. Jr. (ed.), *'Handbook of Statistical Methods for Engineers and Scientists'*, McGraw-Hill, New York (1990).

[28] Weisberg, S., *'Applied Linear Regression'*, 2nd edn., John Wiley & Sons, New York (1985).

Appendix A
Elements of set theory

This appendix is intended as a refresher on a number of fundamental aspects of set theory. For obvious reasons there is no claim of completeness, and we will mainly focus our attention on notion and concepts which are used in the main course of the book.

A.1 Basic definitions and properties

A set is any well-defined collection of objects. These objects are called 'members', 'elements' or sometimes 'points' of the set. Following the customary notation we indicate sets by capital letters and write $x \in A$ when the object x belongs to the set A, that is, x is an element of A. Similarly, $x \notin A$ means that x is not an element of A, that is, it does not belong to the set A.

If A and B are two sets, we say that A is a subset of B – and write $A \subset B$ or, equivalently, $B \supset A$ – if every element of A is an element of B; moreover, we say that A is a proper subset of B if every element of A is an element of B but, in addition, B also contains elements which do not belong to A. Two sets A and B are equal, $A = B$ in symbols, if they contain exactly the same elements; clearly, $A = B$ if and only if $A \subset B$ and $B \subset A$.

In general, a set is defined by specifying its elements. If these elements are few we can list them explicitly between braces: for example, the set O of all possible outcomes of the rolling of a die is defined by writing $O = \{1, 2, 3, 4, 5, 6\}$ where it is understood that any integer between 1 and 6 (1 and 6 included) is a possible outcome of our die-rolling experiment. However, we can also equivalently write $O = \{x \in \mathbb{N} : 1 \leq x \leq 6\}$ which reads 'O is the collection of all elements x belonging to the set of integers $\mathbb{N}$ such that the property $1 \leq x \leq 6$ applies'. This latter, in fact, is often a more convenient way of defining sets: we consider the set of interest A as a subset of a (usually large) set X and specify its elements by means of some property $P(x)$ that these elements must satisfy. Mathematically, this is done by writing the general expression $A = \{x \in X : P(x)\}$ of which the example above is just a special case. As another example, if we write $A = \{x \in \mathbb{N} : x^2 < 7\}$ it is clear that A is the set whose only elements are the positive integers 1 and 2. Also, in this symbolism we can express any open, closed and semi-closed interval of real numbers as, for instance ($\mathbb{R}$ is the set of real numbers),

$$(a, b) = \{x \in \mathbb{R} : a < x < b\}$$
$$[a, b) = \{x \in \mathbb{R} : a \leq x < b\}$$
$$(-\infty, b] = \{x \in \mathbb{R} : x \leq b\}$$
$$(a, \infty) = \{x \in \mathbb{R} : x > a\}$$

etc. The set with no elements is denoted by $\emptyset$ and is called the *empty* or *null* set.

Let us now turn our attention on the fundamental operations on sets. In the following – as it is often the case in applications – we will assume that all sets under consideration are subsets of some fixed 'universal set' W.

The set of all subsets of W is called the *power set* of W and is denoted by $\mathbb{P}(W)$. Clearly, $A \subset W$ if and only if $A \in \mathbb{P}(W)$.

The *union* of two sets A and B, written $A \cup B$, is the set consisting of all elements that belong to either A or B, that is,

$$A \cup B = \{x : x \in A \text{ or } x \in B\}$$

for example, if $A = \{1, 6, 8\}$ and $B = \{1, 5\}$ then $A \cup B = \{1, 5, 6, 8\}$. Moreover, from its definition, it is obvious that commutativity applies for unions of sets, that is $A \cup B = B \cup A$.

The *intersection* of two sets, written $A \cap B$ (however, note that some authors write AB), is the set of all elements that belong to both A and B, that is,

$$A \cap B = \{x : x \in A \text{ and } x \in B\}$$

so that, for example, if A and B are as above, then $A \cap B = \{1\}$. Obviously, also the intersection operation is commutative, that is, $A \cap B = B \cap A$.

If two sets A and B have no elements in common then $A \cap B = \emptyset$ and we say that A and B are *disjoint*.

Given a set $A \in \mathbb{P}(W)$, the *complement* of A – denoted A^C – is the set of all elements of W that do not belong to A, that is,

$$A^C = \{x \in W : x \notin A\}$$

As a consequence, we have that $A \cup A^C = W$, $A \cap A^C = \emptyset$, $W^C = \emptyset$ and $(A^C)^C = A$. Also, if $A \subset B$ then $B^C \subset A^C$.

Intersections and unions of sets satisfy the following *distributive laws* whose proofs are left to the reader: if A, B, C are subsets of a set W, then

$$\begin{aligned} A \cap (B \cup C) &= (A \cap B) \cup (A \cap C) \\ A \cup (B \cap C) &= (A \cup B) \cap (A \cup C) \end{aligned} \tag{A.1}$$

Also, given the sets A, B, C it is immediate to prove the associativity property of unions and intersections, that is $(A \cup B) \cup C = A \cup (B \cup C)$ and $(A \cap B) \cap C = A \cap (B \cap C)$. This implies that we can simply write $A \cup B \cup C$ and $A \cap B \cap C$ without ambiguity.

The operation of complementation, in turn, obeys *De Morgan's laws*: if A and B are subsets of a set W, then

$$\begin{aligned} (A \cup B)^{C} &= A^{C} \cap B^{C} \\ (A \cap B)^{C} &= A^{C} \cup B^{C} \end{aligned} \tag{A.2}$$

We prove here the first of (A.2) and leave the second to the reader. If $x \in (A \cup B)^{C}$ then $x \notin A \cup B$ and therefore $x \notin A$ and $x \notin B$. This implies that $x \in A^{C}$ and $x \in B^{C}$ and therefore $x \in A^{C} \cap B^{C}$ so that – having shown that every element of $(A \cup B)^{C}$ is also an element of $A^{C} \cap B^{C}$ – our first result is that (a) $(A \cup B)^{C} \subset A^{C} \cap B^{C}$. Conversely, if $x \in A^{C} \cap B^{C}$ then $x \in A^{C}$ and $x \in B^{C}$, so that $x \notin A$ and $x \notin B$ and therefore $x \notin A \cup B$, which, in turn, implies $x \in (A \cup B)^{C}$. Thus, our second result is that (b) $A^{C} \cap B^{C} \subset (A \cup B)^{C}$. By putting results (a) and (b) together we finally obtain $(A \cup B)^{C} = A^{C} \cap B^{C}$, which is precisely the first of eq. (A.2).

Other operations on sets are the difference and symmetric difference. The *difference* of two sets, written $A - B$, is the set of all elements of A that do not belong to B. In symbols

$$A - B \equiv \{x : x \in A, x \notin B\}$$

so that, for example, if $A = \{x \in \mathbb{N} : 3 \le x \le 10\}$ and $B = \{8, 9, 10\}$ then $A - B = \{x \in \mathbb{N} : 3 \le x \le 7\}$ and also, since in this case $B \subset A$, we have $B - A = \emptyset$. (Note that the difference $A - B$ is called by some authors the 'complement of B relative to A' and may be written as $A \backslash B$.)

The *symmetric difference* of two sets, in turn, is written $A \Delta B$ and is the set of all elements belonging to either A or B, but not to both. In symbols

$$A \Delta B \equiv \{x : x \in A \text{ or } x \in B, x \notin A \cap B\}$$

So, for example, if $A = \{1, 2, 3, 4\}$ and $B = \{3, 4, 5, 6, 7\}$, then $A \Delta B = \{1, 2, 5, 6, 7\}$. We leave to the reader to show that (a) the symmetric difference is commutative and associative and (b) given two sets A and B, the following relations hold

$$
\begin{aligned}
A\Delta B &= (A \cup B) - (A \cap B) \\
A\Delta B &= (A - B) \cup (B - A)
\end{aligned} \tag{A.3}
$$

where either one of the two eqs (A.3) is often given as the definition of $A\Delta B$.

The last operation we define is the cartesian (or direct) product of two sets: given two sets A and B, their *cartesian product* $A \times B$ is the set of all ordered pairs (x, y) where x is an element of A and y is an element of B; mathematically this is written

$$A \times B \equiv \{(x, y) : x \in A, y \in B\}$$

and it should be noted that $A \times B$ is not the same as $B \times A$ unless $A = B$.

If $A = B$, the product $A \times A$ is usually denoted A^2 and a familiar example from calculus is the ordinary two-dimensional space $\mathbb{R}^2 = \mathbb{R} \times \mathbb{R}$, that is, the set of all pairs (x, y) of real numbers which represent the coordinates of a point in a plane. As an exercise it is left to the reader to show that

$$
\begin{aligned}
A \times (B \cup C) &= (A \times B) \cup (A \times C) \\
A \times (B \cap C) &= (A \times B) \cap (A \times C)
\end{aligned} \tag{A.4}
$$

Given a universal set W, in set theory one often has to deal with finite or infinite collections of subsets (of W) labeled by an index. For our purposes it is sufficient to consider finite and countably infinite collections, which will be denoted by $\{A_n\}_{n=1}^N$ and $\{A_n\}_{n=1}^\infty$, respectively, or simply by $\{A_n\}$ when it is irrelevant whether the sequence is finite or not. Such collections are also called *sequences* of sets. It is immediate to extend the operations of union and intersection as follows

$$
\begin{aligned}
\bigcup_n A_n &= \{x : x \in A_n \text{ for some } n\} \\
\bigcap_n A_n &= \{x : x \in A_n \text{ for all } n\}
\end{aligned}
$$

so that, when B and $\{A_n\}$ are subsets of W, the distributive laws (A.1) can be written in the more general form

$$
\begin{aligned}
B \cap \left(\bigcup_n A_n\right) &= \bigcup_n (B \cap A_n) \\
B \cup \left(\bigcap_n A_n\right) &= \bigcap_n (B \cup A_n)
\end{aligned} \tag{A.5}
$$

and De Morgan's laws (A.2) generalize to

$$\left(\bigcup_n A_n\right)^{C} = \bigcap_n A_n^{C}$$
$$\left(\bigcap_n A_n\right)^{C} = \bigcup_n A_n^{C} \tag{A.6}$$

Given a sequence of subsets $\{A_n\}_{n=1}^{\infty}$ of W, we call it an *increasing sequence* if $A_1 \subset A_2 \subset A_3 \subset \cdots$ or, in other words if $A_n \subset A_{n+1}$ for every index n; similarly, we speak of *decreasing sequence* if $A_n \supset A_{n+1}$ for every n. Both types of sequences are called *monotone*. If $\{A_n\}_{n=1}^{\infty}$ is an increasing sequence as defined above the set

$$A = \bigcup_{n=1}^{\infty} A_n \tag{A.7}$$

is called the limit of the sequence and one often finds the symbols $A = \lim_{n\to\infty} A_n$ or $A_n \uparrow A$. Similarly, if $\{A_n\}_{n=1}^{\infty}$ is decreasing the set

$$A = \bigcap_{n=1}^{\infty} A_n \tag{A.8}$$

is the limit of the sequence and the symbols $A = \lim_{n\to\infty} A_n$ or $A_n \downarrow A$ are frequently encountered. With these definitions it is immediate to show that (a) if $A_n \uparrow A$ then $A_n^{C} \downarrow A^{C}$ and (b) if $A_n \downarrow A$ then $A_n^{C} \uparrow A^{C}$.

More generally, for an arbitrary sequence $\{A_n\}_{n=1}^{\infty}$ of subsets of W we define the set called *limit superior* of the sequence as

$$\limsup_{n\to\infty} A_n = \bigcap_{n=1}^{\infty} \bigcup_{k=n}^{\infty} A_k \tag{A.9}$$

where $a \in \limsup A_n$ if and only if $a \in A_n$ for infinitely many n. Also, we define the *limit inferior* as

$$\liminf_{n\to\infty} A_n = \bigcup_{n=1}^{\infty} \bigcap_{k=n}^{\infty} A_k \tag{A.10}$$

where $a \in \liminf A_n$ if and only if $a \in A_n$ for all but finitely many n. In general $\liminf A_n \subset \limsup A_n$; however, if $\liminf A_n = \limsup A_n = A$ we say that the sequence converges to the set A – or, equivalently, A is the limit of the sequence – and we simply write $A = \lim_{n\to\infty} A_n$. Note that the condition $\liminf A_n = \limsup A_n$ always holds for monotone sequences. As an example consider the sequence of real intervals $A_n = (-\infty, b - 1/n)$; from the definitions (A.9) and (A.10) it is not difficult to determine that $\liminf A_n = \limsup A_n$ and that this common limit is the set $A = (-\infty, b)$. Moreover, since the given sequence is increasing, the result $A = (-\infty, b)$ can also be obtained directly from eq. (A.7).

In many situations one has to deal with *pairwise disjoint sequences of sets*, where this term refers to any indexed collection of sets $\{A_n\}$ such that $A_i \cap A_j = \emptyset$ whenever $i \neq j$. In this regard the concept of *partition* of a set is particularly important: given a set B we call *partition* of B any collection of sets $\{A_n\}$ such that (i) $A_i \cap A_j = \emptyset$ for $i \neq j$ and (ii) $\cup_n A_n = B$.

For our purposes the definition of partition will in general suffice. However, for the interested reader it may be worth noting that there exists a close connection between partitions of a set and the so-called 'equivalence relations' which can be introduced on the same set. In fact – given a set B – it turns out that any equivalence relation on B determines a partition of B and conversely.

Broadly speaking, a (binary) relation is a formalization of the notion that some elements of B may be related to some other elements of B in a special way, thus allowing us to specify the pairs of elements $(a, b) \in B \times B$ that are related. For example, given a set B of 3 people: Mark, aged 36, George, aged 41 and Susy, aged 23, we may define a relation in B by saying '$b_1, b_2 \in B$ are related if b_1 is older than b_2', thus immediately determining the related (and ordered) pairs, that is (George, Susy), (George, Mark) and (Mark, Susy).

In general, given a set B, we speak of a relation *on* B if there exist at least a related pair (a, b) for every $a \in B$ and the fact that two elements a and b are related is indicated $a \approx b$. *Equivalence relations* defined on a set B are special types of relations which obey the three following conditions: for every $a, b, c \in B$

(1) $a \approx a$ (reflexivity),
(2) if $a \approx b$ then $b \approx a$ (symmetry),
(3) if $a \approx b$ and $b \approx c$, then $a \approx c$ (transitivity).

Clearly, the relation 'older than' given above is not an equivalence relation. Conversely, given a set T of triangles $t_1, t_2, \ldots, t_N$, the specification '$t_i \approx t_j$ if t_i has the same area as t_j' defines an equivalence relation (on T) which, in turn, decomposes T in a number of pairwise disjoint subsets (called equivalence classes; in this case the sets made up of triangles with the same area) whose union is T, that is, a partition of T. Since this latter comment on the decomposition of a set is true in general, we note that, in practice, it is often by specifying an equivalence relations on a set that we can construct partitions of this same set.

A.2 Functions and sets, equivalent sets and cardinality

Most readers are familiar with the notion of real-valued function, that is a rule f which associates with each real number x another real number $f(x)$. This, however, is just a special case because the notion of function is essentially a set-theoretic concept and applies in more general contexts.

Now, let X and Y be two sets; a rule f that assigns to every $x \in X$ a unique element $y = f(x) \in Y$ is called a *function* (mapping, transformation or operator) from X to Y. In this case one often writes $f : X \to Y$ and speaks of a 'Y-valued function defined on X' or of a 'function defined on X with values in Y'. Also, the set X is called the *domain* of f – and sometimes indicated $D(f)$ – while the set $R(f) \subset Y$ defined by

$$R(f) = \{y \in Y : y = f(x) \text{ for some } x \in X\}$$

is called the *range* of f. If $x \in X$, the element $y = f(x) \in Y$ is called the *image* of x (under the mapping f) and we can call x a *pre-image* of y. Note that we do not say 'the' pre-image of y because an element y may have more than one pre-image. Moreover it may happen that Y contains elements with no pre-image at all. This implies that $R(f)$ is a proper subset of Y and we say in this case that f maps X *into* Y. Conversely, when every element $y \in Y$ has (at least) a pre-image $x \in X$ (or, in other words, for every $y \in Y$ there is an $x \in X$ such that $y = f(x)$) we say that f maps X *onto* Y or that f is *surjective*. Clearly, this means that $R(f) = Y$.

When $f : X \to Y$ maps distinct elements of X into distinct elements of Y – that is, if $x_1 \neq x_2$ implies $f(x_1) \neq f(x_2)$ – the function f is called *one-to-one* or *injective*.

If a function $f : X \to Y$ is one-to-one and onto (i.e. injective and surjective, or, for short, *bijective*) then for every $y \in Y$ there is a unique pre-image $x \in X$ such that $y = f(x)$. In this case we say that f is *invertible* because we can define the function $f^{-1} : Y \to X$, called the inverse of f. Clearly, $f^{-1}(y) = x$ whenever $y = f(x)$. Note that if $f : X \to Y$ is one-to-one but not onto we can nonetheless define an inverse f^{-1} provided that its domain is $R(f)$, that is, $f^{-1} : R(f) \to X$. In fact, the function $f : X \to R(f)$ is one-to-one and onto.

With the definitions above, given a set $A \subset X$ it should be clear at this point what we mean by $f(A)$, namely

$$f(A) = \{y \in Y : y = f(x) \text{ for some } x \in A\} \subset Y$$

which is called the image of set A under the mapping f. Similarly, given a set $B \subset Y$, we call pre-image of B the subset of X

$$f^{-1}(B) = \{x \in X : f(x) \in B\}$$

Note that this symbolism does not necessarily imply that f is invertible; here we are just focusing our attention on the elements of X which are pre-images of the elements of the set B regardless of the fact whether f is one-to one and/or onto. In fact, if f is not surjective and B is a set whose elements have no pre-image we write $f^{-1}(B) = \emptyset$.

Given three sets X, Y, Z and two functions $f : X \to Y$ and $g : Y \to Z$ we can consider the *composition* of g and f, that is, the function $g \circ f : X \to Z$ defined by

$$(g \circ f)(x) = g(f(x))$$

where we note that this concept is meaningful only if $R(f) \subset D(g)$, otherwise one says that $g \circ f$ does not exist. With this definition at our disposal we note that, broadly speaking, a function $f : X \to Y$ is invertible if there exists a mapping $f^{-1} : Y \to X$ which unravels f. Specifically, $f^{-1}(f(x)) = x$ for every $x \in X$, and since we remarked that a function is invertible if and only if it is one-to-one and onto, the relation $f(f^{-1}(y)) = y$ (for every $y \in Y$) also holds.

At this point we can turn our attention to some important results which will be stated as propositions without proofs. The interested reader can try to work out the proofs or can find them, for instance, in Ref [1] at the end of this appendix.

Proposition A.1 *The pre-image of the union of two sets equals the union of the two individual pre-images, that is,*

$$f^{-1}(A \cup B) = f^{-1}(A) \cup f^{-1}(B) \tag{A.11}$$

Proposition A.2 *The pre-image of the intersection of two sets equals the intersection of the two individual pre-images, that is,*

$$f^{-1}(A \cap B) = f^{-1}(A) \cap f^{-1}(B) \tag{A.12}$$

Proposition A.3 *The image of the union of two sets equals the union of the two individual images, that is,*

$$f(A \cup B) = f(A) \cup f(B) \tag{A.13}$$

Also, these results can be generalized to

$$\begin{aligned} f^{-1}\left(\bigcup_n A_n\right) &= \bigcup_n f^{-1}(A_n) \\ f^{-1}\left(\bigcap_n A_n\right) &= \bigcap_n f^{-1}(A_n) \\ f\left(\bigcup_n A_n\right) &= \bigcup_n f(A_n) \end{aligned} \tag{A.14}$$

Note that nothing has been said about the image of the intersection of sets; this is because, in general

$$f\left(\bigcap_n A_n\right) \subset \bigcap_n f(A_n) \tag{A.15}$$

and the equality sign holds only if f is one-to-one. We close this section with a few short comments on the cardinality (or power) of a set, a term which, broadly speaking, refers to the number of elements in the set. As a preliminary, we say that two sets A and B are *equivalent* if there is a bijective mapping $f : A \to B$ between the elements of A and the elements of B. Two equivalent sets are said to have the same cardinality and, by definition, we say that the empty set $\emptyset$ contains zero elements.

If two sets are finite, that is, contain a finite number of elements, their cardinality is simply the number of their elements and it is easy to determine whether they are equivalent or not. Therefore, in this context, the concepts of cardinality and equivalence have an immediate interpretation. For sets with an infinite number of elements things are not so intuitive and it is here that the notion of equivalence plays a fundamental role. In fact, we say that a set A is countably infinite if there is a bijective mapping between its elements and the set of natural numbers $\mathbb{N} = \{1, 2, 3, \ldots\}$ or, in other words, if A is equivalent to $\mathbb{N}$. Also, as a side comment, note that the term 'countable set' is often used to refer both to finite and countably infinite sets (a better definition is to call 'countable' an infinitely countable set and to use the term 'at most countable' for sets that are either finite or infinitely countable).

In regard to countable sets the following results are worthy of notice:

(1) any subset of a countable set is at most countable;
(2) a finite or countably infinite union of countable sets is itself a countable set;
(3) a countably infinite set is equivalent to one of its proper subsets;
(4) the set $\mathbb{Z} = \{\ldots, -2, -1, 0, 1, 2, \ldots\}$ of all integers is countably infinite;
(5) the set $\mathbb{Q} = \{x : x = p/q, p, q \in \mathbb{Z}, q \neq 0\}$ of rational numbers is countably infinite.

All or most of the five statements above are probably known to the reader, and their proofs can be found on almost any book of mathematical analysis; however, an additional comment on statement (3) may not be out of place. Statement (3) says that it is possible to construct a bijective mapping between an infinite set and any one of its infinite proper subsets. As a matter of fact, this is a distinguishing property of infinite sets and it turns out that a set is infinite if and only if it is equivalent to one of its proper subsets. This is true also for non-countable sets whose existence is guaranteed by a theorem stating that the set of real numbers in the closed interval $[0, 1]$ is uncountable. Furthermore, it can also be shown that the set of all real numbers in any closed or open interval – respectively (a, b) or $[a, b]$, with $a < b$ – and the set of all real numbers (the real line $\mathbb{R}$) are equivalent to $[0, 1]$ and are, therefore, uncountable. Clearly, no uncountable set can be a subset of a countable set because, broadly speaking, countable sets represent the 'smallest' type of infinity while sets equivalent to $[0, 1]$ – which are said to have the power of the continuum – represent a 'larger' type of infinity.

A.3 Systems of sets: algebras and σ-algebras

The general phrase 'system of sets', or collection of sets, is used to indicate a 'set of sets', that is a set whose elements are themselves sets. More specifically, we will be mainly interested in special collections of subsets of a given set W, where the term 'special' refers to the fact that they must satisfy a number of properties in order to qualify.

Let us start with the defining properties of an algebra of sets. A non-empty system R of subsets of a set W is called an *algebra* of sets if

(1) $A \in R$ implies $A^C \in R$,
(2) $A, B \in R$ implies $A \cup B \in R$.

In words, the properties above are expressed by saying that an algebra is closed with respect to the operations of complementation and union. Given a set W, an immediate example of algebra is given by the collection of the two sets $\{\emptyset, W\}$, the so-called trivial algebra on W; another example is the power set $\mathbb{P}(W)$ and a third example is $\{\emptyset, A, A^C, W\}$ where A is a non-empty proper subset of W.

Noteworthy consequences of the defining properties (1) and (2) are that

(i) an algebra is closed under the operation of intersection;
(ii) an algebra is closed under the operations of difference and symmetric difference;
(iii) an algebra is closed under finite unions and intersections, i.e. if $A_1, A_2, \ldots, A_n \in R$ then $\cup_{k=1}^{n} A_k \in R$ and $\cap_{k=1}^{n} A_k \in R$.

The proof of (i) is immediate; in fact, by virtue of De Morgan's laws (A.2) we have $A \cap B = (A^C \cup B^C)^C$, and since the set on the right-hand side belongs to R whenever $A, B \in R$, it follows that $A \cap B \in R$ whenever $A, B \in R$. Moreover, note that property (1) plus closure under intersection imply that $\emptyset \in R$ and $W \in R$. In regard to statement (ii), closure under symmetric difference can be proven by means of the identity $A \Delta B = (A \cap B^C) \cup (B \cap A^C)$ and then closure under difference follows by virtue of the first of eq. (A.3). Finally, the proof of (iii) is trivial.

We can now make a side comment and note that, in the light of (i) and (ii), it should be hardly surprising to find defining properties (of an algebra of sets) other than (1) and (2); for example, one can define [1] an algebra as a system of sets which is closed under the operations of symmetric difference and intersection. Needless to say, this is equivalent to our definition.

Returning to our main discussion, it is useful to point out two theorems which we state in the form of propositions without proof:

Proposition A.4 *Any intersection (finite or not) of algebras of sets is itself an algebra.*

Proposition A.5 *Given a non-empty collection M of subsets of W, there exist a smallest algebra of subsets of W containing M. This algebra is called the* algebra generated *by M and can be denoted by R(M). Also, it can be shown that R(M) is obtained as the intersection of all algebras (on W) containing M (the term 'smaller' above means that if R is an algebra on W containing M, then $R(M) \subset R$).*

We define now another special system of sets. A *σ-algebra* on W is a non-empty collection S of subsets of W satisfying the conditions

(1′) $A \in S$ implies $A^C \in S$;
(2′) $A_1, A_2, \ldots \in S$ imply $\cup_{n=1}^{\infty} A_n \in S$

meaning that a σ-algebra is closed under complementation and countable unions. It is left to the reader to show that a σ-algebra is also closed under countable intersections.

Clearly, every σ-algebra is an algebra, but the converse is not true because the conditions enforced on a σ-algebra are more restrictive. However, two theorems which parallel A.4 and A.5 hold

Proposition A.6 *Any intersection (finite or not) of σ-algebras is itself a σ-algebra.*

Proposition A.7 *Given any non-empty collection M of subsets of W there exist a smallest σ-algebra of subsets of W containing M. This σ-algebra is called the* σ-algebra generated *by M and can be denoted by S(M). Also, it can be shown that S(M) is the intersection of all σ-algebras (on W) containing M.*

If we consider the special case of sets of real numbers, the so-called Borel σ-algebra is particularly important in analysis. Specifically, the *Borel σ-algebra* $\mathbb{B}$ on the real line $\mathbb{R}$ can be defined as the σ-algebra generated by the collection of intervals I of the type $I = \{(-\infty, a] : a \in \mathbb{R}\}$. The members of this σ-algebra are called *Borel sets* (or B-sets) and it can be shown that every open, closed and semi-open set of $\mathbb{R}$ – that is, sets of the type (a, b),

$[a, b]$, $[a, b)$ or $(a, b]$ with $a < b$ – are Borel sets. Also, degenerate intervals of the type $[a, a]$ are Borel sets.

(As a side remark for the more mathematically oriented reader we note that $\mathbb{B}$ is the σ-algebra generated by the natural topology of open sets of $\mathbb{R}$ and that, by extension, the name of Borel sets is used to denote the members of the σ-algebra generated by a topology in a general set X.)

The last results we state in this section consider the way in which systems of sets transform under mappings. Let X, Y be two sets, $f : X \to Y$ a function and let M be a system of subsets of X. We denote by $f(M)$ the system of all images $f(A)$ of sets $A \in M$; similarly, if N is a system of subsets of Y, $f^{-1}(N)$ denotes the system of all pre-images $f^{-1}(B)$ of sets $B \in N$. In this light, the following proposition hold

Proposition A.8

(a) If N is an algebra (in Y) then $f^{-1}(N)$ is an algebra (in X),
(b) If N is a σ-algebra (in Y) then $f^{-1}(N)$ is a σ-algebra (in X).

Reference

[1] Kolmogorov, A.N., Fomin, S.V., *'Introductory Real Analysis'*, Dover Publications, New York (1975).

Further reading

Cafiero, F., Zitarosa, A., *'Elementi di Teoria degli Insiemi'*, Liguori Editore (1977).
Halmos, P.R., *'Naive Set Theory'*, D. Van Nostrand Co. Inc. (1960).
McDonald, J.N., Weiss, N.A., *'A Course in Real Analysis'*, Academic Press (1999).
Rudin, W., *'Principles of Mathematical Analysis'*, McGraw-Hill (1964).

Appendix B

The Lebesgue integral – an overview

Our main purpose here is to introduce and briefly discuss a number of concepts which are relevant to the main subject of the book. For this reason, the exposition will necessarily be concise and limited to the essential aspects of our interest. For a detailed treatment – and for the proof of most theorems – a list of references is given at the end of this appendix.

B.1 Introductory remarks

The reader is probably familiar with the notion and properties of the Riemann integral from fundamental calculus. Unfortunately, this integral suffers from two main limitations: (a) it applies only to functions which are either continuous or 'essentially' continuous and (b) it fails to satisfy certain desirable convergence properties. In fact, in regard to statement (b) it is known, for instance, that if a sequence $\{f_n\}_{n=1}^{\infty}$ of Riemann integrable (R-integrable) functions on an interval $[a, b]$ converges pointwise to a R-integrable function f – that is, if $\lim_{n\to\infty} f_n(x) = f(x)$ – then, in general, it is not true that

$$\lim_{n\to\infty} \int_a^b f_n(x)\,\mathrm{d}x = \int_a^b f(x)\,\mathrm{d}x \tag{B.1}$$

meaning that the operations of limit and integral cannot in general be interchanged. Moreover, this is true even if each f_n and f are continuous (incidentally, we recall that the set of continuous functions is not closed under pointwise limits, that is, the facts that (i) $f_n(x) \to f(x)$ pointwise and (ii) each f_n is continuous do not imply that f is continuous).

If, on the other hand, if the sequence $\{f_n\}_{n=1}^{\infty}$ of R-integrable functions on $[a, b]$ converges uniformly to f, then it can be shown that (i) f is R-integrable and (ii) eq. (B.1) holds. Even so, however, uniform convergence is a sufficient but not necessary condition. Now, since uniform convergence is a rather strong requirement (especially when the common domain is the entire real line $\mathbb{R}$), we are led to the concept of Lebesgue integral which, besides allowing the possibility to integrating a larger class of functions, overcomes many of the difficulties of the Riemann integral.

In addition to all this, the Lebesgue integral can be defined on abstract measure spaces (where the Riemann integral does not make sense) and the Lebesgue integral on the real line is just a special case of this general definition. In probability theory, in fact, we recall that expectations are defined as abstract Lebesgue integral on a probability space. Then, mathematical analysis shows that these integrals turn into Lebesgue–Stieltjes integrals on $\mathbb{R}$ and these, in turn, are calculated as sums or as ordinary Riemann integrals.

B.2 Measure spaces and the Lebesgue measure on the real line

In Chapter 2 we have introduced the notion of probability space. Later in the same chapter (Section 2.2.1) we pointed out that probability spaces are special cases of measure spaces. By this term we mean the following:

Definition B.1 A measure space is a triplet (W, S, ν), where W is a set, S is a σ-algebra of subsets of W and ν is a function $\nu : S \to \mathbb{R}$ satisfying

(M1) $\nu(A) \geq 0$ for all $A \in S$
(M2) σ-additivity: if $A_1, A_2, \ldots \in S$ and $A_i \cap A_j = \emptyset \, (i \neq j)$, then

$$\nu\left(\bigcup_{n=1}^{\infty} A_n\right) = \sum_{n=1}^{\infty} \nu(A_n)$$

The sets belonging to S are called measurable sets and ν is called a measure on S. The peculiar characteristic of probability measures is the additional requirement $\nu(W) = 1$.

In Chapter 2 we also gave some important properties descending from (M1) and (M2) – namely monotonicity, subadditivity and continuity – and noted how a non-negative, σ-additive function defined on a semialgebra G of subsets of a set W can be extended to a σ-algebra of subsets of W. This possibility justifies the fact that, usually, the domain of a measure is a σ-algebra of sets.

More specifically, the extension procedure is accomplished in two steps: (a) for all subsets W, one first defines (in terms of the original set function defined on G) the so-called 'outer measure' and then (b) restricts its domain to all subsets which satisfy a certain 'measurability' criterion (eq. (2.7)). The restriction of the domain is necessary because the outer measure defined in (a) is not σ-additive on its domain (i.e. $\mathbb{P}(W)$; for details see, for instance, Refs [2, 5] or [7]).

Now, the 'construction' of a measure outlined above is a general abstract mathematical procedure which applies to a large number of cases (this means that the set W is not necessarily a set of real numbers). A special important case, however, occurs when the set W is the real line $\mathbb{R}$, a situation which leads to the definition of the Lebesgue measure – usually denoted by μ. This measure generalizes the notion of length (of an interval) to a broad class M – the so-called class of 'Lebesgue measurable' sets – of subsets of $\mathbb{R}$ while at the same time retaining the intuitive concept. Specifically, one starts with the idea of length L of an interval $[a, b] \subset \mathbb{R}$, where $L[a, b] = b - a$, and then extends this notion to the class of 'elementary sets', that is, all sets that can be represented as the finite or countable union of pairwise disjoint intervals. At this point, for any subset A of $\mathbb{R}$, one defines its outer measure $\mu^*(A)$ which, indeed, does generalize the notion of length but lacks the most important property of interest, that is, σ-additivity. In order to retain σ-additivity one must restrict the domain of μ^* to a proper subclass M of $\mathbb{P}(\mathbb{R})$. Then it is shown that (i) M is a σ-algebra and, in particular (ii) contains the σ-algebra $\mathbb{B}$ of all Borel sets of $\mathbb{R}$. Clearly, all open, closed and semiclosed intervals of $\mathbb{R}$, by construction, turn out to be Lebesgue measurable.

Thus, the triplet $(\mathbb{R}, M, \mu)$ – where μ, the Lebesgue measure, denotes the restriction of μ^* to the domain M – is a measure space and, as a matter of fact, it is the measure space which was initially used as a model to develop the abstract concept defined at the beginning of this section. If one ignores, as we did, the historical sequence of facts, the Lebesgue measure is just a particular case of the abstract construction. In fact, it can be shown that our starting point, that is, the collection I of all intervals of $\mathbb{R}$ (including intervals of the form (a, a) and $[a, a]$), is a semialgebra of subsets of $\mathbb{R}$ and the length L, in turn, is a non-negative, σ-additive function $L : I \to \mathbb{R}$. So, for our purposes the point is that we are able to assign a measure to a large number of subsets of $\mathbb{R}$ and that this class of sets – although being a proper subset of $\mathbb{P}(\mathbb{R})$ – is much broader than the class of intervals of $\mathbb{R}$ and even broader than the class $\mathbb{B}$ of Borel sets. As a side remark, it should be noted that, starting from the set $\mathbb{R}$, it is possible to define different measures other than μ. This, in essence, is due to the generality of the basic requirements – that is, properties (M1) and (M2). In their light, in fact, it would be rather surprising if μ were the only possible measure on $\mathbb{R}$.

Closing the parenthesis on the special case $W = \mathbb{R}$, let us return to the abstract setting (W, S, ν) with the final goal of defining an integral on this measure space. The first step in this direction is the definition of measurable function. Before doing this, however, a few further remarks and definitions on measures – and on the Lebesgue measure μ in particular – are in order.

First of all, given a space (W, S, ν), a property is said to hold *almost everywhere* (a.e. or ν-a.e. if the measure needs to be specified) if it holds everywhere except on a set N such that $\nu(N) = 0$. So, for example, if in $\mathbb{R}$ equipped with the Lebesgue measure μ one says that a function $f(x)$ is continuous a.e., it means that the set $A \subset \mathbb{R}$ of all x where f is not continuous is such that $\mu(A) = 0$.

A measure space (W, S, ν) is called *complete* if any subset B of a measurable set A such that $\nu(A) = 0$ is itself measurable, that is, $B \in S$ (and, clearly, $\nu(B) = 0$). In this regard it can be shown that the measure space $(\mathbb{R}, M, \mu)$ is, indeed, complete. If a measure space (W, S, ν) is not complete it can however be completed, as stated in the following proposition:

Proposition B.1 *Let (W, S, ν) be a measure space and let $\bar{S}$ be the class of all sets of the form $B \cup A$ where $B \in S$ and $A \subset C$ for some $C \in S$ such that $\nu(C) = 0$. Define $\bar{\nu}(B \cup A) = \nu(B)$. Then $\bar{S}$ is a σ-algebra, $\bar{\nu}$ is a measure on $\bar{S}$ and $(W, \bar{S}, \bar{\nu})$ is a complete measure space called the* completion *of (W, S, ν). Clearly $S \subset \bar{S}$ and the restriction of $\bar{\nu}$ to S is the original measure ν.*

This result has been given because in the course of this and other books one often considers the Lebesgue measure μ restricted to the domain $\mathbb{B}$ of Borel sets of $\mathbb{R}$; the measure space $(\mathbb{R}, \mathbb{B}, \mu)$ is not complete but it is important to note that $(\mathbb{R}, M, \mu)$ is its completion.

Another definition classifies measures in two classes: a measure ν is called *finite* if $\nu(W) < \infty$ (i.e. $\nu(W)$ is finite) and *σ-finite* if the set W can be represented as the union $W = \cup_{n=1}^{\infty} A_n$ where $A_n \in S$ and $\nu(A_n) < \infty$ for all n. For the cases of our interest it is worth pointing out that

(a) any probability measure is finite,
(b) the Lebesgue measure μ is a σ-finite measure on the real line $\mathbb{R}$ (in fact, consider the sets $I_n = [-n.n]$, $n = 1, 2, \ldots$. Then $\mathbb{R} = \cup_{n=1}^{\infty} I_n$ and $\mu(I_n) < \infty$ for all n. Clearly, $\mu(\mathbb{R}) = \infty$).

B.3 Measurable functions and their properties

The general definition of measurable function is as follows:

Definition B.2(a) Let W_1, W_2 be two sets in which, respectively, we identify two σ-algebras S_1, S_2. Then the function $f : W_1 \to W_2$ is called *measurable* if $f^{-1}(A) \in S_1$ for every set $A \in S_2$.

For our purposes it is often sufficient to consider real-valued set functions, that is functions whose domain is a set $W$and whose range is a subset of $\mathbb{R}$ or $\mathbb{R}$ itself. In this case, the collection $\mathbb{B}$ of Borel sets is considered the 'natural' σ-algebra in the range of the function and the definition is:

Definition B.2(b) Let W be a set, S be a σ-algebra of subsets of W and $f : W \to \mathbb{R}$. The function f is measurable (or, sometimes, S-measurable if the σ-algebra in the domain needs to be indicated) if $f^{-1}(B) \in S$ for every Borel set $B \subset \mathbb{R}$.

In particular, if $f : \mathbb{R} \to \mathbb{R}$ two special cases arise:

(i) if $f^{-1}(B)$ is a Borel set for every Borel set $B \subset \mathbb{R}$, f is called Borel measurable, B-measurable or a Borel function;
(ii) if $f^{-1}(B) \in M$ for every Borel set $B \subset \mathbb{R}$, f is called Lebesgue measurable (L-measurable for short). In general, although for a function $f : \mathbb{R} \to \mathbb{R}$ it is rarely necessary to replace B-measurability by the more general notion of L-measurability, the two concepts need to be distinguished; in fact, every Borel measurable function is L-measurable but not conversely.

Given the definition above (not be confused with the measurability of sets), one should note that:

(a) For the concept of measurable function to make sense we only need to identify two systems of sets, one within its domain (the σ-algebra S_1) and one within its range (the σ-algebra S_2). So, strictly speaking, we do not even need to introduce a measure ν on S_1. However, the concept is frequently used in the context of a measure space (W_1, S_1, ν) because this is the situation of interest in most cases.
(b) In probability terminology (see Chapter 2) a measurable function $f : W \to \mathbb{R}$ is called random variable.

A first useful result is that a function $f : W \to \mathbb{R}$ is measurable if and only if the set $f^{-1}(-\infty, a] = \{w \in W : f(w) \leq a\}$ is measurable – that is, belongs to S – for every $a \in \mathbb{R}$ (see also Proposition 2.9). A second result is that, in general, all ordinary operations of analysis (including limit operations), when applied to measurable functions, lead to measurable functions; in other words all functions that are ordinarily encountered in applications are measurable. This is made more precise in the following propositions:

Proposition B.2 *If f, g are measurable functions, then $f+g, f-g, fg, f/g, \alpha f$ (where $\alpha \in \mathbb{R}$) and $|f|$ are measurable (clearly, when these functions are well-defined, that is, if $f(x)+g(x)$ is never of the form $+\infty - \infty$ and $f(x)/g(x)$ is never of the form ∞/∞ or $\alpha/0$).*

In addition, a measurable function of a measurable function is itself measurable and the limit of measurable functions is measurable, that is,

Proposition B.3 *Let W_1, W_2, W_3 be three sets in which, respectively, we identify the σ-algebras S_1, S_2, S_3. If $f : W_1 \to W_2$ and $g : W_2 \to W_3$ are measurable then the function $h : W_1 \to W_3$ given by $h(x) = g(f(x))$ is measurable.*

Proposition B.4 *If $\{f_n\}_{n=1}^{\infty}$ is a sequence of measurable functions and $\lim_{n\to\infty} f_n(x) = f(x)$, then $f(x)$ is measurable (in other words, the class of measurable functions is closed with respect to pointwise convergence).*

This last result is important because, as we noted in the introductory remarks, the class of continuous functions is not closed under pointwise limits. It is then evident (and it can be proven rigorously) that the class of continuous functions is a proper subclass of measurable functions, that is, that every continuous function is measurable.

We noted above that the notion of measurability of a function does not require a measure to be defined; nevertheless, this is by far the most frequent situation and the domain is generally a set with the structure of a measure space, that is, a triplet (W, S, ν) which, by virtue of Proposition B.1, can be considered a complete measure space without loss of generality (in fact, we can always set $\nu(A') = 0$ for any subset A' of a set A such that $\nu(A) = 0$). Unless otherwise stated, we will always assume completeness.

So, in all cases where a measure is defined the values taken on by a measurable function on sets of zero measure can often be neglected by bringing into play the concept of 'almost everywhere' properties introduced in the previous section. In this light, two functions f, g defined on the same set are called 'equal almost everywhere' (equal a.e. or, for some authors, equivalent) if the set of values where $f(x) \neq g(x)$ has zero measure, that is if $\nu\{x : f(x) \neq g(x)\} = 0$. From this it follows that a function f equal a.e. to a measurable function g is itself measurable. The next definition proceeds along this line of reasoning:

Definition B.3 A sequence of measurable functions $\{f_n\}_{n=1}^{\infty}$ is said to converge a.e. to a function f if $\lim_{n\to\infty} f_n(x) = f(x)$ except on a set of measure zero. In this case one often writes $f_n \to f[\text{a.e.}]$ or $f_n \to f[\nu - \text{a.e.}]$ if the measure needs to be specified.

In this regard, since pointwise limits of measurable functions are measurable functions (Proposition B.4), it can be asked if this remains true for convergence a.e. We have the following result:

Proposition B.5 *If the sequence of measurable functions $\{f_n\}_{n=1}^{\infty}$ converges a.e. to f, f itself is measurable.*

A remark, however, is in order here. Proposition B.5 is not, in general, true if the measure space (W, S, ν) is not complete. Therefore, in particular, if $f_n : \mathbb{R} \to \mathbb{R}$ are L-measurable functions and $f_n \to f$ a.e., then f is L-measurable but if the functions f_n are B-measurable we cannot conclude that f is B-measurable.

Another type of convergence used in probability theory is the so-called 'convergence in probability' which, in the context of measure theory, is called by mathematicians 'convergence in measure'.

Definition B.4 A sequence $\{f_n\}_{n=1}^{\infty}$ of measurable functions converges in measure to the measurable function f if for each $\varepsilon > 0$ we have

$$\lim_{n\to\infty} \nu\{x : |f(x) - f_n(x)| \geq \varepsilon\} = 0$$

which, in words, means that the measure of the set where f_n differs from f by more than any prescribed positive number tends to zero as $n \to \infty$. To indicate convergence in measure one often finds the symbol $f_n \to f[\nu]$.

In general, there is no relationship between convergence a.e. and convergence in measure. However, for finite measure spaces (hence for probability spaces) convergence a.e. implies convergence in measure, i.e. the following proposition holds:

Proposition B.6 *Let (W, S, ν) be a finite measure space and let the sequence of measurable functions $\{f_n\}_{n=1}^{\infty}$ converge a.e. to f. Then, $f_n \to f[\nu]$.*

The converse statement is not true and convergence in measure does not imply convergence a.e. There exists, however, a partial converse

Proposition B.7 *Let the sequence of measurable functions $\{f_n\}_{n=1}^{\infty}$ converge in measure to the measurable function f. Then, there is a subsequence $\{f_{n_k}\}_{k=1}^{\infty}$ such that $f_{n_k} \to f$ a.e.*

B.4 The abstract Lebesgue integral

Among measurable functions it is useful to distinguish the class of simple functions, that is functions whose range is a finite set. So, let (W, S, ν) be a measure space:

Definition B.5 A measurable function $s : W \to \mathbb{R}$ is called simple if it takes on only finitely many distinct values $a_1, a_2, \ldots, a_n$. Equivalently, σ is simple if and only if it can be expressed as the sum

$$s = \sum_{k=1}^{n} a_k I_{A_k} \tag{B.2}$$

where I_{A_k} is the indicator function (defined by eq. (2.21)) of the set $A_k \equiv \{w \in W : s(w) = a_k\}$. The sets A_k, in turn, can be assumed to be disjoint without loss of generality and, clearly, the function s is measurable if and only if the A_k are measurable. The Lebesgue integral is first defined for simple functions:

Definition B.6 Let s be a measurable simple function on W of the form (B.2). Then the abstract Lebesgue integral of s over W (with respect to ν) is defined by

$$\int_W s \, d\nu \equiv \sum_{k=1}^{n} a_k \nu(A_k) \tag{B.3a}$$

as long as $+\infty$ and $-\infty$ do not both appear in the sum on the r.h.s.; if they do we say that the integral does not exist. Also, one adopts the usual convention $0 \cdot \infty = 0$ if $a_k = 0$ and $\nu(A_k) = \infty$ for some index k. As incidental remarks to notation and terminology we note that (i) other frequently encountered symbols for $\int_W s \, d\nu$ are $\int_W s(w) \, d\nu(w)$ or $\int_W s(w)\nu(dw)$ and (ii) one sometimes speaks of 'abstract Lebesgue integral' if the setting is a general measure space (W, S, ν) and of 'Lebesgue integral' in the special case $(W, S, \nu) = (\mathbb{R}, M, \mu)$.

If $A \in S$ the abstract Lebesgue integral of s over A is defined by

$$\int_A s \, d\nu \equiv \int_W I_A s \, d\nu \tag{B.3b}$$

where the definition makes sense because if s is measurable and $A \in S$ then the product $I_A s$ is a measurable function.

With the integral of simple functions, we can now define the integral of non-negative functions. In fact, by virtue of the following Proposition [9] we determine that the non-negative, measurable functions can be approximated by non-negative, measurable simple functions:

Proposition B.8 *Let f be a measurable, real-valued function $f : W \to [0, +\infty]$. Then there exist non-negative, measurable simple functions s_n such that*

(i) $0 \le s_1 \le s_2 \le \cdots \le f$
(ii) the sequence $\{s_n\}_{n=1}^{\infty}$ converges pointwise to f.

Definition B.7 If f is a non-negative, measurable, real-valued function, its (abstract) Lebesgue integral over W is defined by

$$\int_W f \, d\nu \equiv \lim_{n \to \infty} \int_W s_n \, d\nu \tag{B.4a}$$

where the sequence of simple functions s_n approximate f in the sense of Proposition B.8. Equivalently, (B.4a) can be replaced by

$$\int_W f \, d\nu \equiv \sup \int_W s \, d\nu \tag{B.4b}$$

where the supremum is taken over all measurable simple functions such that $0 \le s \le f$.

Again, if $A \in S$, the Lebesgue integral over A is

$$\int_A f \, d\nu = \int_W I_A f \, d\nu \tag{B.4c}$$

It should be noted that the integral may have the value $+\infty$.

Some properties of the integral of non-negative functions are as follows.

Proposition B.9 *Let f, g be two non-negative, measurable, real-valued functions on $W, a \geq 0$ and $A \in S$. Then*

(a) $f \leq g$ *a.e. implies* $\int_A f \, d\nu \leq \int_A g \, d\nu$;
(b) $B \in S, B \subset A$ *implies* $\int_B f \, d\nu \leq \int_A f \, d\nu$;
(c) *if* $f(w) = 0$ *for all* $w \in A$ *then* $\int_A f \, d\nu = 0$ *even if* $\nu(A) = \infty$;
(d) *if* $\nu(A) = 0$ *then* $\int_A f \, d\nu = 0$ *even if* $f(w) = \infty$ *for all* $w \in A$;
(e) $\int_A af \, d\nu = a \int_A f \, d\nu$.

In addition to the above properties we can now state some of the key results of Lebesgue integration.

Proposition B.10 (Monotone Convergence Theorem) *Let $\{f_n\}_{n=1}^{\infty}$ be a monotone nondecreasing sequence of measurable, non-negative functions converging pointwise to f – that is, $0 \leq f_1 \leq f_2 \leq \cdots$ and $\lim_{n\to\infty} f_n(w) = f(w)$. Then f is measurable and*

$$\int_A f \, d\nu = \lim_{n\to\infty} \int_A f_n \, d\nu \tag{B.5}$$

meaning that – under the assumptions of the theorem – the operations of limit and integral can be interchanged.

Three corollaries to Proposition B.10 are worthy of mention:

Corollary 1 Under the assumptions of Proposition B.9 on f, g and A we have

$$\int_A (f+g) \, d\nu = \int_A f \, d\nu + \int_A g \, d\nu \tag{B.6}$$

Corollary 2 (B. Levi) If f_n ($n = 1, 2, \ldots$) are non-negative, measurable functions and $A \in S$ then

$$\int_A \left(\sum_{n=1}^{\infty} f_n \right) d\nu = \sum_{n=1}^{\infty} \int_A f_n \, d\nu \tag{B.7}$$

Corollary 3 Let f be a measurable function and $A_n \in S$ such that $A_i \cap A_j = \emptyset$ for $i \neq j$. Then

$$\int_{\bigcup_n A_n} f\,\mathrm{d}\nu = \sum_n \int_{A_n} f\,\mathrm{d}\nu \tag{B.8}$$

Proposition B.11 *Let $f : W \to [0,\infty]$ be measurable. The set function defined by*

$$\varphi(A) \equiv \int_A f\,\mathrm{d}\nu \tag{B.9a}$$

for all $A \in S$ is a σ-additive (see Corollary 3) measure on the σ-algebra S. Moreover, for all measurable functions $g : W \to [0,\infty]$

$$\int_W g\,\mathrm{d}\varphi = \int_W fg\,\mathrm{d}\nu \tag{B.9b}$$

As remarks to Proposition B.11 we note that:

(i) Equation (B.9b) is often expressed by writing $\mathrm{d}\varphi = f\,\mathrm{d}\nu$, where this equation simply indicates that (B.9b) holds for all measurable functions $g \geq 0$;
(ii) the inverse of Proposition B.11 – which will be considered later – is an important result of mathematical analysis known as Radon–Nikodym theorem.

Proposition B.12 (Fatou's lemma) *Let $f_n : W \to [0,\infty]$ be measurable for all $n = 1, 2, \ldots$. Then, for all $A \in S$*

$$\int_A \left(\liminf_{n\to\infty} f_n\right) \mathrm{d}\nu \leq \liminf_{n\to\infty} \int_A f_n\,\mathrm{d}\nu \tag{B.10}$$

where, we recall from basic analysis, the limit inferior of a sequence of real numbers $\{x_n\}_{n=1}^{\infty}$ is defined as

$$\liminf_{n\to\infty} x_n = \sup_n(\inf_{k\geq n} x_k)$$

and clearly for a sequence of functions f_n we mean the function defined by $(\liminf_{n\to\infty} f_n)(x) = \liminf_{n\to\infty}(f_n(x))$.

At this point we can relax the restriction of non-negativity. Let f be a measurable real-valued function on W, its positive and negative part, denoted by f^+, f^- respectively, and defined by

$$f^+ \equiv \max(f, 0)$$
$$f^- \equiv -\min(f, 0)$$

are two measurable, non-negative real-valued functions. Clearly, $f = f^+ - f^-$ and $|f| = f^+ + f^-$. We define the integral of f as

$$\int_W f \, \mathrm{d}\nu \equiv \int_W f^+ \mathrm{d}\nu - \int_W f^- \mathrm{d}\nu \tag{B.11}$$

if at least one of the integrals on the r.h.s. is finite; if both integrals are finite then (B.11) is finite (one often writes $\int_W f \, \mathrm{d}\nu < \infty$ in this case) and we say that f is *L-integrable*, or summable, on W. Also, if f is L-integrable over W, then it is L-integrable over every $A \in S$.

Similarly, we have

$$\int_W |f| \mathrm{d}\nu = \int_W f^+ \mathrm{d}\nu + \int_W f^- \mathrm{d}\nu \tag{B.12}$$

and it is evident that f is L-integrable if and only if $|f|$ is L-integrable.

Now, since real-valued functions are special cases of complex-valued functions – that is, they are complex-valued functions whose imaginary part is zero – it is only a small step to turn to this more general setting and define the integral of a measurable function $f : W \to \mathbb{C}$ (where $\mathbb{C}$ is the set of complex numbers) as

$$\int_W f \, \mathrm{d}\nu \equiv \int_W (\operatorname{Re} f) \, \mathrm{d}\nu + i \int_W (\operatorname{Im} f) \, \mathrm{d}\nu \tag{B.13}$$

Clearly, f is measurable if and only if its real and imaginary part $\operatorname{Re} f$ and $\operatorname{Im} f$ – which, in turn, are real-valued functions – are measurable.

If $\int_W |f| d\nu < \infty$ (where $|f| = \sqrt{(\operatorname{Re} f)^2 + (\operatorname{Im} f)^2}$) then, as above, f is said to be L-integrable and this is also stated in mathematical terms by writing $f \in L^1(W, S, \nu)$ where, by definition, $L^1(W, S, \nu)$ is the set of all complex-valued, measurable functions such that $\int_W |f| \mathrm{d}\nu < \infty$ and one does not distinguish between functions that are a.e. equal (in fact, if $f = g$ a.e., their integrals are the same).

Proposition B.13 *If f, g are two L-integrable functions and $a, b \in \mathbb{C}$ then $af + bg$ is L-integrable and*

$$\int_W (af + bg) \, \mathrm{d}\nu = a \int_W f \, \mathrm{d}\nu + b \int_W g \, \mathrm{d}\nu \tag{B.14}$$

which, in other words, means that $L^1(W, S, \nu)$ is a linear space.

Corollary 4 If $f : W \to \mathbb{C}$ is L-integrable then

$$\left|\int_W f \, d\nu\right| \leq \int_W |f| \, d\nu \tag{B.15}$$

Another fundamental result of Lebesgue integration is the so-called Lebesgue dominated convergence theorem. We state the theorem in the setting of complex-valued functions but it is evident that it remains valid in the particular case of real-valued functions:

Proposition B.14 (Dominated convergence theorem) *Let $\{f_n\}_{n=1}^{\infty}$ be a sequence of measurable, complex-valued functions converging a.e. to a function f. If there exists a non-negative L-integrable function g such that $|f_n| \leq g$ a.e. for all $n = 1, 2, \ldots$ then $f \in L^1(W, S, \nu)$ and*

$$\int_W f \, d\nu = \lim_{n\to\infty} \int_W f_n \, d\nu \tag{B.16}$$

Corollary 5 Let $\{f_n\}_{n=1}^{\infty}$ be a sequence of measurable, complex-valued functions such that $\sum_{n=1}^{\infty} \int_W |f_n| d\nu < \infty$. Then the series $\sum_n f_n$ converges a.e. to a L-integrable function and

$$\int_W \sum_{n=1}^{\infty} f_n \, d\nu = \sum_{n=1}^{\infty} \int_W f_n \, d\nu \tag{B.17}$$

which means that, under the assumptions of the corollary, a series of measurable functions can be integrated term by term.

Another result worthy of notice is that the dominated convergence theorem remains true when a.e. convergence is replaced by convergence in measure, in fact, we have

Proposition B.15 *Let $\{f_n\}_{n=1}^{\infty}$ be a sequence of measurable, complex-valued functions converging in measure to a function f. If there exists a non-negative L-integrable function g such that $|f_n| \leq g$ a.e. for all $n = 1, 2, \ldots$ then $f \in L^1(W, S, \nu)$ and*

$$\int_W f \, d\nu = \lim_{n\to\infty} \int_W f_n \, d\nu \tag{B.18}$$

which, as in Propositions B.10 and B.14, means that the operations of limit and integral can be interchanged under reasonably mild conditions.

All the above results are important in probability theory because of the strict relation between the notion of abstract Lebesgue integral of a measurable function and the so-called expected value of a random variable. In fact – as it is shown in Chapter 2 and in the next section – if (W, S, P) is a probability space and X is a random variable on W, then the *expectation* (or mean, see Section 2.3.2) of X, denoted by $E(X)$ or $E[X]$, is given by

$$E(X) = \int_W X \, \mathrm{d}P \tag{B.19}$$

which is the abstract Lebesgue integral of X (over W) with respect to the probability measure P. So, for the main purpose of the book, we note that saying that 'f is a measurable, L-integrable function' translates in probability terminology into the sentence 'f is a random variable with finite expectation', or, in symbols, $f \in L^1(W, S, P)$.

As a final result in this section we may consider functions defined on the real line and note that, as a matter of fact, the Lebesgue integral is a generalization of the Riemann integral of basic calculus. This fact is formally stated in the following proposition:

Proposition B.16 *Suppose that f is a Riemann integrable function on an interval $[a, b] \subset R$. Then f is Lebesgue integrable – that is, integrable with respect to the Lebesgue measure μ – on $[a, b]$ and the two integrals are equal.*

The converse, however, is not true and there exist Lebesgue integrable functions that are not Riemann integrable.

B.5 Further results in integration and measure theory and their relation to probability

Let $F : \mathbb{R} \to \mathbb{R}$ be a finite, non-decreasing function with the property that it is right-continuous at every point (from Chapter 2 we know that distribution functions of random variables are functions of this type). If, on the collection of half-open intervals $(a, b] \subset \mathbb{R}$ we define the set function $m(a, b] = F(b) - F(a)$ it can be shown that this set function has the characteristics of a measure and can therefore be extended (Section 2.3) to a finite measure μ_F defined on the σ-algebra of Borel sets of the real line. This measure is called the *Lebesgue–Stieltjes measure* corresponding to F and it is finite because the assumptions on F (monotonicity and finiteness) imply

$$\mu_F(\mathbb{R}) = \lim_{x \to \infty} F(x) - \lim_{x \to -\infty} F(x)$$

and both limits on the r.h.s. exist and are finite. The function F, in turn, is called a generating function for μ_F. Note that we do not say 'the' generating function because we may add any real constant to F to obtain a new generating function which induces the same measure on the real line. If one of the two limits at infinity is assigned a specific value (as it is the case in probability) then the uniqueness of the generating function follows and there exists a one-to-one correspondence between Lebesgue–Stieltjes measures and generating functions.

The connection with probability is evident if one notes that a random variable X defined on a probability space (W, S, P) induces a probability measure P_X on the Borel sets by means of the relation $P_X(B) = P(X^{-1}(B))$ (Chapter 2, eq. (2.13)). The measure P_X – which is called in mathematical terms the *image measure* of P by X and often denoted by $P \circ X^{-1}$ – in turn defines a distribution function F_X through the relation $F_X(x) = P_X(-\infty, x]$; this function is finite, non-decreasing and right-continuous at every point. In addition, since P_X is a probability measure, F_X is subjected to the conditions $\lim_{n\to-\infty} F(x) = 0$ and $\lim_{n\to\infty} F(x) = 1$ and these requirements make it unique (in the sense above). In probability theory, therefore, P_X is the Lebesgue–Sieltjes measure corresponding to F_X and F_X, in turn, is called distribution function (of X) instead of generating function. So, while in analysis one generally starts from a function F with certain characteristics and then obtains the measure μ_F, probability theory proceeds, so to speak, backwards – that is, from the measure P_X to the function F_X – with the random variable X being the starting point of the chain.

In any case, once we have a Lebesgue–Stieltjes measure μ_F we can integrate with respect to this measure.

Definition B.8 Let μ_F be a Lebesgue–Stieltjes measure corresponding to the function F, f a Borel measurable function. Its integral with respect to μ_F is called Lebesgue–Stieltjes integral and is generally denoted by $\int f \, \mathrm{d}F$.

The connection with probability theory is immediate because one can consider integrals with respect to the measure P_X, which – according to the definition above – are written as integrals in $\mathrm{d}F_X$. In this regard, consider the following problem: let (W, S, P) be a probability space, X a random variable on W and $g : \mathbb{R} \to \mathbb{R}$ a Borel function. Then the composite function $Z : W \to R$ defined by $Z(w) \equiv g(X(w))$ is itself a random variable (i.e. measurable) and, by definition, we know that its expectation $E(Z)$ is given by the abstract Lebesgue integral

$$E(Z) = \int_W Z \, \mathrm{d}P$$

The question is, can we express $E(Z)$ in terms of the measure P_X, that is, as an integral on the real line? The answer is yes and we have the result

Proposition B.17 *Let X, g and Z as above, then*

$$\int_W Z\,\mathrm{d}P = \int_{\mathbb{R}} g(x)\,\mathrm{d}F_X \tag{B.20a}$$

in the sense that if either of the two sides exists, so does the other and they are equal (in other words $Z \in L^1(W, S, P)$ if and only if $g \in L^1(\mathbb{R}, \mathbb{B}(\mathbb{R}), P_X)$).

So, in particular, if $X(w) = x$ we get

$$E(X) \equiv \int_W X\,\mathrm{d}P = \int_{\mathbb{R}} x\,\mathrm{d}F_X \tag{B.20b}$$

Pushing the argument further, it is clear that eq. (B.20b) applies to any random variable and therefore, if Z is as above – that is, $Z(w) \equiv g(X(w))$ – then we have the possibility of calculating $E(Z)$ as

$$E(Z) = \int_{\mathbb{R}} z\,\mathrm{d}F_Z \tag{B.21}$$

so that, owing to (B.21) and (B.20a), we have

$$\int_R z\,\mathrm{d}F_Z = \int_{\mathbb{R}} g(x)\,\mathrm{d}F_X \tag{B.22}$$

Which integral to use to calculate $E(Z)$ is generally a matter of convenience.

Given a random variable X, the expectation (B.20b) is a special case of a number of quantities called *moments* of X. In fact, if $n = 1, 2, \ldots$ we define the nth moment of X as $E(X^n)$, that is,

$$E(X^n) \equiv \int_W X^n\,\mathrm{d}P = \int_{\mathbb{R}} x^n\,\mathrm{d}F_X \tag{B.23}$$

where the second equality is due to eq. (B.22). So, the expectation is just the first moment of X. More about moments is in Chapter 2; here we limit ourselves to the following proposition:

Proposition B.18 *If $n > 1$ and the nth moment of the random variable X is finite – that is, $E(X^n) < \infty$ – then $E(X^k) < \infty$ for $1 \le k \le n$. (Note that this result, for convenience, is also given in Chapter 2 as Proposition 2.12.)*

Let us now return to analysis and give some more definitions:

Definition B.9 Let (W, S) be a measurable space and ν_1, ν_2 two measures on the σ-algebra S. We say that ν_2 is *absolutely continuous* with respect to ν_1 – and write $\nu_2 \ll \nu_1$ – if $\nu_1(A) = 0$ implies $\nu_2(A) = 0$ or, in other words, $\nu_2(A) = 0$ whenever $\nu_1(A) = 0$. On the other hand the two measures are *mutually singular* – denoted $\nu_1 \perp \nu_2$ – if there is a set $A \in S$ such that $\nu_1(A) = 0$ and $\nu_2(A^C) = 0$.

Owing to Propositions B.11 and B.9 (property (d)) we note that the measure φ defined by eq. (B.9a) is such that $\varphi(A) = 0$ whenever $\nu(A) = 0$, so that $\varphi \ll \nu$. This means that $\varphi \ll \nu$ is a necessary condition in order for the 'integral representation' of eq. (B.9a) to be possible. An important result known as Radon–Nikodym theorem shows that (in σ-finite measure spaces) this condition is also sufficient.

Proposition B.19 (Radon–Nikodym theorem) *Let (W, S, ν) be a σ-finite measure space and φ a measure on S. If $\varphi \ll \nu$ then there exists a non-negative, measurable function $f : W \to [0, \infty]$ such that*

$$\varphi(A) = \int_A f \, \mathrm{d}\nu \tag{B.24}$$

for all $A \in S$. Moreover, f is unique in the sense that if $\varphi(A) = \int_A g \, \mathrm{d}\nu$ for all $A \in S$, then $f = g$ a.e. (i.e. $\nu\{w : f(w) \neq g(w)\} = 0$).

The function f of eq. (B.19) is called the Radon–Nikodym derivative of φ with respect to ν and one often finds the symbols $\mathrm{d}\varphi = f \, \mathrm{d}\nu$ or $f = \mathrm{d}\varphi / \mathrm{d}\nu$.

Another result is the so-called Lebesgue decomposition of a measure φ with respect to a given measure ν:

Proposition B.20 (Lebesgue decomposition theorem) *Let (W, S, ν) be a σ-finite measure space and φ a σ-finite measure on S. Then there exists a unique decomposition $\varphi = \varphi_1 + \varphi_2$ where φ_1, φ_2 are two measures such that $\varphi_1 \ll \nu$ and $\varphi_2 \perp n$.*

This result is very general and it is the first step leading to further decomposition of measures. Without entering into details – which can be found, for instance, in Chapter 6 of Ref. [7] – it is sufficient for our purposes to recall that (i) P_X is a finite measure defined on the Borel sets of $\mathbb{R}$ and (ii) the Lebesgue measure μ is the 'natural' measure in $\mathbb{R}$. In this light, an important result is a theorem stating that any finite Borel measure on $\mathbb{R}$ can be uniquely decomposed as the sum of three finite Borel measures $m_{\mathrm{ac}}, m_{\mathrm{sc}}$ and m_{d}, where m_{ac} is absolutely continuous with respect to μ, m_{sc} is continuous and singular with respect to μ and m_{d} is discrete. This result is given as

Proposition 2.11 in Chapter 2 where it was also noted that this decomposition reflects a decomposition of a general PDF into the sum of three parts (eq. (2.28)). There, however, although we introduced the notions of continuous and discrete PDFs in Section 2.3.1, we did not specify the meaning of the terms 'absolutely continuous' and 'singular' functions. We give here the appropriate definitions:

(i) F is called a singular PDF if $F' = 0$ a.e. (with respect to the Lebesgue measure μ);
(ii) F is an absolutely continuous PDF if F' exists a.e. and it is Lebesgue integrable on $\mathbb{R}$, and

$$F(x) = \int_{-\infty}^{x} F'(t)\,dt \tag{B.25}$$

for $-\infty < x < \infty$. Clearly, in this case F' is the pdf associated to F.

By virtue of these definitions (plus the definition of discrete and continuous PDFs), the decomposition (2.28) is the PDF counterpart of the decomposition (2.26) in the light of the fact that a finite Borel measure (on $\mathbb{R}$) is singular if and only if its distribution function is singular, discrete if and only if its distribution function is a jump (discrete) function, etc.

In regard to the concept of absolute continuity – which, it should be noted, is not limited to PDFs but applies to real-valued functions in general – some important considerations are in order. If, as a particular case of the definition above, a function f is defined on an interval $[a, b]$ and its derivative f' exists a.e. and is Lebesgue integrable on $[a, b]$, then we say that f is absolutely continuous on $[a, b]$ if

$$f(x) = f(a) + \int_{a}^{x} f'(t)\,dt \tag{B.26}$$

for $a \le x \le b$. Equation (B.26), in turn, brings back to mind the second fundamental theorem of calculus – and rightly so, because the class of functions for which this theorem holds is precisely the class of absolutely continuous functions. In most books of analysis, in fact (see, for instance, [2, 5] or [7]), one generally finds the following definition of absolute continuity:

Definition B.10 A function f defined on $[a, b]$ is absolutely continuous on $[a, b]$ if for each $\varepsilon > 0$ there is a $\delta > 0$ such that for any finite sequence (a_i, b_i) of disjoint subintervals of $[a, b]$ with $\sum_i (b_i - a_i) < \delta$, then $\sum_i |f(b_i) - f(a_i)| < \varepsilon$. (Absolute continuity on the entire real line $\mathbb{R}$ in terms of this definition is a bit more involved. However, a necessary and sufficient condition is that f is absolutely continuous on every finite closed interval, is of bounded variation and $\lim_{x \to -\infty} f(x) = 0$. For more details the reader is referred to Refs [5] or [7].)

Then, starting from this definition, it is then shown that eq. (B.26) or, equivalently, the relation

$$\int_a^x f'(t)\,\mathrm{d}t = f(x) - f(a)$$

holds for all $x \in [a, b]$ if and only if f is absolutely continuous on $[a, b]$.

The above considerations suffice for our purposes but one final key result in Lebesgue integration deserves to be mentioned for its importance in both theory and applications. This is Fubini's theorem which, under rather mild conditions, allows to calculate multiple integrals as iterated integrals. In the general setting of complex-valued measurable functions, the theorem can be stated as follows:

Proposition B.21 *Let (W, S, ν) and (U, T, τ) be σ-finite measure spaces. Let f be a complex-values $S \times T$-measurable function on $W \times U$ such that at least one of the quantities*

(i) $\int_{W \times U} |f(x, y)|\,\mathrm{d}(\nu \times \tau)$
(ii) $\int_W \left\{ \int_U |f(x, y)|\,\mathrm{d}\tau \right\} \mathrm{d}\nu$
(iii) $\int_U \left\{ \int_W |f(x, y)|\,\mathrm{d}\nu \right\} \mathrm{d}\tau$

is finite. Then

$$\begin{aligned} \int_{W \times U} |f(x, y)|\,\mathrm{d}(\nu x \tau) &= \int_W \left\{ \int_U |f(x, y)|\,\mathrm{d}\tau \right\} \mathrm{d}\nu \\ &= \int_U \left\{ \int_W |f(x, y)|\,\mathrm{d}\nu \right\} \mathrm{d}\tau \end{aligned} \tag{B.27}$$

Clearly, in order to arrive at Proposition B.21, a number of preliminary results are needed. In fact, one must, for instance, introduce the concept of product measure space, show that we can appropriately define a measure on a σ-algebra of subsets of this space in terms of the individual measures of the one-dimensional spaces entering the product etc. All this is beyond our scope although, in general, it should be noted that – besides some minor complications due to the higher dimensionality – the main ideas are direct generalizations of the one-dimensional case. The details, however, are worthy of study in their own right and the interested reader can refer, for instance, to [2, 3, 5, 6, 7] or [9].

References and further reading

[1] Cramer, H., '*Mathematical Methods of Statistics*', Princeton Landmarks in Mathematics, Princeton University Press, 18th printing (1991).
[2] Friedman, A., '*Foundations of Modern Analysis*', Dover Publications, New York (1982).
[3] Haaser, N.B., Sullivan, J.A., '*Real Analysis*', Dover Publications, New York (1991).
[4] Kirillov, A.A., Gvishiani, A.D., '*Theorems and Problems in Functional Analysis*', Springer-Verlag, New York (1982).
[5] Kolmogorov, A.N., Fomin, S.V., '*Introductory Real Analysis*', Dover Publications, New York (1970).
[6] Kolmogorov, A.N., Fomin, S.V., '*Elements of the Theory of Functions and Functional Analysis*', Dover Publ., New York (1999).
[7] McDonald, J.N., Weiss, N.A., '*A Course in Real Analysis*', Academic Press, San Diego (1999).
[8] Rudin, W., '*Principles of Mathematical Analysis*', 3rd edn., McGraw-Hill, New York (1976).
[9] Rudin, W., '*Real and Complex Analysis*', McGraw-Hill, New York (1966).
[10] Wilcox, H.J., Myers, D.L., '*An Introduction to Lebesgue Integration and Fourier Series*', Dover Publications, New York (1978).

Appendix C

As a complement to the main text, we give in this appendix a number of miscellaneous definitions, results and remarks. Without claim of completeness, we consider only a few selected topics which may be helpful in reading this and/or other texts on probability and statistics. Theorems, in general, will not be proven although we will make some occasional comments on some of the proofs.

In regard to probability distributions we use the following notation: (a) we write $X \approx PL(\cdot)$ to mean that the r.v. X is distributed according to the probability law $PL(\cdot)$, where we specify within parenthesis any parameter(s) pertinent to PL. So, for instance, $X \approx N(0,1)$ means that X is standard normal (zero mean and unit variance) and $X \approx \chi^2(\nu)$ means that X is distributed according to a Chi-square law (see below) with ν degrees of freedom.

The expression $X \approx \text{As} - PL(\cdot)$, on the other hand, indicates that X is asymptotically distributed according to the probability law $PL(\cdot)$.

C.1 The Gamma Function $\Gamma(x)$

The definition is

$$\Gamma(x) = \int_0^\infty t^{x-1} e^{-t}\, dt \quad (x > 0) \tag{C.1}$$

The function is continuous and has continuous derivatives of all orders given by

$$\Gamma^{(m)}(x) = \int_0^\infty t^{x-1} (\ln t)^m e^{-t}\, dt \quad (x > 0) \tag{C.2}$$

and $\Gamma(x) \to \infty$ as $x \to 0$ or $x \to \infty$. It has a minimum at $x_0 \cong 1.4616$ and $\Gamma(x_0) = 0.8856$. Its main properties are as follows: for any $x > 0$

$$\Gamma(x+1) = x\Gamma(x) \tag{C.3}$$

which can be obtained integrating (C.1) by parts. If x is equal to a positive integer n we get from (C.3)

$$\Gamma(n+1) = n! \tag{C.4}$$

since $\Gamma(1) = 1$. Also, $\Gamma(2) = \Gamma(1) = 1$, $\Gamma(1/2) = \sqrt{\pi}$ and the well-known Stirling's formula

$$n! \approx n^n \mathrm{e}^{-n}\sqrt{2\pi n} \tag{C.5}$$

is just a special case of the approximation

$$\Gamma(x) \cong \sqrt{2\pi}\, x^{x-1/2}\mathrm{e}^{-x} \tag{C.6}$$

which is valid for large values of x. Moreover, for any $x > 0$ it can be shown that

$$\Gamma(x) = \lim_{n\to\infty} \frac{n!n^x}{x(x+1)\cdots(x+n)} \tag{C.7}$$

C.2 Gamma distribution

$X \approx \Gamma(a, b)$ if its pdf is

$$f(x; a, b) = \Gamma(a, b) = \frac{x^{b-1}}{a^b\Gamma(b)}\mathrm{e}^{-x/a} \tag{C.8}$$

for $x > 0$ (and zero otherwise). The quantities a, b are two (positive) parameters. The CF of X is

$$\varphi(u) = \frac{1}{(1 - iau)^b} \tag{C.9}$$

from which it is easy to determine $\mu \equiv E(X) = ab$, $\alpha_2 \equiv E(X^2) = a^2b(1+b)$ and $\mathrm{Var}(X) = a^2b$. Also, we have the recursion formulas for ordinary and central moments

$$\begin{aligned} \alpha_k &= a(k-1+b)\alpha_{k-1} \\ \mu_{k+1} &= ak(\mu_k + ab\,\mu_{k-1}) \end{aligned} \tag{C.10}$$

respectively. An important property of the gamma distribution is easily obtained from the CF (C.9): the sum of two independent gamma r.v.s with parameters a, b_1 and a, b_2 is a gamma variable with parameters $a, b_1 + b_2$.

C.3 The χ^2 distribution

A r.v. distributed according to a χ^2 distribution with ν degrees of freedom (where ν is a positive integer) has a pdf

$$f(x) = \frac{x^{(\nu/2)-1}}{2^{\nu/2}\Gamma(\nu/2)} e^{-x/2} \tag{C.11}$$

for $x > 0$ (and zero otherwise). For $\nu \le 2$ the function is monotonically decreasing while for $\nu > 2$ it has a maximum at $x = \nu - 2$.

The CF of a χ^2 random variable is

$$\varphi(u) = \frac{1}{(1-2iu)^{\nu/2}} \tag{C.12}$$

and taking its derivatives at $u = 0$ we get immediately $E(X) = \nu$ and $\alpha_2 \equiv E(X^2) = \nu(\nu+2)$. More generally, the kth order (ordinary) moment can be written in the form of product of k terms as

$$\alpha_k \equiv E(X^k) = \nu(\nu+2)\cdots(n+2k-2) \tag{C.13}$$

and therefore $\alpha_3 = \nu(\nu+2)(\nu+4)$, $\alpha_4 = \nu(\nu+2)(\nu+4)(\nu+6)$, etc. For the central moments we have $\mu_2 \equiv \mathrm{Var}(X) = 2\nu$, $\mu_3 = 8\nu$ and $\mu_4 = 12\nu(4+\nu)$ so that, consequently, the coefficients of skewness and kurtosis are $\kappa_3 \equiv \mu_3/\mu_2^{3/2} = 2\sqrt{2/\nu}$ and $\kappa_4 \equiv \mu_4/\mu_2^2 = 3 + 12/\nu$, respectively.

Note that $\Gamma(2, \nu/2) = \chi^2(\nu)$ and also $\chi^2(2) = \exp(2)$ where the exponential distribution is defined below (eq. (C.33)).

The χ^2 variables are strictly related to standard normal r.v.s. In fact

Proposition C.1 *If* $X \approx N(0,1)$ *then* $X^2 \approx \chi^2(1)$.

This theorem can be proven by using the considerations of Section 2.5.3 where, right before Example 2.11, the reader was invited to show that if f_X is the pdf of X and $Y = X^2$ then (for $y > 0$)

$$f_Y(y) = \frac{f_X(-\sqrt{y}) + f_X(\sqrt{y})}{2\sqrt{y}}$$

In the special case in which f_X is even then $f_Y(y) = y^{-1/2} f_X(\sqrt{y})$; if, moreover, f_X is the standard normal pdf we get $f_Y(y) = (2\pi y)^{-1/2}\exp(-y/2)$ which, owing to $\Gamma(1/2) = \sqrt{\pi}$, is precisely the distribution $\chi^2(1)$.

Also, an important 'reproducibility' property is given by the following result

Proposition C.2 *If $X_1, X_2, \ldots, X_n$ are independent and χ^2-distributed with $\nu_1, \nu_2, \ldots, \nu_n$ degrees of freedom, respectively, then*

$$\sum_i X_i \approx \chi^2(\nu_1 + \nu_2 + \cdots + \nu_n).$$

The proof is immediate if one notes that independence implies that the CF of the r.v. $\sum_i X_i$ is the product of the individual CFs of the X_i (see eq. (3.29) and Example 3.3).

From Propositions C.1 and C.2 it follows

Proposition C.3 *If $X_1, X_2, \ldots, X_n$ are independent standard normal r.v.s, then $\sum_i X_i^2 \approx \chi^2(n)$.*

Another result connecting normal and χ^2 r.v.s is

Proposition C.4 *Let $\mathbf{X} = (X_1, X_2, \ldots, X_n)$ be a sample of size n from $N(\mu, \sigma^2)$ and let $M, \bar{S}^2$ be the sample mean and the unbiased sample variance (see eq. (5.29)). Then*

(i) *M and $\bar{S}^2$ are independent;*
(ii) $\sqrt{n}(M - \mu)/\sigma \approx N(0, 1)$;
(iii) $(n - 1)\bar{S}^2/\sigma^2 \approx \chi^2(n - 1)$.

As an immediate remark, note that by choosing the (biased) estimator S^2 for the variance, result (iii) can be equivalently stated as $nS^2/\sigma^2 \approx \chi^2(n - 1)$.

Finally, we consider an asymptotic result

Proposition C.5 *Let $X \approx \chi^2(\nu)$. Then, as $n \to \infty$ the random variables $(X - \nu)/\sqrt{2\nu}$ and $\sqrt{2X} - \sqrt{2\nu}$ tend in distribution to a standard normal r.v.*

C.4 Student's distribution

First introduced in 1908 in a paper by Gosset writing under the pen-name of 'Student', it is also known as the t-distribution. A r.v. X is distributed according to Student's distribution with ν degrees of freedom – and we write $X \approx St(\nu)$ – if its pdf is

$$f(x) = \frac{\Gamma((\nu + 1)/2)}{\sqrt{\pi\nu}\,\Gamma(\nu/2)} \left(1 + \frac{x^2}{\nu}\right)^{-(\nu+1)/2} \tag{C.14}$$

where ν is a positive integer and $-\infty < x < \infty$. The mean $E(X)$ is finite for $\nu \geq 2$ and the variance is finite for $\nu \geq 3$ and we have

$$E(X) = 0$$
$$\operatorname{Var}(X) = \frac{\nu}{\nu - 2} \tag{C.15}$$

where the first result is obvious because the distribution is symmetric about $x = 0$. Consequently, all existing moments of odd order are zero; note that we speak of 'existing moments' because the k-th order moment is finite for $k < \nu$. Provided that $2r < \nu$, setting $k = 2r$ gives

$$\mu_{2r} = \alpha_{2r} = \frac{1 \cdot 3 \cdots (2r-1)\nu^r}{(\nu-2)(\nu-4)\cdots(\nu-2r)} \tag{C.16}$$

so that (for $\nu \geq 5$) the coefficient of kurtosis is $\kappa_4 = 3 + 6/(\nu - 4)$.

Student's distribution is also related to the standard normal distribution; in fact

Proposition C.6 *As $\nu \to \infty$, Student's distribution tends to $N(0, 1)$.*

As a remark of practical nature, it turns out that the approximation to $N(0, 1)$ can be considered quite satisfactory for $\nu > 30$; this is why tabulated values of $\text{St}(\nu)$ are generally given only for $\nu \leq 30$ or $\nu \leq 40$.

Proposition C.7 *Let X, Y be two independent r.v.s such that $X \approx N(0, 1)$ and $Y \approx \chi^2(n)$. Then the r.v. $T \equiv X/\sqrt{Y/\nu}$ has a Student distribution with $\nu - 1$ degrees of freedom (i.e. $T \approx \text{St}(\nu - 1)$).*

A typical case involving the Student distribution follows directly from Propositions C.4 and C.7:

Proposition C.8 *Let $\mathbf{X} = (X_1, X_2, \ldots, X_n)$ be a sample from $N(\mu, \sigma^2)$. Then $\sqrt{n}(M - \mu)/\bar{S} \approx St(n-1)$ or, equivalently, $\sqrt{n-1}(M - \mu)/S \approx \text{St}(n-1)$.*

With two independent samples, on the other hand, we have:

Proposition C.9 *Let $\mathbf{X} = (X_1, \ldots, X_n)$ and $\mathbf{Y} = (Y_1, \ldots, Y_m)$ be two independent samples from the distribution $N(\mu, \sigma^2)$. Let M_1, S_1^2 be the sampling mean and variance of the first sample and M_2, S_2^2 the sampling mean and variance of the second sample. Then*

$$\sqrt{\frac{mn(m+n-2)}{m+n}}\,\frac{M_1 - M_2}{\sqrt{nS_1^2 + mS_2^2}} \approx \text{St}(m+n-2) \tag{C.17}$$

C.5 Fisher's distribution

This distribution – sometimes also called Snedecor's distribution – is characterized by two (integer) parameters ν_1, ν_2 and we denote it by the symbol $\text{Fsh}(\nu_1, \nu_2)$. A continuous r.v. X is such that $X \approx \text{Fsh}(\nu_1, \nu_2)$ when its pdf is

$$f(x) = \frac{\Gamma((\nu_1 + \nu_2)/2)}{\Gamma(\nu_1/2)\Gamma(\nu_2/2)} \left(\frac{\nu_1}{\nu_2}\right)^{\nu_1/2} \frac{x^{(\nu_1/2)-1}}{(1 + \nu_1 x/\nu_2)^{(\nu_1+\nu_2)/2}} \tag{C.18}$$

for $x > 0$ and zero otherwise. The two parameters ν_1, ν_2 are generally referred to as the degrees of freedom of the Fisher distribution.

The mean and variance are defined for $\nu_2 > 2$ and $\nu_2 > 4$, respectively, and they are

$$\begin{aligned} E(X) &= \frac{\nu_2}{\nu_2 - 2} \\ \text{Var}(X) &= \frac{2\nu_2^2(\nu_1 + \nu_2 - 2)}{\nu_1(\nu_2 - 2)^2(\nu_2 - 4)} \end{aligned} \tag{C.19}$$

while the ordinary k-th order moment, defined for $2k < \nu_2$, is

$$\alpha_k = E(X^k) = \left(\frac{\nu_2}{\nu_1}\right)^k \frac{\Gamma(\nu_1/2 + k)\Gamma(\nu_2/2 - k)}{\Gamma(\nu_1/2)\Gamma(\nu_2/2)} \tag{C.20a}$$

so, that, in particular, owing to (C.4) we get the first of eqs (C.19) and

$$\alpha_2 = \frac{\nu_2^2(\nu_1 + 2)}{\nu_1(\nu_2 - 2)(\nu_2 - 4)} \tag{C.20b}$$

for $\nu_2 > 4$. The mode (i.e. the x-value where $f(x)$ has a maximum) of Fisher's distribution is at the point $x = [\nu_2(\nu_1 - 2)]/[\nu_1(\nu_2 + 2)]$ which is always less than 1 (in this regard note that the mean $E(X)$, when it exists, is always greater than 1).

A connection between Fisher's r.v.s and χ^2 variables is given by the following theorem:

Proposition C.10 *Let X_1, X_2 be two independent, χ^2-distributed r.v.s with ν_1 and ν_2 degrees of freedom, respectively. Then*

$$\frac{X_1/\nu_1}{X_2/\nu_2} \approx \text{Fsh}(\nu_1, \nu_2)$$

In statistics, moreover, a typical situation involving Fisher's distribution is expressed by the following result:

Proposition C.11 *Let* $\mathbf{X} = (X_1, \ldots, X_n)$ *and* $\mathbf{Y} = (Y_1, \ldots, Y_m)$ *be independent samples from the distributions* $N(\mu_1, \sigma_1^2)$ *and* $N(\mu_2, \sigma_2^2)$*, respectively and let* S_1^2, S_2^2 *be the corresponding sample variances. Then for any* μ_1, μ_2

$$\frac{n(m-1)\sigma_2^2 S_1^2}{m(n-1)\sigma_1^2 S_2^2} \approx \text{Fsh}(n-1, m-1) \tag{C.21a}$$

If, instead of the estimators S_1^2, S_2^2 *we consider the unbiased estimators* $\bar{S}_1^2, \bar{S}_2^2$*, eq. (C.21) turns into the equivalent result*

$$\frac{\bar{S}_1^2/\sigma_1^2}{\bar{S}_2^2/\sigma_2^2} \approx \text{Fsh}(n-1, m-1) \tag{C.21b}$$

Noteworthy properties of Fisher's distribution are given in Propositions C.12, C.13 and C.14.

Proposition C.12 *If* $X \approx \text{Fsh}(\nu_1, \nu_2)$ *then* $1/X \approx \text{Fsh}(\nu_2, \nu_1)$.

This result is, in essence, a consequence of Proposition C.10. In fact, if $X \approx \text{Fsh}(\nu_1, \nu_2)$ then there exist two independent, χ^2-distributed r.v.s X_1, X_2 with ν_1 and ν_2 degrees of freedom, respectively, such that $X = \nu_2 X_1/\nu_1 X_2$. Consequently, $X^{-1} = \nu_1 X_2/\nu_2 X_1$ and the theorem follows.

As a corollary, let $F_{1-\alpha;\,\nu_1,\nu_2}$ denote the (upper or lower) $1-\alpha$ quantile of $\text{Fsh}(\nu_1, \nu_2)$; then

$$F_{1-\alpha;\,\nu_1,\nu_2} = \frac{1}{F_{\alpha;\,\nu_2,\nu_1}} \tag{C.22}$$

where, clearly, $F_{\alpha;\,\nu_2,\nu_1}$ is the (upper or lower, in agreement with above) α quantile of $\text{Fsh}(\nu_2, \nu_1)$. This property is used, for instance, in Chapter 5, Example 5.11(b).

Proposition C.13 *Let* $X \approx \text{Fsh}(\nu_1, \nu_2)$*. Then* $\nu_1 X \to \chi^2(\nu_1)[D]$ *as* $\nu_2 \to \infty$ *(the meaning of limit in distribution, that is, D-limit, is explained in Chapter 2, Section 2.4, and Chapter 4, Section 4.2).*

Proposition C.14 *Let* $X \approx \text{St}(\nu)$*; then* $X^2 \approx \text{Fsh}(1, \nu)$.

As in the proof of Proposition C.1, by setting $Y = X^2$ we have $f_Y(y) = y^{-1/2} f_X(\sqrt{y})$ for $y > 0$ (and $f_Y(y) = 0$ otherwise) because Student's pdf is even. Substituting eq. (C.14) into this relation and noting that $\Gamma(1/2) = \sqrt{\pi}$, we immediately get the pdf of the distribution Fsh$(1, \nu)$.

Also, this last proposition implies a connection between Student's and Fisher's quantiles. The relation between lower quantiles is $F_{\gamma;1,\nu} = t^2_{(1+\gamma)/2;\nu}$ and has been used in Chapter 7 to obtain eq. (7.41b) from eq. (7.41). In fact, by temporarily omitting, for brevity, the degrees of freedom in the symbols of quantiles, we have

$$\gamma = P(X^2 \le F_\gamma) = P\left(-\sqrt{F_\gamma} < X \le \sqrt{F_\gamma}\right) = \int_{-\sqrt{F_\gamma}}^{\sqrt{F_\gamma}} S(x)\,dx \qquad \text{(C.23)}$$

where we denoted by $S(x)$ the pdf of St(ν). Owing to the symmetry (about $x = 0$) of $S(x)$, eq. (C.23) implies that the area to the right of $\sqrt{F_\gamma}$ is $(1-\gamma)/2$ which, in other words, means that $\sqrt{F_\gamma}$ is the upper $(1 - \gamma)/2$ quantile of St (ν) or, equivalently, the lower $1 - (1 - \gamma)/2 = (1 + \gamma)/2$ quantile. The consequence, as we set out to prove, is that $F_{\gamma;1,\nu} = t^2_{(1+\gamma)/2;\nu}$.

C.6 Some other probability distributions

The distributions considered in the preceding sections of this appendix are frequently used in statistical applications. In addition, some other fundamental probability distributions have been introduced and discussed in the main text and among these, just to name a few, we recall the uniform distribution, the binomial, the multinomial, Poisson's and the Gaussian (normal) distribution. However, since it is reasonable to expect that there exist other important distributions, we mention here a few more among the most common.

Considering a sequence of Bernoulli trials (see Example 2.8(a)), let us focus our attention on the number of failures before the first success. Recalling that the probability of success in each trial is usually denoted by $p(0 < p < 1)$ while $q = 1-p$ is the probability of failure, the relevant (discrete) distribution in this case is the so-called *geometric distribution* whose pmf is given by

$$p_X(x) = p(1-p)^x = pq^x \qquad \text{(C.24)}$$

where $x = 0, 1, 2, \ldots$ represents the number of failures before the first success (in other words, $p_X(x)$ is the probability of obtaining the first success at the $(x + 1)$th trial). It is left to the reader to determine that

(i) $\varphi(u) = p/(1 - qe^{iu})$ is the CF corresponding to (C.24);
(ii) the mean and variance of a geometric r.v. are $E(X) = q/p$ and $\mathrm{Var}(X) = q/p^2$, respectively.

Note that some authors call geometric the distribution with pmf $p_X(x) = pq^{x-1}$, where $x = 1, 2, \ldots$ is the number of trials at which the first success occurs. With this definition $p_X(x)$ is the probability of obtaining the first success at the xth trial after $x - 1$ failures.

An important property of the geometric distribution which distinguishes it from other discrete distributions is its 'lack of memory', mathematically expressed by the relation

$$P\{X = n + k | X \geq n\} = P\{X = k\} \tag{C.25}$$

which, in words, means that the fact of having observed n failures has no influence whatsoever on the number of (future) trials that we still have to perform before obtaining the first success. In terms of waiting time before the first success, the 'memoryless property' (C.25) states that the probability of having to perform other k trials – resulting in failures – given the fact that we have already observed n failures is the same as the initial probability that we had of observing k failures in the first k trials. At first sight, this property may appear counterintuitive because it seems that a long waiting time without success should somehow reduce the remaining time left before success (incidentally, the habit of playing 'late numbers' in a lottery is based on this mistaken belief). If, however, one considers the fact that the various Bernoulli trials are independent, the property is not surprising at all.

On the other hand, a r.v. with pmf

$$p_X(x) = \binom{x + n - 1}{n - 1} p^n (1 - p)^x = \binom{x + n - 1}{x} p^n q^x \tag{C.26a}$$

where $x = 0, 1, 2, \ldots$ is called *Pascal's* (or *negative binomial*) r.v. In a scheme of Bernoulli trials the r.v. X represents the number of failures before obtaining the nth success. In other words $X = x$ when we obtain a success at the $(x + n)$th trial and we have observed $n - 1$ successes – in whatever order – in the preceding $(x + n - 1)$ trials. Clearly, (C.24) is the special case $n = 1$ of (C.26). The CF of a Pascal r.v. is

$$\varphi(u) = \left(\frac{p}{1 - (1 - p)e^{iu}} \right)^n = \left(\frac{p}{1 - qe^{iu}} \right)^n \tag{C.27}$$

and its mean and variance are, respectively, $E(X) = nq/p$ and $\mathrm{Var}(X) = nq/p^2$. Moreover, by virtue of eq. (3.29) it is immediate to see that the sum of two independent Pascal r.v.s with the same parameter p and indexes n_1 and n_2, respectively, is a Pascal r.v. with parameter p and index $n_1 + n_2$.

As a word of caution, it should be noted that some authors call Pascal's (or negative binomial) a discrete r.v. Y with pmf

$$p_Y(y) = \binom{y-1}{n-1} p^n (1-p)^{y-n}, \quad y = n, n+1, \ldots \tag{C.26b}$$

where Y represents the total number of trials at which we achieve the nth success. Stated differently, $Y = y$ if we obtain the nth success at the yth trial after having observed $(n-1)$ successes in the preceding $(y-1)$ trials. According to this definition it is clear that (i) $y \geq n$ because we need at least n trials to have n successes and (ii) being $X = Y - n$ the number of failures before the nth success, the pmf of X is given by (C.26a).

In the course of the main text we often considered sampling problems. The typical situation is an urn containing N objects, m with a given desired property and $N - m$ without this property. If, without replacement, we draw at random n objects ($n \leq N$), the probability $p_X(x)$ of having x objects with the desired property is given by the *hypergeometric* distribution

$$p_X(x) = \frac{\binom{m}{x}\binom{N-m}{n-x}}{\binom{N}{n}}, \quad x = 0, 1, \ldots, k \tag{C.28}$$

The pmf of (C.28) is easily justified if one considers that the numerator is the number of ways in which we can get – in n draws – x successes (object with the property) and $n - x$ failures (object without the property) while the denominator is simply the total number of equally likely samples of size n. Starting from (C.28), some rather cumbersome calculations lead to

$$\begin{aligned} E(X) &= n\frac{m}{N} \\ \mathrm{Var}(X) &= \frac{nm(N-m)(N-k)}{N^2(N-1)} \end{aligned} \tag{C.29}$$

Moreover, by indefinitely increasing both m and N in such a way that the ratio $p \equiv m/N$ remains constant, we have $m = Np$ and $N - m = N(1-p)$ and it is not difficult to show that

$$\lim_{N \to \infty} p_X(x) = \binom{n}{x} p^x (1-p)^{n-x} \tag{C.30}$$

that is, the hypergeometric distribution tends to a binomial distribution. In the light of the fact that the binomial distribution can be associated to a sampling scheme with replacement, eq. (C.30) is not surprising. In fact, if the number of objects in the urn is very high the difference between sampling with replacement and sampling without replacement is negligible. In the limit of $N \to \infty$, clearly, there is no difference at all.

Turning to continuous distributions, let X be a normal r.v. with mean m and variance σ^2. If we let $Y = e^X$, Y is a so-called *lognormal* r.v. and its pdf is

$$f_Y(y) = \frac{1}{y\sigma\sqrt{2\pi}} \exp\left(-\frac{(\ln y - \mu)^2}{2\sigma^2}\right) \tag{C.31}$$

for $y > 0$ and zero otherwise. The kth order ordinary moment is

$$E(Y^k) = \exp\left(k\mu + \frac{k^2\sigma^2}{2}\right) \tag{C.32a}$$

from which we get the mean and variance as

$$\begin{aligned} E(Y) &= \exp\left(\mu + \frac{\sigma^2}{2}\right) \\ \mathrm{Var}(Y) &= \exp(2\mu + 2\sigma^2) - \exp(2\mu + \sigma^2) = e^{2\mu+\sigma^2}(e^{\sigma^2} - 1) \end{aligned} \tag{C.32b}$$

Going back to the Gamma distribution (C.8), the particular case $b = 1$ deserves special attention because the pdf

$$f(x) = \frac{1}{a}\exp(-x/a), \quad x > 0 \tag{C.33}$$

($f(x) = 0$ otherwise) is frequently encountered in applications. A r.v. X with pdf (C.33) is called *exponential* of parameter a and in this case one often writes $X \approx \mathrm{Exp}(a)$. In the light of the results of Section C.2 we have the CF $\varphi(u) = (1 - iau)^{-1}$ and the moments $E(X^k) = k!a^k$ $(k = 1, 2, \ldots)$. In particular

$$\begin{aligned} E(X) &= a \\ \mathrm{Var}(X) &= a^2 \end{aligned} \tag{C.34}$$

and also $\mu_3 = 2a^3$, $\mu_4 = 9a^4$, etc.

Noting that the exponential PDF (for $x > 0$) is $F(x) = 1 - \exp(-x/a)$ and therefore $P\{X > x\} = 1 - F(x) = \exp(-x/a)$, the equalities

$$P\{X > t + s | X > s\} = \frac{P\{X > t + s\}}{P\{X > s\}} = \mathrm{e}^{-t/a} = P\{X > t\}$$

$(s, t > 0)$ show that the exponential distribution – like the geometric – is memoryless. This, in words, can be expressed as follows: if, in order for an event E to occur, I have already waited for a time s, the probability of having to wait for a further time t is the same as if I started to wait right now.

As mentioned above, the distribution (C.33) is often used in applications, the typical case being to model events that occur in sequence with random (and independent) 'arrival times'. These arrival times, in fact – for instance, in the disintegration of radioactive particles, the sequence of seismic phenomena, etc. – are adequately represented by means of an exponential model. The consequence is, as it is generally observed in practice, that close events are more likely than distant events (the adjectives 'close' and 'distant' refer here to the distance in time on an appropriate scale) and we have clusters of events separated by long waiting times. As a matter of fact, some rare events like serious accidents or natural disasters do fit into this description and therefore it is quite likely that two or more such events may be close together. This is a law of nature and there is no need – as it is often heard – to invoke mysterious correlations between calamities.

A final remark on the exponential distribution: we noted above that $\Gamma(a, 1) = \exp(a)$; since, however, $\Gamma(2, \nu/2) = \chi^2(\nu)$ it follows that $\chi^2(2) = \exp(2)$. This fact had already been pointed out in Section C.3.

In Section 4.6.1 we introduced the *Cauchy* distribution which, in the form of eq. (4.35), is a special case of the Student distribution (C.14) when $\nu = 1$. A more general form of Cauchy pdf is

$$f(x) = \frac{b}{\pi[b^2 + (x - a)^2]} \tag{C.35}$$

for $-\infty < x < \infty$ and $b > 0$. The parameters a, b are called the 'location' parameter and the 'scale' parameter, respectively. Since the CF of (C.35) is $\varphi(u) = \exp(iua - b|u|)$ and this function is not differentiable in $u = 0$, the Cauchy distribution has no finite moments of any order. Finally, in regard to this distribution, a noteworthy result (whose proof can be found at the end of Section 3.5) is as follows:

Proposition C.15 *Let X, Y be two independent r.v.s such that $X \approx N(0, 1)$ and $Y \approx N(0, 1)$. Then, the r.v. X/Y is distributed according to a Cauchy distribution with location parameter $a = 0$ and scale parameter $b = 1$.*

The last continuous distribution that we consider in this appendix is the so-called *beta* distribution. Before this, however, we must introduce the beta function $B(p, q)$ defined as

$$B(p, q) = \int_0^1 x^{p-1}(1-x)^{q-1}\,dx \tag{C.36}$$

where p, q are two positive real constants (which have nothing to do with the probabilities appearing in Bernoulli trials). A first property is $B(p, q) = B(q, p)$, which can easily be obtained by performing the change of variable $y = 1 - x$. A second property is the relation with the the gamma function; in fact it can be shown that

$$B(p, q) = \frac{\Gamma(p)\Gamma(q)}{\Gamma(p+q)} \tag{C.37}$$

A beta r.v. with parameters p and q is a variable with the pdf

$$f(x) = \frac{1}{B(p, q)} x^{p-1}(1-x)^{q-1} \tag{C.38}$$

for $0 < x < 1$ and zero otherwise. The kth order moment of this distribution is

$$\begin{aligned} E(X^k) &= \frac{B(p+k, q)}{B(p, q)} = \frac{\Gamma(p+k)\,\Gamma(p+q)}{\Gamma(p)\Gamma(p+q+k)} \\ &= \frac{p+k-1}{p+q+k-1} E(X^{k-1}) \end{aligned} \tag{C.39a}$$

where the last expression is a recursion formula. In particular, we have

$$\begin{aligned} E(X) &= \frac{p}{p+q} \\ \mathrm{Var}(X) &= \frac{pq}{(p+q)^2(p+q+1)} \end{aligned} \tag{C.39b}$$

Depending on the values of the two parameters, the pdf (C.38) varies; more specifically

(a) if $p > 1, q > 1$ it is kind of bell-shaped with a maximum at $x = (p-1)/(p+q-2)$;
(b) if $p < 1, q < 1$ it is U-shaped with a minimum at $x = (p-1)/(p+q-2)$;
(c) if $p > 1, q \le 1$ it is monotone increasing;
(d) if $p \le 1, q \ge 1$ it is monotone decreasing with the special case $p = q = 1$ in which it becomes the uniform distribution on the interval $(0, 1)$.

C.7 A few final results

A first result of interest is of practical use in calculations and concerns the standard normal PDF, often denoted by the symbol $\Phi(x)$. It is well known that there is no explicit expression for

$$F(x) = \frac{1}{\sqrt{2\pi}} \int_{-\infty}^{x} \exp(-z^2/2)\, dz$$

and this is why its values are extensively tabulated. Nonetheless, with $a = 1.5976$ and $b = 0.044715$, a sufficiently accurate approximation for most applications is

$$\Phi(x) \cong \frac{1}{1 + \exp\{-ax(1 + bx^2)\}} \tag{C.40}$$

which leads to a maximum absolute error < 0.0002.

Now, let $\mathbf{A}$ be a square $n \times n$ symmetric matrix, that is, $\mathbf{A} = \mathbf{A}^T$. It is known from matrix theory that $\mathbf{x}^T\mathbf{A}\,\mathbf{x}$ is the quadratic form associated to $\mathbf{A}$ and one says that (i) $\mathbf{A}$ is positive semidefinite if $\mathbf{x}^T\mathbf{A}\mathbf{x} \geq 0$ for all (non-zero) n-dimensional vectors $\mathbf{x}$ and (ii) $\mathbf{A}$ is positive definite if $\mathbf{x}^T\mathbf{A}\mathbf{x} > 0$ for all (non-zero) n-dimensional vectors $\mathbf{x}$. For our purposes, let us consider a sample $\mathbf{X} = (X_1, \ldots, X_n)^T$ from $N(0, 1)$, the quadratic form $Q \equiv \mathbf{X}^T\mathbf{A}\mathbf{X}$ and the m linear forms

$$t_i = \sum_{j=1}^{n} b_{ij} X_j, \quad j = 1, 2, \ldots, m$$

which, in turn, can also be written in matrix notation as $\mathbf{t} = \mathbf{B}\mathbf{X}$ where $\mathbf{t}$ is a $m \times 1$ column vector and $\mathbf{B}$ is the $m \times n$ matrix of coefficients. Then, denoting by $\mathbf{0}$ the null matrix of the appropriate dimensions relevant to the theorem being stated, we have

Proposition C.16 *If* $\mathbf{B}\mathbf{A} = \mathbf{0}$ *then the functions* Q *and* $\mathbf{t}$ *are independent.*

If now, on the other hand, we consider two quadratic forms $Q_1 = \mathbf{X}^T\mathbf{A}_1\mathbf{X}$ and $Q_2 = \mathbf{X}^T\mathbf{A}_2\mathbf{X}$, the following result applies

Proposition C.17 *If* $\mathbf{A}\mathbf{B} = \mathbf{B}\mathbf{A} = \mathbf{0}$ *then* Q_1 *and* Q_2 *are independent.*

Finally, the following two theorems concern the distribution of quadratic forms of normal variables. Denoting, as it is customary, by tr$\mathbf{A}$ the trace of $\mathbf{A}$ and by rk$\mathbf{A}$ its rank, the first theorem is

Proposition C.18 *Let* $Q = \mathbf{X}^{\mathrm{T}}\mathbf{A}\mathbf{X}$ *and let* $rk\mathbf{A} = r \leq n$. *If* $\mathbf{A}$ *is idempotent (i.e.* $\mathbf{A} = \mathbf{A}^2$*) then*

(i) $r = \mathrm{tr}\,\mathbf{A}$ *and*
(ii) $Q \approx \chi^2(r)$

while the second asserts.

Proposition C.19 *Let* $\mathbf{Y}$ *be a n-dimensional random vector distributed according to the multivariate, non-degenerate, normal distribution with mean* $\mathbf{m}$ *and covariance matrix* $\mathbf{K}$ *(in this regard, recall Section 3.3.2). Then*

$$(\mathbf{Y} - \mathbf{m})^{\mathrm{T}}\mathbf{K}^{-1}(\mathbf{Y} - \mathbf{m}) \approx \chi^2(n)$$

For the interested reader, the proofs of Propositions C.16–C.19 can be found in Ref [6]. All of them, however, are based on the fact that a for a real symmetric matrix it is always possible to find an orthogonal matrix $\mathbf{U}$ such that $\mathbf{D} \equiv \mathbf{U}^{\mathrm{T}}\mathbf{A}\mathbf{U}$ is diagonal and its (of $\mathbf{D}$) only non-zero elements are the eigenvalues of $\mathbf{A}$. We recall here that a matrix $\mathbf{U}$ is called orthogonal if $\mathbf{U}^{\mathrm{T}} = \mathbf{U}^{-1}$. Moreover, it turns out that the columns $\mathbf{u}_1, \ldots, \mathbf{u}_n$ of the matrix $\mathbf{U}$ which 'diagonalizes' $\mathbf{A}$ are the eigenvectors of $\mathbf{A}$. A detailed treatment of these fundamental aspects of matrix analysis can be found in [2, 5, 10].

References and further reading

[1] Azzalini, A., *'Inferenza Statistica: una Presentazione Basata sul Concetto di Verosimiglianza'*, Springer-Verlag Italia, Milano (2001).
[2] Bickley, W.G., Thompson, R.S.H.G., *'Matrices: Their Meaning and Manipulation'*, The English Universities Press, London (1964).
[3] Cramér, H., *'Mathematical Methods of Statistics'*, Princeton University Press, Princeton, Princeton, 19th printing (1999).
[4] Di Crescenzo, A., Ricciardi, L.M., *'Elementi di Statistica'*, Liguori Editore, Napoli (2000).
[5] Horn, R.A., Johnson, C.R., *'Matrix Analysis'*, Cambridge University Press, Cambridge (1985).
[6] Ivchenko, G., Medvedev, Yu., *'Mathematical Statistics'*, Mir Publishers, Moscow (1990).
[7] Keeping, E.S. *'Introduction to Statistical Inference'*, Dover Publications, New York (1995).
[8] Rinne, H., *'Taschenbuch der Statistik'*, Verlag Harri Deutsch, Frankfurt (2003).
[9] Suhir, E., *'Applied Probability for Engineers and Scientists'*, McGraw-Hill, New York (1997).
[10] Wilkinson, J.H., *'The Algebraic Eigenvalue Problem'*, Clarendon Press, Oxford (1965).

Index

For Product Safety Concerns and Information please contact our EU representative GPSR@taylorandfrancis.com
Taylor & Francis Verlag GmbH, Kaufingerstraße 24, 80331 München, Germany

www.ingramcontent.com/pod-product-compliance
Ingram Content Group UK Ltd.
Pitfield, Milton Keynes, MK11 3LW, UK
UKHW022008180926
475701UK00003B/256

9780367446529